Tourism Case Studies
Asia-Pacific Region

Tourism Case Studies Asia-Pacific Region

Edited by

Taufik Abdullah and Reni Polus

CABI

CABI is a trading name of CAB International

CABI
Nosworthy Way
Wallingford
Oxfordshire OX10 8DE
UK

Tel: +44 (0)1491 832111
E-mail: info@cabi.org
Website: www.cabi.org

CABI
200 Portland Street
Boston
MA 02114
USA

T: +1 (617)682-9015
E-mail: cabi-nao@cabi.org

A catalogue record for this book is available from the British Library, London, UK.

ISBN-13: 9781800628663 (hardback)
9781800628670 (ePDF)
9781800628687 (ePub)

DOI: 10.1079/9781800628687.0000

Commissioning Editor: Victoria Taylor
Editorial Assistant: Emma McCann
Production Editor: Rosie Hayden

Typeset by Exeter Premedia Services Pvt Ltd, Chennai, India

Contents

Supplementary Content

https://www.cabidigitallibrary.org/doi/book/10.1079/9781800628687.0000

Three videos are available for Chapter 27, Haunting Memory of Forgotten History: A Case Study of WWII 'Comfort Stations' in the Philippines as a Tourism Resource Featured in Selected Filipino Films (Aileen C. Ani):

Video 1: Short tour of the Bahay na Pula

Video 2: Interior of the Bahay na Pula

Video 3: Interior of the Bahay na Pula

Contributors

Abhisek Porya – Department of Tourism and Hospitality Management, Jamia Millia Islamia (A Central University), New Delhi, India

Ade Suherlan – University of Western Sydney, Penrith, Australia

Aileen C. Ani – College of Asia Pacific Studies, Ritsumeikan Asia Pacific University, Beppu, Oita, Japan

Aindrila Chakraborty – Department of Anthropology, College of Arts and Sciences, University at Albany, SUNY, New York, USA

Alexandra Coghlan – Universitas Pendidikan Indonesia, Bandung, Indonesia

Alka Maheshwari – Amity Institute of Travel & Tourism, Amity University, Noida, Uttar Pradesh, India

Amanda Natasanti – Universitas Gadjah Mada, Yogyakarta, Indonesia

Amran Hamzah – Department of Urban and Regional Planning, Faculty of Built Environment and Surveying, Universiti Teknologi Malaysia, Johor Bahru, Johor, Malaysia

Ana Delevska – Business and Construction, Melbourne Polytechnic, Melbourne, Australia

Anjusha P.P. – School of Tourism Studies, Mahatma Gandhi University, Kottayam, India

Bobbie Chew Bigby – University of Waterloo, Canada

Brigita Talentiana Mahen – Universitas Gadjah Mada, Yogyakarta, Indonesia

Carolina I. Garcia – School of Management and Marketing (Tourism Research Cluster), Faculty of Business and Law, Curtin University, Western Australia, Australia

Christof Pforr – School of Management and Marketing (Tourism Research Cluster), Faculty of Business and Law, Curtin University, Western Australia, Australia

David N. Nguyen – International Research Institute for Disaster Science, Tohoku University, Sendai, Japan

Della Ade Saputri – Universitas Gadjah Mada, Yogyakarta, Indonesia

Ding Xu – School of Tourism Management, Sun Yat-sen University, China

E. Joshu Ajoon – Department of Tourism, University of Otago, Dunedin, New Zealand

Eiji Ito – School of Health and Sport Sciences, Chukyo University, Toyota, Aichi, Japan

Fahmi Prihantoro – Universitas Gadjah Mada, Yogyakarta, Indonesia

Gabriel Fuentes Cordoba – Faculty of Liberal Arts, Sophia University, Tokyo, Japan

Guiqing Li – School of Tourism Management, Sun Yat-sen University, China

Heike Schänzel – School of Hospitality & Tourism, Auckland University of Technology, Auckland, New Zealand

I. Putu Gede Eka Praptika – Universitas Gadjah Mada, Yogyakarta, Indonesia

Isabel Paulino – Department of Organisation, Business Management and Product Design, Faculty of Tourism, University of Girona, Girona, Spain

Jennifer Kim Lian Chan – Centre for Economic Development and Policy, Faculty of Business, Economics and Accountancy, Universiti Malaysia Sabah, Kota Kinabalu, Sabah, Malaysia

Joo-Ee Gan – School of Hospitality and Service Management, Sunway University, Bandar Sunway, Malaysia

Kadek Sumaja – I Gusti Ngurah Rai Meteorological Office, Indonesian Agency for Meteorology Climatology and Geophysics (BMKG), Bali, Indonesia; School of Aviation, Faculty of Science, University of New South Wales (UNSW Sydney), Australia

Karen E. McNamara – School of the Environment, The University of Queensland, St Lucia, Brisbane, Australia

Krittawit Krittayaruangroj – College of Management, Mahidol University, Bangkok, Thailand

Lan Wang – School of Art, Design and Media, Nanyang Technological University, Singapore

Longrui Han – The Development Planning Unit, University College London, London, UK

Madura Thivanka Pathirana – Edith Cowan University, Sri Lanka Campus, School of Business and Law, Colombo, Sri Lanka

Michael Volgger – School of Management and Marketing (Tourism Research Cluster), Faculty of Business and Law, Curtin University, Western Australia, Australia

Michelle Szydlowski – University of Exeter, St. Luke's Campus, Exeter, UK

Mohamed Shenaan – School of Hospitality & Tourism, Auckland University of Technology, Auckland, New Zealand

Morakot Ditta-Apichai – College of Innovation and Management, Songkhla Rajabhat University, Songkhla Province, Thailand

Muhammad Ibnu Azzulfa – Universitas Gadjah Mada, Yogyakarta, Indonesia

N. Scott – University of the Sunshine Coast, Australia

Nadzirah Hosen – Department of Urban and Regional Planning, Faculty of Built Environment and Surveying, Universiti Teknologi Malaysia, Johor Bahru, Johor, Malaysia

Nimit Chowdhary – Jamia Millia Islamia (A Central University), New Delhi, India

Parishmita Gogoi – LEHO (Ladakh Environment & Health Organization), Leh, Ladakh, India

Paul Gemmell – School of Business and Hospitality, Algonquin College, Ottawa, Canada

Rachel Clissold – International Centre for Environmental Management, Hanoi, Vietnam; School of the Environment, The University of Queensland, St Lucia, Brisbane, Australia

Ramon Benedicto A. Alampay – Asian Institute of Tourism, University of the Philippines Diliman, Quezon City, Philippines

Ranjana Tiwari – Department of Commerce & Management Studies, Jain (Deemed to be University), Bengaluru, Karnataka, India

Reiza Miftah Wirakusuma – Department of Tourism, Sport and Hotel Management, Griffith University, Gold Coast, Australia; Universitas Pendidikan Indonesia, Bandung, Indonesia

Reni Polus – Business School at Edinburgh Napier University, Scotland

Ross Westoby – Griffith Institute for Tourism, Griffith University, Nathan, Brisbane, Australia

Sabiha Yeasmin Rosy – Department of Women and Gender Studies, University of Dhaka, Dhaka, Bangladesh

Sandra Cherro Osorio – Business and Construction, Melbourne Polytechnic, Melbourne, Australia

Sarah Gardiner – Institute of Tourism; Department of Tourism, Sport and Hotel Management, Griffith University, Gold Coast, Australia

Sarin Chhuon – Cambodia Living Arts, Phnom Penh, Cambodia

Stu Hayes – Department of Tourism, University of Otago, Dunedin, New Zealand

Sunayana – Department of Management Studies, FMS, Jamia Millia Islamia (A Central University), New Delhi, India

Suparak Suriyankietkaew – College of Management, Mahidol University, Bangkok, Thailand

Sweety Mishra – Department of Tourism and Hospitality Management, Jamia Millia Islamia (A Central University), New Delhi, India

Tay T.R. Koo – School of Aviation, Faculty of Science, University of New South Wales (UNSW Sydney), Australia

Taufik Abdullah – Faculty of Social Science Education at Universitas Pendidikan Indonesia

Toney K. Thomas – School of Tourism Studies, Mahatma Gandhi University, Kottayam, India

Van K. Nguyen – Business and Construction, Melbourne Polytechnic, Melbourne, Australia

Viviane Obed – Further Arts, Port Vila, Vanuatu

Yana Wengel – HNU-ASU Joint International Tourism College, Hainan University, Haikou, China; Borneo Tourism Research Centre, Faculty of Business, Economics and Accountancy, Universiti Malaysia Sabah, Kota Kinabalu, Malaysia

Yu Li – College of Ethnology, Guangdong Polytechnic Normal University, Guangzhou, Guangdong, China

Yunxin Hu – HNU-ASU Joint International Tourism College, Hainan University, Haikou, China

Zainub Ibrahim – School of Business and Hospitality, Algonquin College, Ottawa, Canada

Foreword

The collection of case studies from the Asia-Pacific region in this book is important in its own right. Yet it also stands as an integral component of a series of special issues of *Tourism Cases* that have focused on tourism outside of the global north/western context and perspective. These special issues have evolved to provide both a venue for authors from beyond the global north to tell their stories to the world and for those from beyond the Asia-Pacific region to develop their understanding of tourism in the region and to potentially be exposed to different conceptualizations of tourism and its operation. Consequently, this book, as a representation of the special issue of *Tourism Cases* dedicated to the Asia-Pacific offers the reader a chance to expand their knowledge and ways of knowing.

Case studies may not have always had the best press; they may not always have been loved. Instead, they have sometimes been seen as somehow inferior to *real* academic pieces of work; book chapters, books, and, at the pinnacle, double-blind peer-reviewed journal articles. Yet in an era when universities, and the academics that reside within them, are being reminded of the need to ensure the relevance and accessibility of their work to the world outside their walls case studies offer a potentially important tool to help ensure this. Stripped of the academic pageantry, case studies are able to focus on the story that lies at their heart instead of filling space with long and detailed methodological discourses, complex data analysis processes, and literature reviews that extoll the links of a study to *everything* that has gone before and ground the work in various philosophical debates. This is not to say that such work is without value, far from it, but that it can make impenetrable important stories that can help not just other academics but, in the context of tourism, students, industry operators, governments, and simply interested others who want to understand what is going on and how the situation (whatever that may be) can be improved. The case studies in this book not only offer the reader the chance to engage with real-world stories of tourism in an accessible and informative way, but also offer a template for how researchers can help to make their work accessible to others.

The content of the book can be read as a whole, offering insights into a wide range of issues spread across virtually all of the Asia-Pacific region. Alternatively, in the best traditions of all edited books, it can be dipped in and out of, focusing on whichever particular issue the reader is interested in. Both uses of the book are equally valid though for anyone with the time in our fast-moving world I do urge the former approach for the unexpected learning opportunities it has to offer. When we know what we want the latter approach is clearly the way to go. All I can say is that journeying through the book in its totality will not be a waste of your time.

Thanks must go to all the authors of the case studies, both in this book and the wider special issue. There simply is no space here to go into detail about each of the cases but their diversity, both in terms of the stories they tell and the places they are set in, is awesome. They talk not just of what is happening and has happened in the region but of numerous different ways of knowing and being. Thanks must also go to the work of the editors, Taufik Abdullah and Reni Polus, who oversaw the almost 50 case studies that were submitted to *Tourism Cases* for the Asia-Pacific special issue. They may see themselves as early career academics but the truth is that they are awe-inspiring, to be able to take on a mammoth task and make an excellent job of it. It just goes to show it is not the length of time that you have had a PhD for that matters but your thoughts and actions. Finally, thanks go to Claire Parfitt and everyone at CABI for having the vision to give a home to the idea that became *Tourism Cases*. We all too often live in a fast-paced world and it is nice to know that within this there is CABI, a publisher who clearly cares for all those they interact with, treating authors and editors with respect and compassion.

Professor Neil Carr, University of Otago
Editor-in-Chief for CABI's *Tourism Cases*

Introduction

Taufik Abdullah[1] and Reni Polus[2]
[1]*Faculty of Social Science Education at Universitas Pendidikan Indonesia;*
[2]*Business School at Edinburgh Napier University, Scotland*

Setting the Stage: Key Themes and Contexts in Asia-Pacific Tourism

Welcome to an exploration of the Asia-Pacific region, where tourism not only flourishes but profoundly influences the diverse cultural and economic landscapes. In *Tourism Case Studies from the Asia-Pacific Region*, we invite you to embark on a journey through a selection of compelling case studies that illustrate the dynamic world of tourism in some of the fastest growing destinations on the globe. This region, renowned for its vibrant tourism hotspots and a significant outflow of travellers from countries such as China, Japan, Korea, Singapore, Malaysia and the burgeoning market of India, plays an important role in shaping global tourism trends.

As we venture through this exploration, Leung and Baloglu (2013) illuminate the region's unique competitive advantages. Hong Kong and Singapore shine with their robust policy frameworks and tourist appeal, while Japan, Korea and Taiwan excel in cultural resources and advanced infrastructure. China, India and Indonesia offer rich cultural and natural treasures, while Australia and New Zealand stand out for their superior air transportation and infrastructure prowess. Malaysia, Mongolia, the Philippines and Thailand are celebrated for their impressive price competitiveness.

However, the global COVID-19 pandemic casts a long shadow over the tourism industry. According to the APEC Tourism Working Group (2024), the pandemic slashed tourism's contribution to global GDP by 49.1% and cut tourism-related jobs by 18.5%. As we emerge from this crisis, the tourism landscape is shifting towards sustainability and regeneration, with a focus on planetary health, human well-being, and community empowerment (Becken and Loehr, 2023). This volume captures the transformation through a series of insightful case studies which highlight how the Asia-Pacific region is managing and reinventing tourism.

In this collection, we present practical and engaging case studies that offer real-world insights into the current opportunities and challenges within the tourism industry. Through these case studies, you will gain a deeper appreciation of how tourism serves as both a catalyst for growth and a reflection of the diverse cultures across the Asia-Pacific region. Developed by experts from academia and industry, this volume features 32 case studies organized into seven sections. These sections address critical topics such as sustainable practices, cultural preservation, empowerment, resilience, management, policy and emerging trends. By exploring these themes, the volume provides valuable lessons and practical strategies for fostering a more ethical and inclusive tourism industry.

Section 1: Environmentally sustainable practices for tourism

Our first section explores the essential theme of environmental sustainability within tourism, spotlighting four distinct case studies that illustrate innovative practices across various destinations in the Asia-Pacific region. From national parks to mountain trekking zones, this section explores how sustainable tourism is being implemented and the role of stakeholder engagement in these efforts.

Our journey begins in India with the case study **Sustainable Hiking Through the Green Trails: Examples of Exceptional Responsible Practices** by Anjusha P.P. and Toney K. Thomas. We explore the transformative 'green trails' initiative by Indiahikes, a prominent trekking organization. Set against the backdrop of India's sensitive mountain ecosystems, Indiahikes has redefined mountain tourism through sustainable practices that engage trekkers, local villages, and the trekking team. This case reveals how these practices can foster environmental stewardship and integrate sustainability into daily life through the lenses of stakeholder theory, transformative learning and place-building.

Next, we travel to Indonesia with **Sustainable Tourism and Stakeholders' Satisfaction in Batulayang Tourism Village, Bogor, Indonesia** by Ade Suherlan. This case reveals the dynamic between sustainable tourism development and stakeholder satisfaction in Batulayang, West Java. By emphasizing a bottom-up approach, this case illustrates how stakeholder involvement drives the success of tourism villages and highlights the importance of collective effort in achieving sustainability.

Still in Indonesia, our next stop is Komodo National Park, where **Overtourism and Tourism Sustainable Management in the Komodo National Park, Indonesia** by Reiza Miftah Wirakusuma, Sarah Gardiner and Alexandra Coghlan examines the management of overtourism in this UNESCO World Heritage site. This case explores the challenges of managing tourism in fragile ecosystems and provides insights into collaborative management practices that can help mitigate the impacts of overtourism.

Finally, we venture to Malaysia with **Adopting Sustainable Practices at an Island Ecotourism Destination in Sabah Through a Sea Turtle Conservation Project** by Jennifer Kim Lian Chan. Here, we explore the award-winning sea turtle conservation project at Walai Penyu Conservation Park. This case highlights how ecotourism initiatives, supported by local communities and various organizations, exemplify practical sustainability and the harmonious coexistence of conservation and tourism.

Together, these case studies highlight the diverse approaches to sustainable tourism and the importance of tailored strategies and stakeholder collaboration. They demonstrate that sustainability in tourism is not a one-size-fits-all approach but requires tailored strategies that consider the unique characteristics and needs of each destination.

Section 2: Cultural preservation and identity through tourism

Our exploration continues with a focus on cultural sustainability, featuring six case studies from Indonesia, Singapore, Cambodia, Japan, India and China. This section showcases how tourism can both preserve and promote cultural heritage while shaping destination identity.

Our first case, **Improving Cultural Sustainable Tourism: Evidence of Saba Budaya Baduy from the Baduy Tribe, Indonesia** by I. Putu Gede Eka Praptika, Amanda Natasanti, Della Ade Saputri, Muhammad Ibnu Azzulfa, Brigita Talentiana Mahen and Fahmi Prihantoro explores the Baduy people's efforts to preserve their cultural identity through the 'Saba Budaya Baduy' initiative. This case illuminates how adherence to cultural principles and customary provisions helps the Baduy community manage tourism impacts and uphold their cultural resilience.

Next, we travel to Singapore with **Hawker Centres, Cultural Heritage and Tourism – A Case Study of Lau Pa Sat, Singapore** by Lan Wang. This case examines how Lau Pa Sat, a historic hawker centre, integrates cultural heritage within a modern urban setting. It highlights Lau Pa Sat's role in fostering community identity and balancing historical preservation with contemporary needs.

From Singapore, we journey to Cambodia with **Experiencing Cambodia's Living Arts Through Tourism** by Bobbie Chew Bigby, Madura Thivanka Pathirana and Sarin Chhuon. This case presents the Cambodian Living Arts' (CLA) nightly performances and their impact on heritage preservation and economic development. It offers a model for using cultural tourism to foster national identity and generate economic benefits.

Our next stop is Japan with **Aikido Tourism Development in Tanabe, Wakayama, Japan** by Eiji Ito. This case explores how Japan's traditional martial art, aikido, is leveraged to create niche tourism experiences. Ito's case illustrates how cultural heritage can be harnessed to engage tourists while preserving traditional practices.

In India, we explore **Forging Social Identity: The Making of a Destination – A Case of the Bullet Baba Shrine in Marwar, Rajasthan, India** by Abhisek Porya, Sweety Mishra, Sunayana and Nimit Chowdhary. This case examines how the Bullet Baba temple has become a cultural landmark through its unique rituals and storytelling, offering a vivid example of how cultural tourism can create a distinct destination identity.

Finally, we travel to China with **Lived Religious Practice and Tourism Perceptions of Southern Buddhist Monks in an Ethnic Tourism Community in Xishuangbanna, China** by Yu Li. This case explores the complexities of Buddhist tourism in Menghan Town, highlighting the challenges and strategies for managing religious tourism in a way that respects cultural and spiritual practices.

This section provides a multifaceted view of how tourism can support cultural preservation and identity, balancing heritage with modernization while fostering cultural exchange. Each case highlights the importance of integrating cultural heritage with tourism development to ensure sustainable and meaningful travel experiences.

Section 3: Empowerment through tourism

Our journey now takes us to the theme of empowerment through community-based tourism initiatives. This section highlights six case studies from the Philippines, Maldives, India and Thailand, demonstrating how tourism can drive economic empowerment and social change.

We start in the Philippines with **Community-Based Ecotourism and Conservation in Tboli, Philippines** by Ramon Benedicto A. Alampay. This case explores how the Tboli Indigenous people have embraced tourism to balance economic, environmental and cultural goals. Through local planning and stakeholder engagement, the Tboli have crafted a unique tourism identity that promotes cultural and environmental sustainability.

Next, we head to the Maldives with **The Guesthouse Phenomenon in the Maldives – Development and Issues** by Mohamed Shenaan and Heike Schänzel. This case explores the shift from the traditional 'one island, one resort' model to community-based tourism through guesthouses on inhabited islands. It examines the rapid growth, challenges, and benefits of integrating community-based accommodation, emphasizing how these changes encourage more direct interactions between tourists and local communities.

Our journey then takes us to India with two compelling cases. First, we explore the picturesque village of Munsiyari in the Himalayan foothills through **Breaking Barriers, Building Futures: The Transformative Power of Tourism in Empowering Women in Munsiyari, India** by Ranjana Tiwari. This case showcases how tourism can empower women and promote social entrepreneurship. It illustrates the transformative effects of tourism on local communities and offers insights into achieving long-term sustainability and empowerment.

Next, we move to Himachal Pradesh with **Community-Based Tourism as a Sustainable Approach to Tourism Development in Jibhi Valley, Himachal Pradesh, India** by Parishmita

Gogoi and Alka Maheshwari, which highlights how tourism in Jibhi Valley supports local livelihoods and preserves cultural heritage. This case explores how community-based tourism can enhance local economies and foster inclusiveness, particularly among women and marginalized groups.

Our journey continues to Thailand, starting in Satun Province with **Business Strategy and Development Towards Sustainability: A Story of Ban Bo Chet Luk Community Entrepreneurship, Thailand** by Morakot Ditta-Apichai. This case examines how Ban Bo Chet Luk village's community-based tourism initiatives promote sustainability and empower local entrepreneurs. It provides insights into capacity-building programmes and marketing strategies that drive sustainable tourism growth.

Finally, we visit Trang Province with **Baan Khoa Lak: A Case Study on Sustainable Rural Development Through Community-Based Tourism and Natural Resource Conservation** by Krittawit Krittayaruangroj and Suparak Suriyankietkaew. This case demonstrates how Baan Khoa Lak has successfully implemented a tourism programme that boosts income while preserving the environment and cultural heritage. It showcases the effectiveness of community engagement and natural resource conservation in rural tourism development.

This section offers an insightful view of how community-based tourism can drive empowerment and sustainable development, emphasizing the role of local participation and collaboration in achieving transformative impacts. Each case highlights the diverse benefits and challenges of integrating tourism with community and environmental goals.

Section 4: Resilience and recovery through tourism

In this section, we embark on a journey across the Asia-Pacific region where communities transform adversity into opportunity through the resilient spirit of tourism. From Malaysia's verdant forests to the pristine islands of Vanuatu, this section offers a journey through seven compelling case studies that illustrate how tourism fosters recovery and resilience amidst natural disasters and global crises like the COVID-19 pandemic.

Our first stop is Malaysia with **Green Recovery and Regenerative Tourism: The Success Story of Miso Walai Homestay in Building a Resilient Rural Tourism Destination**, by Nadzirah Hosen, Isabel Paulino and Amran Hamzah, where we visit the Miso Walai Homestay in the lush Kinabatangan region. Amidst towering trees and rich wildlife, the community faced the withdrawal of timber companies, leading to economic uncertainty. However, they turned this challenge into an opportunity by pioneering a community-based tourism model centred on forest restoration. When the COVID-19 pandemic struck, this green recovery initiative provided a lifeline, consolidating income streams until international tourists returned. This case details the critical success factors that made this green recovery programme a beacon of resilience.

Next, we sail to Vanuatu, an archipelago in the southwestern Pacific Ocean. Here, in **The Realities of Peripheral Handicraft Livelihoods in the Face of Disasters in Vanuatu**, Rachel Clissold, Ross Westoby, Karen E. McNamara and Viviane Obed explore how female handicraft vendors navigate a multihazard environment. From tropical cyclones to the COVID-19 pandemic, these women face significant challenges. Yet, their stories are also about resilience, community support and the pivotal role of local institutions in recovery.

Our journey continues to Kerala's enchanting backwaters in India. **Resilience and Loyalty in the Face of Adversity: A Case of Local Employees in Kerala's Backwater Tourism**, by E. Joshu Ajoon and Stu Hayes, reveals the steadfastness of local employees in the backwater tourism sector. Despite rising costs and the COVID-19 pandemic's pressures, their intrinsic motivation and community spirit ensure that tourism businesses remain resilient.

Next, we travel to the northern coasts of Japan, still echoing with the memories of the 2011 Great East Japan Earthquake. David N. Nguyen and Gabriel Fuentes Cordoba, in **Responses from Businesses in Rural Communities Recovering from the 2011 Tohoku Disasters During the 2020 Tokyo Olympics**, show us how local businesses hoped the 2020 Tokyo Olympics would catalyse economic revival. Despite the challenges of the COVID-19 pandemic, these businesses exemplify resilience and adaptation, reflecting on varied approaches to spectator policies in Miyagi and Fukushima.

Our next destination is Australia, where the excitement of hosting the 2026 Commonwealth Games turned into disappointment with its cancellation. Ana Delevska, Sandra Cherro Osorio and Van K. Nguyen, in **'Not All Fun and Games': The Case of the Victoria 2026 Commonwealth Games**, dissect the logistical and financial challenges that led to this decision. This case highlights the delicate balance between hosting mega-events and ensuring sustainable, inclusive benefits for the host communities.

From the sporting arenas of Australia, we move to the Philippines where **The Bojo Algouinsan Ecotourism Initiative: Crisis Mitigation Through Community Tourism Development**, by Zainub Ibrahim and Paul Gemmell, showcases how tourism can protect against environmental and economic threats. This case takes us through Bojo's journey from vulnerability to becoming an award-winning ecotourism destination, driven by community spirit and innovative tourism practices.

Finally, we land in Sri Lanka, exploring the post-war tourism boom in Jaffna. M.T. Pathirana, B.C. Bigby, N. Scott and S. Gardiner, in **Tourism Development Opportunities in Post War: A Case of Jaffna, Sri Lanka**, guide us through the transformation of this once conflict-ridden region into a burgeoning tourist hotspot. Infrastructure upgrades and cultural heritage play key roles in this revival, offering strategies for equitable development in post-conflict zones.

This section illustrates how tourism can serve as a formidable force for recovery and resilience across the Asia-Pacific region. The diverse case studies presented demonstrate how communities, faced with adversity, have harnessed the power of tourism to foster renewal and growth.

Section 5: Critical studies in tourism

Our journey now takes a deeper turn as we explore tourism's role in addressing and exacerbating societal power imbalances. From the Indigenous communities of Bangladesh to the colonial legacies in India, this section presents five critical studies that reveal the complexities and ethical challenges in tourism.

We begin in the rugged terrain of the Chittagong Hill Tracts in Bangladesh. In **The Growth of Tourism and 'Development' in an Indigenous Setting: Exploring a Case of the Chittagong Hill Tracts, Bangladesh**, Sabiha Yeasmin Rosy examines the economic interventions favouring the Bengali population at the expense of Indigenous communities. This case challenges the notion of 'development' in a post-conflict setting, revealing issues of militarization, resource control and environmental impacts.

Next, we traverse Malaysia's diverse rural landscapes. **Equal Opportunities in Rural Tourism Entrepreneurship: Challenges in Malaysia** by Joo-Ee Gan explores how government initiatives aimed at uplifting socio-economic status often bypasses the Orang Asli, Malaysia's Indigenous people. This case highlights the need for tailored capacity-building efforts to ensure true equality in tourism opportunities.

From human marginalization, we shift to the ethical dilemmas of animal exploitation in Nepal. In **Elephant-Based Tourism in Nepal: The Key to Reducing Human Poverty or a Tool to Commodify Endangered Species?**, Michelle Szydlowski raises questions about the socio-economic gaps exacerbated by tourism practices. This case highlights the ethical implications of using captive elephants for tourism, balancing human poverty alleviation against the commodification of endangered species.

We then travel to the Philippines, where the echoes of World War II still linger. Aileen C. Ani, in **Haunting Memory of Forgotten History: A Case Study of World War II 'Comfort Stations' in the Philippines as a Tourism Resource Featured in Selected Filipino Films**, invites us to uncover the untold stories of 'comfort women' and 'comfort gays' through Filipino films and potential tourism routes. This case takes an interdisciplinary perspective that integrates history (memory), media (representation) and tourism (sites) to propose a tour as a form of remembering the war victims. Three videos are available for this case study - see details below.

Our final stop in this section is the vibrant cultural tapestry of Puducherry, India. Aindrila Chakraborty's **'We Invite You for Lunch': Post/Colonial Subjectivities and Performances of Tourism in Puducherry, India** explores how hosts in Puducherry use cultural performances to share their heritage with tourists. This case explores how colonial legacies shape contemporary tourism encounters, fostering a deeper understanding of collective identity.

This section reveals tourism's dual nature as a powerful driver of development and a complex source of ethical dilemmas. It challenges us to rethink our approach to tourism, urging a move towards more inclusive and culturally attuned practices.

Section 6: Management and policy

Our exploration of the Asia-Pacific region's tourism landscapes now brings us to the strategic and policy dimensions that guide this industry. This section presents two insightful case studies from Indonesia and Australia, illustrating the impact of data and evidence on tourism management and policy making.

Our first stop is Bali, Indonesia. Kadek Sumaja and Tay T.R. Koo, in **Meteorological Trends and the Travel Sector in Bali 1991–2020**, analyse how changing weather patterns influence Bali's aviation and travel sectors. This case emphasizes the importance of meteorological data in shaping regulatory, physical and economic strategies, providing a model for evidence-based policy making.

We then travel to Western Australia, where Carolina I. Garcia, Christof Pforr, and Michael Volgger present **Evidence Use in Policymaking: The Case of a Hotel Policy in Western Australia**. This case explores a policy launched in 2011 to incentivize hotel investments, examining its development, implementation and outcomes. It highlights how evidence-based approaches can inform effective tourism strategies and foster sustainable growth.

This section highlights the critical role of data-driven approaches in guiding tourism management and policy decisions, emphasizing the need for rigorous analysis to support strategic planning and regulatory frameworks.

Section 7: Emerging trends in tourism

As our journey through the Asia-Pacific region draws to a close, we explore the exciting new trends shaping the future of tourism. Two captivating cases from China highlight the evolving nature of tourism, showcasing surfing tourism in Hainan and educational tourism in Mount Danxiashan.

We first ride the waves to Hainan Island. In **Surfing in Hainan Island: An Emerging Surfing Tourism Destination in China**, Yana Wengel, Longrui Han and Yunxin Hu explore Hainan's burgeoning surfing scene. From surf schools to cultural transformations, this case illustrates how surfing tourism is putting Hainan on the global map, bringing economic opportunities and cultural changes to the island.

Our final destination is the serene landscapes of Mount Danxiashan. Guiqing Li and Ding Xu, in **Learning by Doing: Educational Tourism at Mount Danxiashan**, explore how educational tourism balances conservation and community involvement. This case demonstrates how study tours and local engagement can drive sustainable tourism development, providing economic benefits and personal growth opportunities for locals.

This final section highlights the dynamic and adaptive nature of tourism as it embraces new trends and opportunities. These final two cases highlight how innovation and thoughtful engagement can lead to positive outcomes for both local economies and cultures. As we reflect on these examples, it becomes evident that the future of tourism lies in embracing these emerging trends while staying committed to sustainability and community well-being.

The cases presented in this volume vary in length but follow a uniform structure. Most begin with a summary that provides essential background information and features a brief section titled 'The Value and Interest of the Case Study', which succinctly highlights the case's relevance. The main content typically starts with a 'Background' or 'Context' section, followed by in-depth discussion in the central sections, and concludes with a summary of key findings and lessons learned. Each case study tells a compelling story, providing enough context to understand why certain actions were taken and their outcomes. These narratives not only inspire but also enhance your comprehension of the underlying principles illustrated. The volume concludes with overarching remarks that synthesize the main themes and observations from the case studies, and it also includes a section outlining ongoing and future research plans by the contributors, detailing their upcoming research initiatives and objectives. These plans are intended to advance understanding and address emerging challenges in the field of study.

References

APEC Tourism Working Group (2024) *Annual Report on Asia-Pacific Tourism (2021)*. Asia-Pacific Economic Cooperation. Available at: https://www.apec.org/publications/2024/01/annual-report-on-asia-pacific-tourism-(2021)#:~:text=As%20a%20result%20of%20the,2019%2C%20reducing%20by%2018.5%25

Becken, S. and Loehr, J. (2023) Asia-Pacific tourism futures emerging from COVID-19 recovery responses and implications for sustainability. *Journal of Tourism Futures* 9(1), 35–48.

Leung, X.Y. and Baloglu, S. (2013) Tourism competitiveness of Asia Pacific destinations. *Tourism Analysis* 18(4), 371–384.

Supplementary content

https://www.cabidigitallibrary.org/doi/book/10.1079/9781800628687.0000

Three videos are available for Chapter 27, Haunting Memory of Forgotten History: A Case Study of WWII 'Comfort Stations' in the Philippines as a Tourism Resource Featured in Selected Filipino Films (Aileen C. Ani):

Video 1: Short tour of the Bahay na Pula

Video 2: Interior of the Bahay na Pula

Video 3: Interior of the Bahay na Pula

Section 1: Environmentally Sustainable Practices for Tourism

This section explores sustainable practices within tourist destinations like national parks, conservation areas, and mountain trekking zones. Through cases from India, Indonesia and Malaysia, it illustrates how stakeholder involvement and community engagement are important in implementing effective environmental conservation efforts.

1 Sustainable Hiking Through the Green Trails: Examples of Exceptional Responsible Practices

Anjusha P. P. and Toney K. Thomas
School of Tourism Studies, Mahatma Gandhi University, Kottayam, India

Summary
'Indiahikes' is a leading trek team in India, leaving a thoughtful impact on the trekkers as well as the mountains. Through different initiatives for sustainability, and their iconic eco bags, the company leaves the mountains better than they found them. The mission titled Green Trail Initiative focuses on green ways of trekking. Over the years, the programme collected over 100 tonnes of litter from the trails and disposed of it through proper segregation methods. Working with sustainability at its core, the green trail focuses on the six golden practices of green sweeping,[1] segregation at source, food composting, bio-toilets, and saving water and energy. This study addresses this exceptional case as an example of stakeholder initiative towards effective sustainable practices of destination management. The trekkers are an important aspect of the study, they collect waste and follow eco-sensitive practices at the camp sites. The act has a significant impact on the travel behaviour of trekkers by transforming them into environmentally responsible travellers. As an exceptional initiative for the mountain travel system, the act of creating green trails sets ideal guidelines for various trek companies to follow.

Methodological Process Flow

The information for the case study is primarily based on the participant observation and first-hand experience of the author who participated in a 7-day trek conducted by Indiahikes to *Kuari Pass* in December 2022. This made it easier to understand the significance they place on sustainability measures right away after making a reservation. Various theoretical and conceptual models of sustainability and conservation relate to the author's observations. Interactions with the fellow trekking participants, the trek leader, technical team, and support team from Indiahikes also contribute to this reflective account. Also, the first-hand experience of these sustainability awareness initiatives contributed to the reflection on transformative experiences. The author also recollects the thoughts and experiences of other fellow trekkers and technical team members from the group activities during the trek, especially from the feedback session. Other information regarding the background and success of these initiatives was gathered from the official website of the organization and multiple informative vlogs and blog posts of the organization.[2]

The Value and Interest of the Case Study

This case study addresses an example of the sustainability efforts of a trekking company in

India transforming the lives of the trekkers, trek leaders, and the villages through its eco-friendly initiative called 'green trails'. In this way, the case is an example of better management of mountain tourism. Details of the case study also explain the interconnection of different theories and models in sustainability with the functioning of the organization. The case also details the relation to the United Nations Sustainable Development Goals (SDGs).

Context

The mountains and sustainability

Emphasizing the importance of protecting terrestrial ecosystems and prioritizing mountain ecosystems, SDG 15 explicitly mentions the systematic process of mountain conservation (United Nations, 2022). There are 671 million people living in high mountain regions (10% of the global population). The current trend in tourism indicates the increasing popularity of mountain tourism, which contributes 9–16% of all international tourist arrivals (UNWTO, 2023). Since mountain tourism activities take place in a defined and limited geographical space, the concern of carrying capacity leading to adverse environmental impacts is a notable issue. In general, mountain tourism is considered a rural tourism activity. Mountain tourism has the advantage over many other types of tourism destinations of attracting year-round visitors with a range of activities (Steiger *et al.*, 2022), including mountaineering, biking, hiking, biking, and snow sports. Its social and economic impact is important for the livelihood of local people and is of paramount importance as mountains are highly vulnerable to human and natural ecological balance (United Nations Environmental Programme, 2008). Technology has revolutionized the experience of trekking in the mountains. People can now climb faster and with less effort due to significant improvements in clothing, equipment, ropes, and ice axes. All these advancements pave the way for mass tourist engagement. This significant influx of tourists, especially in the Himalayas, increases the risk of environmental degradation (Biswas *et al.*, 2022). Hence an increasing number of trekkers and their activities will have adverse impacts on mountain tourism. Traveller's interest in digital detox holidays, their increasing affinity and appreciation for nature, and their interest in attaining more personal experience indicate the increasing importance of mountain tourism (Ozdemir and Goktas, 2021). In this growing sector of tourism, attaining sustainability is a specific responsibility of tourists. Due to the high level of dependence on natural resources and the presence of rare, endemic, and endangered species of flora and fauna, mountain landscapes and communities are highly sensitive and vulnerable to climate change (Steiger *et al.*, 2022), which emphasises the need for sustainable initiatives in the mountains to reduce the impact of the tourists.

Trekking in India

Mountain ranges cover about 30% of the land in India. India's mountain ranges are some of the most popular tourist destinations in the World. The Himalayan ranges, which include the world's tallest mountain, have been a place of expedition since 1953, when Edmund Hillary and his climbing partner, Tenzing Norgay, became the first to reach the highest point on Earth, the summit of Mount Everest (National Geographic Education, n.d.). India has a number of peaks above 7,000m which are still unexplored (Indian Mountaineering Federation, 2018). Many of India's mountain ranges occupy a prominent place in the pride of Indian culture and religion. Mountains such as Kailash, Vaikuntha, Meru, Govardhan, Kishkintha, and Kedarnath are replete within Hindu religious texts. The mountain ranges of the Himalayas, such as the Karakoram and Pir Panjal ranges; the eastern mountains like Purvanchal, the Satpura, and Vindhya ranges; and the Aravalli ranges of the Western Ghats, offer abundant opportunities for the country to promote and practice mountain adventure tourism (Kohli, 2002). Some of the world's best trekking spots are also located in the famous mountain ranges of India. Peaks such as Hampta Pass Trek in Himachal Pradesh, Beas Kund Trek, Chandrakhani Pass Trek, and Markha Valley Trek are some of the major trekking destinations. India's share of the global mountain-based trekking market is estimated at around US$0.08 billion from the global market of US$89 billion. The growth of the trekking industry in India is expected to be 10–20% annually in the coming years (Shah, 2020). Due to the steady growth of trekking-related tourist activities and increasing tourist numbers, it needs the utmost consideration to promote environmental, economic, and socio-cultural sustainability and reduce the risks associated with adventure activities (Ziegler *et al.*, 2021). This indicates the urgency of a well-structured, proper management of mountain tourism resources and the socio-economic development of the mountain region through a proper policy framework that helps in attaining economically productive, socially responsible, and ecologically sound development strategies. Also, according to Nepal (2002), it is important to emphasize the need for sustainable eco-tourism practices in the mountains not only to conserve nature but to also foster responsible tourists, acknowledge the local cultures and ways of living in the mountains, and build sustainable livelihood for people who are living in the villages.

Indiahikes

A stakeholder initiative for sustainable practices of destination management in the mountains in India is a trekking company called Indiahikes (n.d.). The organization focuses on shaping travel behaviour of trekkers in India and moulds them into environmentally responsible travellers. The organization was established more than a decade ago by Arjun Majumdar and Sandhya U C with the goal of making hiking accessible to everybody. The firm, which first served as a forum for documenting various treks, quickly saw an opportunity and expanded to become India's biggest trekking organization, serving more than 20,000 trekkers every year without sacrificing its basic ideals of sustainability. Not only did they provide treks but also have a huge database of trek related information. The trek team is known for its efforts to leave the mountains better, safety, making trekkers independent, and the transformative experiences they provide for the trekkers. Through their initiatives to make trekking accessible to all, the vision of Indiahikes is that 'Everyone must trek because it transforms life'. As a trekking organization, the company has highlighted the importance of sustainability in its core values and principles. Protecting the environment and blending with nature are among the main principles of the firm. Figure 1.1 shows an example of experiencing and appreciating nature during the trek. Being a 'sustainability champion' is a core value at the company. It has taken efforts to leave the mountains better than they have found them. The efforts of Indiahikes stands as a transformation of various theories of sustainability into practice.

As a trekking team, they are known for their iconic 'eco-bags' as shown in Fig. 1.2, which they tie around their waists to collect garbage on the trails. This clearly distinguishes them from fellow trekkers. The trekkers are prompted to pick up any waste materials they come across during the trails, which are then segregated and properly disposed of at the campsites. Through exploring these possibilities of creating a transformative experience for the trekkers, Indiahikes put forward awareness to protect the sensitive ecosystem.

Green trails – leaving the mountains better

Green trails is a major sustainability drive of the organization, focusing on six golden rules to be followed during the trekking and beyond. It aims at leaving the mountains better. The initiative was started to inculcate a sustainable trekking culture in India. The concept of 'green trails' revolves around a holistic approach to addressing the challenge of garbage and waste management. It focuses on two main strategies: increasing the collection of rubbish and reducing overall trash production to prevent it from ending up in landfills. Over 32 major treks in India, the green trails have collected over 160,000 kg of waste, helping to divert 80% of it from landfills. The programme also manages 100% of the wet waste that the company generates at the source itself. In addition, the initiative realizes the resolution of the waste crisis goes beyond merely establishing appropriate infrastructure. Instead, it necessitates inducing a

Fig. 1.1. Deoriatal Chandrashila trek by Indiahikes (Courtesy of Jothiranjan. Available at: https://indiahikes.com/category/photo_story).

Fig. 1.2. Iconic Eco bag (courtesy of India Hikes. Available at: https://store.indiahikes.com/product/indiahikes-eco-bag/).

shift in behaviour to bring about a solution. From the initial stage of booking, trekkers are constantly communicated with regarding the green trails policies as well as green alternatives to be followed during the treks. At the very core is the aim to stop generating waste.

As shown in Fig. 1.3, every trek by Indiahikes follows the six golden rules of green sweeping, segregation at source, food composting, bio-toilets, and saving water and energy. Through the utilization of eco bags, individuals who engage in trekking or hiking activities take on a proactive role in addressing the issue of litter and waste left behind on the trails. Waste segregation involves sorting discarded materials into distinct categories based on their type and composition. This practice promotes effective recycling, composting, and disposal

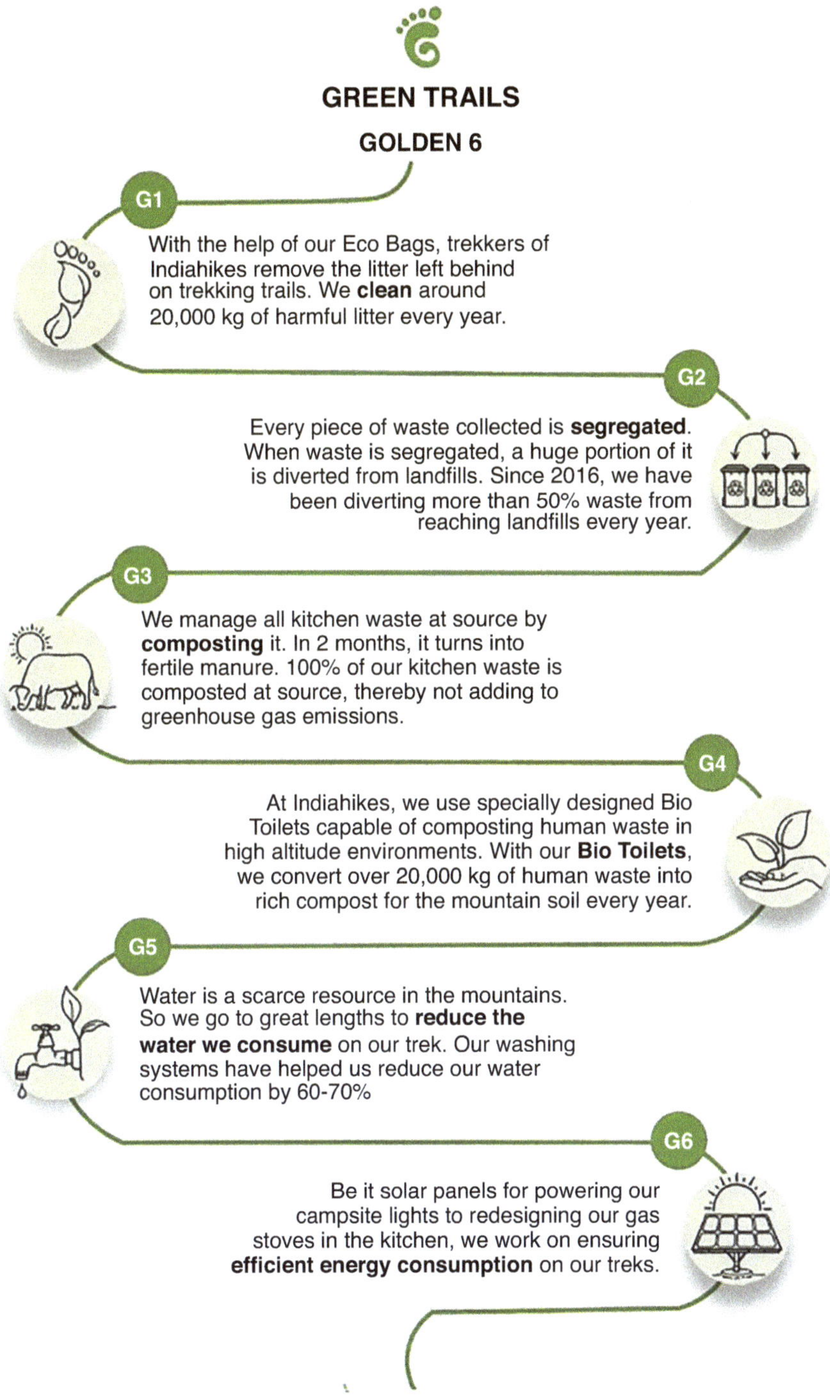

Fig. 1.3. Golden Gs of Green Trails (courtesy of Indiahikes: https://indiahikes.com/green-trails).

methods. By separating waste at its source, valuable resources are conserved, pollution is reduced, and landfill pressure is alleviated. Effective management of organic waste is done at the campsite through the management of wet waste. The biggest issue of managing human waste at campsites is done with the help of deep-pit bio-toilets. Human waste if not treated properly will adversely impact the mountain ecosystem, even contaminating the water sources. Also, high altitudes do not usually provide a suitable environment for the decomposition of human waste due to the thin air and low temperatures. Large volumes of human waste, lack of infrastructure, and social stigma attached to human waste management make the task even harder. Deep-pit bio-toilets with coco peat[3] offer a sustainable solution rather than open defecation and catholes for larger groups of trekkers. The benefits of bio-toilets are trifold. They protect over 80% of the campsite from human waste due to a lesser number of pits, speed up the decomposition process with the heat in deep pits, and protect topsoil from permanent damage due to the need to only remove some topsoil in the process of digging pits. Green trails also take efficient measures to conserve water and energy resources as well through minimal consumption.

Green Trails Impacts

The green trails project has made a significant impact on the trekking trails and the trekking community. Green trails is about much more than just cleaning trails and using sustainable products for trekking. It also has an impact on the remote villages of the Himalayas through its programmes, which include teaching sustainable living skills in schools and villages, organizing clean-ups, establishing eco-villages through proper waste management, turning waste into useful materials, and planting new trees. It also shows examples of upcycling from the waste materials which may otherwise end in landfill. The initiative focuses on transformation towards sustainability and community engagement. Green trails' website of Indiahikes, as seen in Fig. 1.4, showcases different small steps which can be adopted by other trekking companies in the mountains. The large database of information on the site regarding the green trails impacts and sustainability resources create awareness among the readers. The visible impacts of green trails is a clear example of engagement of different stakeholders in achieving the sustainability goals of the organization. The case portrays the importance of stakeholder theory (Freeman, 1984), where the organization values the involvement of various stakeholders in bringing about change. It also values the interactions and relationships between those who have a stake in the organization.

Stories of transformation – Creating sustainability champions

There are appreciable changes in the perspectives and actions of trekkers, the villages and campsites as well as the people working in Indiahikes through the green trails project. Trekkers with their eco bags, as they journey along the trails, conscientiously pick up and gather any discarded items or debris they come across. It encourages trekkers to actively engage with the surroundings and fosters a heightened sense of awareness regarding the impact of human activities on nature. Additionally, the use of eco bags and the maintenance of the paths provide a concrete example for others who could be motivated to imitate the actions. It encourages a culture of collective responsibility and changes the way individuals approach outdoor activities. Figure 1.5 highlights a brochure which is distributed to the trekkers, reminding them to pick up garbage on the trails. Also, the segregation and minimal generation of waste are practices which many trekkers take home, which leaves a positive impact on the eco-sensitive practices of multiple societies. As more people adopt comparable practices and work together to preserve the beauty of natural settings, this might eventually have a beneficial ripple effect.

Through regular training and sustainability challenges, the company is creating sustainability champions in the organization to work closely with the community. The trek leaders and different employees of the company lead different initiatives of sustainable living. Some of the examples include composting, reducing inorganic waste, and DIY projects to upcycle

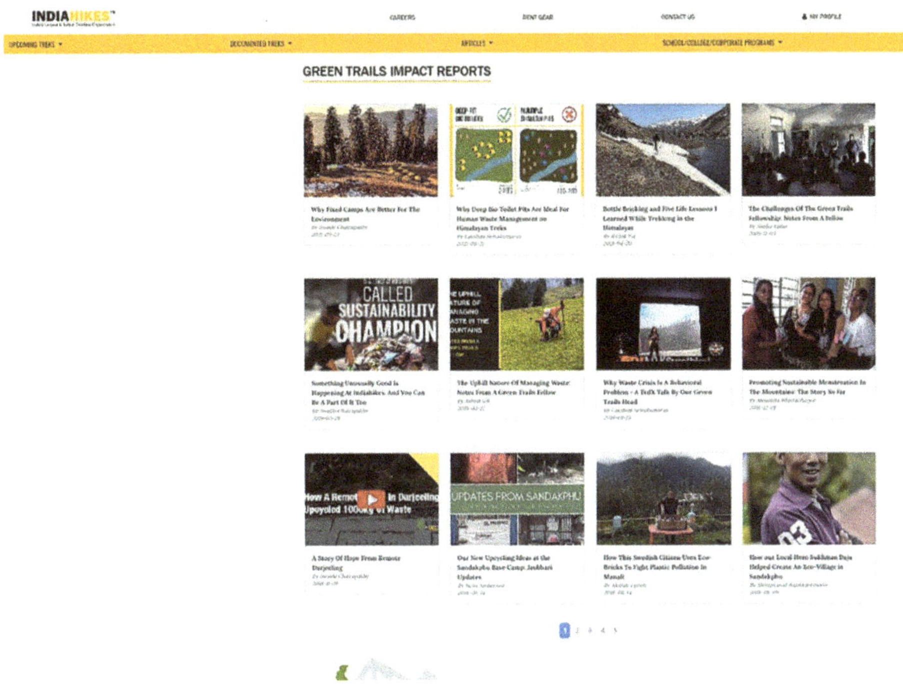

Fig. 1.4. Green trails database (courtesy of India Hikes: https://indiahikes.com/category/green_trails_impact_reports).

waste materials. During treks, the examples shown by the trek leaders motivate the fellow trekkers to engage in sustainable practices. This immersive experience of the trek team sheds light on the transformative learning theory (Mezirow, 1991) which reflects on the changes in perspectives adults go through when having an experience. The sustainable initiatives of Indiahikes lead to significant changes in the everyday life of fellow trekkers, the community in the villages where the camps are set up, and also the employees who work for the firm.

The impacts of transformation are seen in the remote villages, where the team engages in awareness programmes to develop eco-sensitive living because of the difficulty in managing waste on mountains. In the villages, the project has developed sustainable menstruation,[4] upcycling products, building bottle bricks to make infrastructure, and encouraging local vendors to reduce the amount of waste generation. Lohajung in Uttarakhand and Jaubhari in Sikkim are examples of villages where the project has created substantial impacts which others can learn from.

Community engagement

As one of their major goals, green trails put a lot of emphasis on encouraging community involvement. Community forums are used as an effective way to educate people about sustainable trekking practices and foster a strong sustainable trekking community. These forums act as venues for information exchange, raising awareness, and the instillation of a feeling of responsibility for protecting natural habitats.

The influence of the project extends beyond the trekkers through eco clubs at the schools in remote villages, encouraging students to develop a lifelong commitment to eco-friendly habits. The project also taps the potential of rainwater harvesting, which is impactful in the mountains where water resources are scarce.

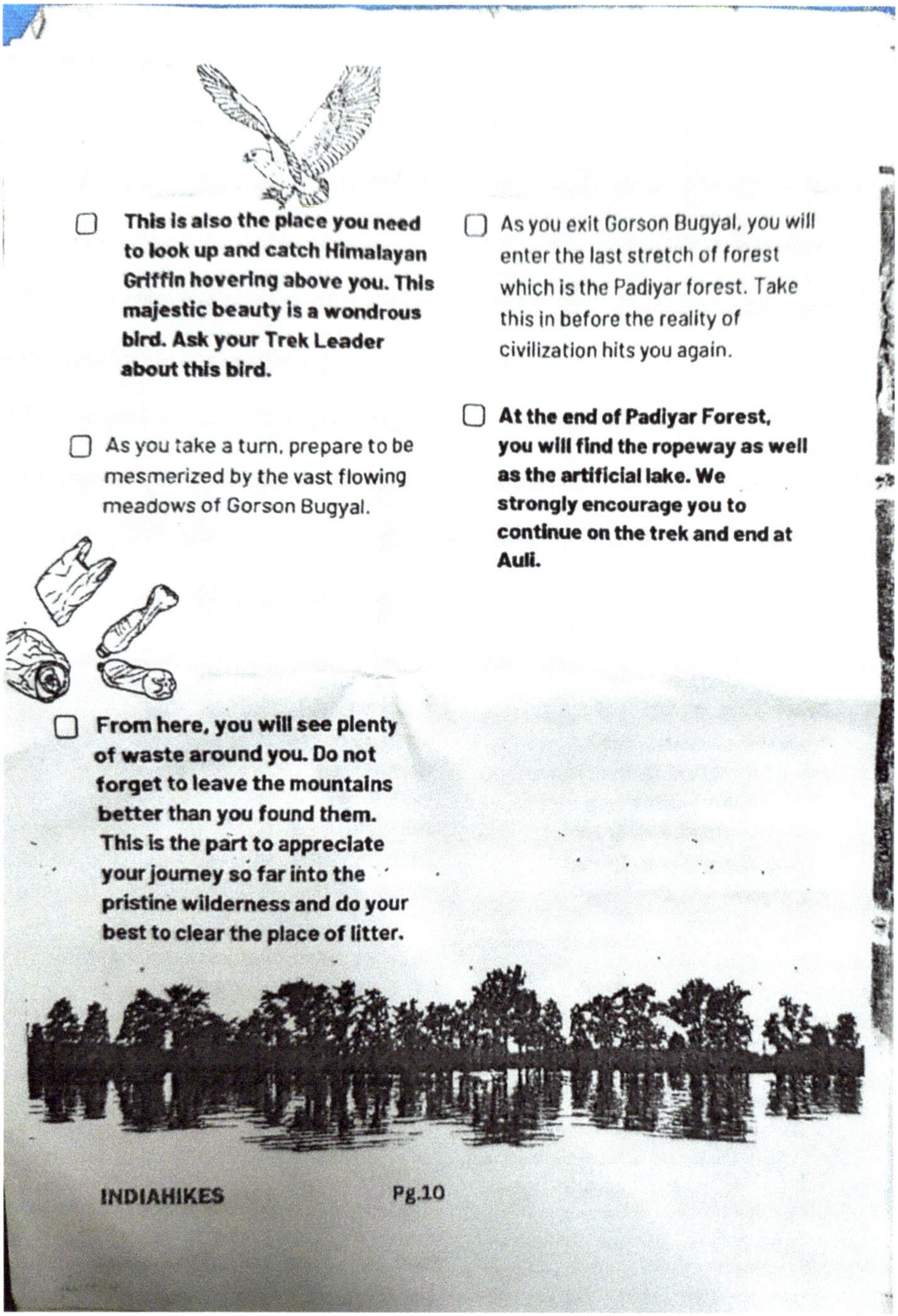

- [ ] **This is also the place you need to look up and catch Himalayan Griffin hovering above you. This majestic beauty is a wondrous bird. Ask your Trek Leader about this bird.**
- [ ] As you take a turn, prepare to be mesmerized by the vast flowing meadows of Gorson Bugyal.
- [ ] As you exit Gorson Bugyal, you will enter the last stretch of forest which is the Padiyar forest. Take this in before the reality of civilization hits you again.
- [ ] **At the end of Padiyar Forest, you will find the ropeway as well as the artificial lake. We strongly encourage you to continue on the trek and end at Auli.**
- [ ] **From here, you will see plenty of waste around you. Do not forget to leave the mountains better than you found them. This is the part to appreciate your journey so far into the pristine wilderness and do your best to clear the place of litter.**

INDIAHIKES Pg.10

Fig. 1.5. Highlighting sustainability while trekking – A page from brochure for Kuari pass trek of Indiahikes, Courtesy of Indiahikes.

Equally impactful is the outdoor leadership programme for schools and colleges, which helps to create strong and responsible connections with nature. The various sustainability initiatives of Indiahikes can also be looked upon in relation to place-building theory, which was originally put forward to assess corporate social responsibility (Thomas, 2004). The theory is concerned with how much an organization appreciates and contributes to its local community. By working towards achieving the goals of sustainability, the organization also undertakes to make its social and geographical location better. Within the context of place building theory, this case provides an example of a contributive organization focused on giving back to the community. Also through the efforts of the green trails internship and green trails fellowship schemes, trekking enthusiasts are trained in sustainable trekking practices. This helps individuals who are willing to work in the field of sustainability, where they get hands-on experience and understand the associated challenges.

Conclusions

Indiahikes, as a leading trek team in India, stands out through its efforts to promote sustainability through their trekking endeavours. Their initiatives focus not only on enhancing trekking experiences but also prioritize the preservation of nature, empowerment of local communities, and upliftment of villages. Their efforts revolve around showcasing improved trekking methods that leave minimal impact on the environment. By adhering to eco-friendly practices, they actively contribute to the conservation of the natural beauty of the mountains and surrounding areas, in line with the United Nations Sustainable Development Goals.

The organization gives an excellent example of stakeholder initiative for sustainable tourism. Additionally, their approach is characterized by a sense of responsibility, aiming for trekkers not just to enjoy the mountains but to leave them in a better state than they were found. Empowering everyone associated with the treks leaves the individuals with a transformative experience which translates to their everyday lives as well.

Notes

[1] The term used by the organization for the activity of cleaning the trail of visible litter.
[2] The research has completed all ethical approvals and received permission from the organization to be referenced in the case study.
[3] Coco peat refers to a product derived from coconut husks, which is widely used as a growing medium or substrate for plants. Composting will happen faster because of its high carbon content.
[4] The initiative is to create awareness about menstrual cups and encourage the villagers to switch from sanitary pads to cups, which are more economical and environmentally friendly.

References

Biswas, C., Roy, K., Dutta, R. and Gayen, S.K. (2022) An exploratory analysis of mountaineering risk estimation among the mountaineers in the Indian Himalaya. In: Mandal, S., Maiti, R., Nones, M. and Beckedahl, H.R. (eds) *Applied Geomorphology and Contemporary Issues*. Springer Link, Cham.

Freeman, R.E. (1984) *Strategic Management: A Stakeholder Approach*. Cambridge University Press, Cambridge.

Indiahikes (n.d.) Indiahikes – India's Safest & Largest Trekking Organisation. Available at: https://indiahikes.com/ (accessed 13 June 2023).

Indian Mountaineering Federation (2018) Indian Mountaineering Foundation. Available at: https://www.indmount.org/IMF/mountaineering (accessed 17 October 2023).

Kohli, M.S. (2002) *Mountains of India*. Indus Publishing, New Delhi.

Mezirow, J. (1991) *Transformative Dimensions of Adult Learning*. Ed.gov. Jossey-Bass, 350, Sansome Street, San Francisco, CA. Available at: https://eric.ed.gov/?id=ED353469

National Geographic Education (n.d.) Mount Everest. Available at: https://education.nationalgeographic.org/resource/mount-everest/#:~:text=The%20first%20ever%20recorded%20people (accessed 17 October 2023).

Nepal, S.K. (2002) Mountain ecotourism and sustainable development. *Mountain Research and Development* 22(2), 104–109. DOI: 10.1659/0276-4741(2002)022[0104:MEASD]2.0.CO;2.

Ozdemir, M.A. and Goktas, L.S. (2021) Research trends on digital detox holidays: A bibliometric analysis, 2012–2020. *Tourism Management Studies* 17(3), 21–35. DOI: 10.18089/tms.2021.170302.

Shah, R. (2020) How India's hiking industry is changing? *Medium*. Available at: https://medium.com/@revant248/how-indias-hiking-industry-is-changing-9b4776cceda3#:~:text=In%20fact%2C%20according%20to%20a (accessed 17 October 2023).

Steiger, R., Knowles, N., Pöll, K. and Rutty, M. (2022) Impacts of climate change on mountain tourism: A review. *Journal of Sustainable Tourism* 30, 1–34. DOI: 10.1080/09669582.2022.2112204.

Thomas, D.F. (2004) Toward an Understanding of Organization Place Building in Communities., [Dissertation] Colorado State University. Available at: https://www.proquest.com/openview/b4fd379afb7d99dd8335ee4782a96281/1?pq-origsite=gscholar&cbl=18750&diss=y (accessed 13 June 2023).

United Nations (2022) Goal 15. United Nations. Available at: https://sdgs.un.org/goals/goal15 (accessed 5 July 2023).

United Nations Environmental Programme (2008) UNEP 2007 Annual Report. Available at: https://wedocs.unep.org/20.500.11822/7647 (accessed 1 July 2023).

UNWTO (2023) *Understanding and Quantifying Mountain Tourism*. World Tourism Organization. DOI: 10.18111/9789284424023.

Ziegler, A.D., Wasson, R.J., Sundriyal, Y., Srivastava, P., Sasges, G. *et al.* (2021) A call for reducing tourism risk to environmental hazards in the Himalaya. *Environmental Hazards* 22(1), 1–28. DOI: 10.1080/17477891.2021.1984196.

2 Sustainable Tourism and Stakeholders' Satisfaction in Batulayang Tourism Village, Bogor, Indonesia

Ade Suherlan
University of Western Sydney, Penrith, Australia

Summary

Desa Wisata Batulayang (Batulayang Tourism Village) is an emerging tourist destination in Bogor Regency, West Java, Indonesia, offering a unique experience of rural life, traditions, and nature. Developing such tourism villages is vital for the community, providing economic benefits, alleviating poverty, and unemployment, and preserving the environment and culture. Enhancing the availability of job opportunities in rural areas by creating sustainable tourism villages will expand the range of services such as healthcare and education. Consequently, this is expected to reduce the tendency of rural communities to migrate to urban areas.

Desa Wisata Batulayang adheres to sustainable tourism principles to ensure the village's long-term sustainability. The success of sustainable tourism hinges on active involvement, satisfaction, and support from various stakeholders. Involving stakeholders in decision making promotes the village's viability, community development, resource preservation, and economic prosperity.

The Value and Interest of the Case Study

This case study sheds light on the complexities of sustainable tourism development in Batulayang tourist village. By exploring sustainable tourism principles, stakeholder involvement, and factors influencing success, readers gain valuable insights into effective tourist village development in Batulayang. Prioritizing stakeholder involvement, support, and collaboration throughout the planning process is crucial for achieving success in sustainable tourism villages.

A Brief Methodology

The data in this case study was acquired by systematically searching diverse online sources, encompassing websites, social media platforms, and YouTube. In addition, primary data were obtained through personal communication with Rusmana, A., the head of the Tourism Awareness Group, or *Pokdarwis,* an acronym for *Kelompok Sadar Wisata* (means Tourism Awareness Group), conducted from 16 July to 24 July 2023. The data was analysed through content analysis (Krippendorff, 2004; Kohlbacher, 2005).

The Tourism Trends and Tourism Villages Policy in Indonesia

After the COVID-19 pandemic, the Indonesian government implemented various policies to bolster tourism recovery. These included measures like the policy stimulus of the National Economic

Recovery Program (NERP), the Proud Tourism Movement in Indonesia, support for micro, small, and medium enterprises (MSMEs) and the local tourism sector, technological innovations in e-tourism, and CHSE Certification (Cleanliness, Health, Safety & Environment Sustainability). These programmes are being implemented in Tourism Indonesia, including developing tourism villages in Bogor Regency, West Java.

Tourism Village represents a prominent facet of Indonesia's strategic agenda for tourism sector development, underlined by an array of governmental policies and regulatory frameworks (Government Regulation, 2011; RPJMN, 2020–2024; Permenparekraf, 2021, Permenparekraf, 2022; Permendesa PDTT, 2022). In the new normal era, tourist villages are seen as a flagship programme to accelerate tourism revival and stimulate economic growth in Indonesia. Globally, tourism trends are shifting towards nature and local culture for unique experiences (Arora and Sharma, 2021), including in Indonesia. Village tourism aligns with these trends and has significant potential to meet tourists' needs and cater to global trends. Aside from resilience, the tourism industry is further expected to transform into a sustainable industry capable of fostering community welfare, addressing climate change, and engaging local communities as agents of transformation (Sharma *et al.*, 2021).

Village tourism was selected as a priority in Indonesia's development strategy, employing the 'building from the village' approach. This approach represents a cohesive endeavour to attain national sustainable development goals (SDGs) (Bakti, 2020). Employing a people-centric development strategy and fostering community groups, development is directed towards the village itself. Under this framework, villages are no longer regarded as passive recipients of development but are designated as active agents and vanguards of progress in enhancing community well-being. Villages possess the autonomy to oversee their development in alignment with the specific circumstances prevailing in each village (Law No. 6/2014 (Law, 2016)).

The global and national SDGs do not explicitly encompass preserving local customs and traditions within villages. To address this, the Village SDGs introduce an additional dimension (Village SDG 18: Dynamic Village Institutions and Adaptive Village Culture), which focuses on regulating and preserving local wisdom. This empowers village authorities to nurture their communities harmoniously within the rich tapestry of local traditions. This decision holds significant political implications to improve overall well-being. It seeks to achieve this by fostering fair income distribution, creating employment opportunities, alleviating poverty, enhancing human resource quality, ensuring equitable access to social services, boosting village competitiveness, and ultimately enhancing the quality of life for rural communities. This focus arises from the notable disparities between urban and rural regions within Indonesia.

Community groups initiate the development of tourist villages in Indonesia through village meetings *or musyawarah desa*. The outcomes of these meetings are then proposed to the local government agencies responsible for tourism (Tourism Department) and the agencies responsible for governance and community empowerment through proposals for tourist village development. Subsequently, if these proposals successfully meet the established criteria to attain tourist village status, they are officially conferred with such recognition through a decision made by the regent or mayor.

Rural tourism is rising, with tourists seeking safe, comfortable, clean, and sustainable villages. Developing tourism villages requires collaboration (McComb *et al.*, 2016), particularly with the community (Bramwell and Lane, 1993). Community-based tourism empowers communities to manage tourism for economic, social, and environmental welfare (Ngo and Creutz, 2022).

In a bottom-up collaboration, the village takes the lead in development. Active community participation, understanding of the village's potential, and local wisdom are crucial for tourist village development (Vitasurya *et al.*, 2018). Community participation can be defined as the involvement of all elements in the decision making process regarding matters to be addressed and their implementation methods (Maryati, 2005). The community must actively design, plan, manage, innovate, evaluate, and continuously improve (Stone, 2015). Each tourist village should have a Tourism Awareness Group or *Pokdarwis* to bridge policies, enhance tourism awareness, improve human resources quality, and create job opportunities. *Pokdarwis* is a community-level institution comprising tourism stakeholders that actively drive the creation of a favourable environment for tourism growth and development, aligning with the *Sapta Pesona* (seven charms) principles (the Guidelines of Tourism Awareness Group or Pedoman Kelompok Sadar Wisata, 2012, p. 16). This entity plays a crucial role in advancing regional development through tourism, enhancing the well-being of the local community, and realizing tourist villages.

The Ministry of Tourism and Creative Economics (*Kemenparekraf*) of the Republic of Indonesia supports this bottom-up collaboration through various policies and programmes, including roadshows to 50 tourist village locations in 2022, launching the *'Beti Dewi'* (Buy Creative Village Tourism) programme, *Pak Wisnu (Paket Wisata Nusantara)* or Nusantara Tourism Package offering around 100 destinations and the Indonesia Tourism Village Award (*Anugerah Desa Wisata Indonesia – ADWI*). This collaboration aims to integrate regional and central programmes for equitable regional tourism development. These programmes were followed by several government regulations, such as accelerating COVID-19 vaccination, easing travel regulations, and the CHSE programme. Collaboration, mainly via partnerships involving multiple stakeholders, has been recognized as a successful approach to promoting initiatives related to the advancement of tourism (Graci, 2013). Effective collaboration and synergy among all stakeholders are essential for creating beneficial and well-timed policies. Successful tourism village development relies on strategic planning of central and regional policies (Liu *et al.*, 2023), with active community involvement (Nagarjuna, 2015). Adaptation, innovation, and cooperation should be customized to the village's unique characteristics and potential, ensuring resilience in the new normal. Tourism village development is expected to attract both domestic and foreign tourists.

The Growth of Tourism Village in Bogor Regency

Village tourism represents an emerging trend that hinges on the synergy between nature, culture, and local communities. Currently, tourism villages in Bogor Regency are progressively establishing greater autonomy and resilience. The Bogor Regency Government is actively pushing the tourism sector forward through the *'Karsa Bogor Maju'* programme (Diskominfo Kabupaten Bogor, 2021). In 2019, there were 25 tourist villages. This number increased to 35 in 2020, expanded to 40 in 2021, and reached 41 in 2022 (Diskominfo Kabupaten Bogor, 2021). This growth reflects a prevailing trend and a targeted effort by the Bogor Regency Government to invigorate the rural economies. This upward trajectory is set to continue in 2023.

The development of tourist villages in Bogor Regency is categorized into six regions: the southern: (1) *Tugu Selatan* Village to *Ciawi*; (2) *Ciawi* to *Sukabumi* district border; the western: (3) *Tamansari* to *Pamijahan*; (4) *Ciampea-Nanggung-Sukajaya*; and (5) *Iwul* and *Jampang* Tourism Villages. Additionally, the eastern region (6) includes *Pasir Mukti, Tarikolot, Linggar Mukti,* and *Bojong Kulur* (Widyanti, 2022). The Bogor Regency Culture and Tourism Office has been actively promoting the empowerment of tourist villages since 2009. From 2012 until today, tourism villages in Bogor Regency have started to gain popularity and recognition.

The administration of tourist villages in Bogor Regency operates dynamically, initiated autonomously by the local community. Subsequently, the Bogor Regency Tourism and Culture Office steps in to help and deliver training. From then on, the Bogor District Tourism and Culture Office provides community groups with instruction in the principles of tourism villages, destination management, and human resource management.

Fig. 2.1 shows that the number of tourist visits to Batulayang tourist village steadily increased until 2019. However, there was a decline in 2020 and 2021 due to the COVID-19 pandemic. From 2022 to the present, there has been a gradual increase as the village recovers after the end of the pandemic. The number of tourists is predicted to continue to increase in 2023. In relation to this, Humaedi (2023) predicts that the number of tourist visits to Bogor Regency is expected to increase by 20% compared to the preceding year (in 2022, 8,150,000 people visited). It is expected that

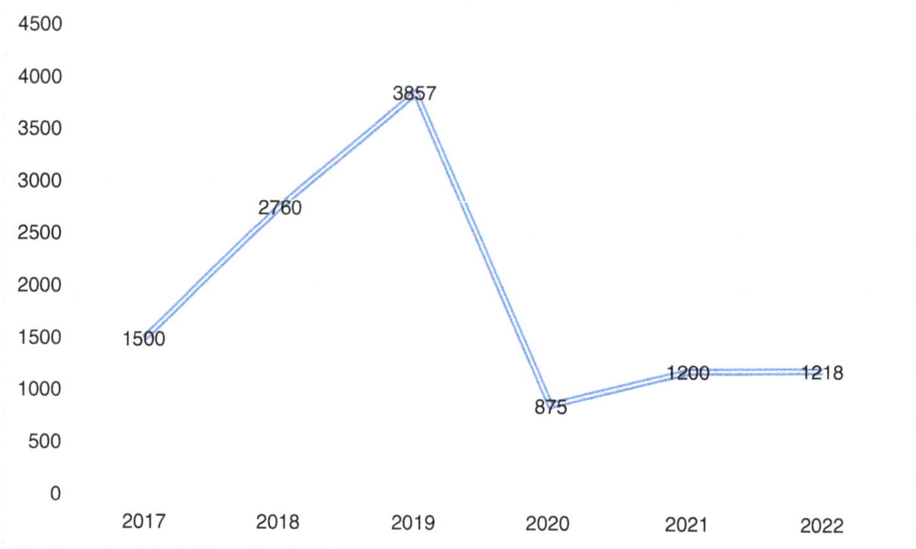

Fig. 2.1. The number of Batulayang visitors (© Bahri *et al*., 2023; Rusmana, A., Bogor, 16–24 July 2023, personal communication).

50% of these visitors will travel to the Puncak area, where Batulayang tourist village is located (Rama, 2023). Batulayang Tourism Village is considered one of the most exemplary sustainable tourism villages in Bogor Regency. In 2023, Batulayang Tourism Village earned the distinction of being one of the eight tourism villages chosen to represent Indonesia at the UNWTO Best Tourism Villages (Setiawan, 2023).

Profile of Batulayang Tourism Village

Batulayang tourist village is situated in *Kampung Pasir Manggis*, Batulayang Village, Cisarua District, Bogor Regency, West Java, Indonesia. Nestled in the Bogor highlands, it is approximately 72 km from Jakarta (Jakarta is the capital city of Indonesia), accessible in about 2.5 hour by car (Fig. 2.2).

Most of Batulayang Village's land consists of dry lands (moorlands, fields, hills, and yards) (Sianturi, 2007). Buildings and structures cover around 66 ha, while rice fields span roughly 26 ha. The village features wetlands covering about 4 ha and approximately 0.40 ha of land for public facilities (Sianturi, 2007).

Batulayang Village is adjacent to *Megamendung* Village in the north, *Cibereum* Village in the south, *Cisarua* Village in the west, and *Tugu Utara* Village in the east. This village is 800–1,000m above sea level. The terrain in this area exhibits a diverse topography, ranging from flat to undulating, with a 40% slope. Rainfall is experienced approximately 90 days a year, with an average of 33.36 mm/day or 3,002 mm/year (Sianturi, 2007). According to the Schmidt-Ferguson climate classification, Batulayang encounters an annual precipitation of 4,000 mm, peaking in February at 550 mm and reaching its lowest point in August at 200 mm. The air temperature in this area typically spans from 18 to 24°C (Sianturi, 2007). Situated adjacent to a protected forest, Batulayang Village preserves its innate natural splendour.

The Batulayang Tourism Area is adorned with diverse vegetation, encompassing various species of forests, greatly enhancing its allure. With a population of approximately 9,500 residents (RBG, 2023), Batulayang Village is subdivided into two hamlets, comprising 4 *RWs* (neighbourhood associations) and 23 *RTs* (residential units) (Rusmana, A., Bogor, 16–24 July 2023, personal communication). It is

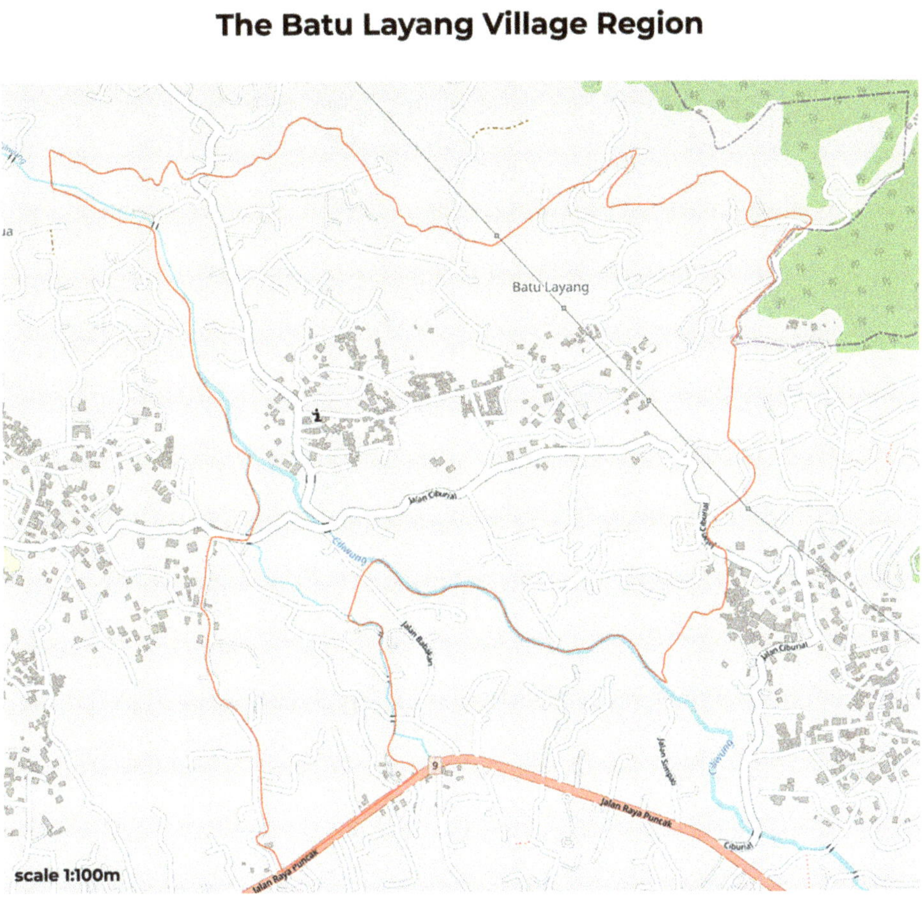

Fig. 2.2. The map of the Batulayang Region, Cisarua, Bogor, Indonesia (courtesy of Dinas PUPR Kota Depok (Dinas Pekerjaan Umum dan Perumahan Rakyat Kota Depok; Public Works and Housing Office of Depok City), 2023 , modified from basic map PUPR by Fatchan Ridho Fadillah, 2023).

noteworthy that the Ministry of Tourism and Creative Economy has classified its ecotourism potential as 'Advanced.' (Anon, n.d). The village derives substantial advantages from its inherent natural conditions. Therefore, what is imperative is the proficient management of these resources by all elements of the local community.

Batulayang offers attractions like waterfalls, the Cimandala River, and scenic landscapes (Anon, 2022). From hiking and glamping to cultural experiences and craft-making, Batulayang Tourism Village is a versatile destination, offering jungle tracks, glamping, outbound activities, off-roading, cultural education, *jaipong* dance, *angklung* music, craft workshops, *pencak silat*, and experiences of Sundanese traditional cuisines and beverages (Intan, 2021; Pamela, 2022; Bahri *et al.*, 2023). The preservation of nature and culture and the oversight of tourist attractions within Batulayang tourism village falls upon the collective responsibility of all relevant stakeholders, specifically the village government, *Pokdarwis*, NGOs (ecovillage community), the local populace, visitors, and *Perhutani*. *Perhutani* plays a pivotal role in the context of Batulayang tourist village, which represents a village forest area under the ownership of *Perum Perhutani* (Anon, 2022). *Perhutani*, a state-owned enterprise, manages Java and Madura islands' state forest resources to sustain ecological, social, and economic systems, upholding constitutional principles (the 1945 State Constitution of the Republic of Indonesia) (Anon, 2023). Batulayang Village boasts abundant natural beauty and cultural riches. In this context, it becomes imperative for all major stakeholders to exhibit creativity and innovation, thereby guaranteeing the long-term sustainability of tourism in this village.

The Development of Batulayang as a Tourism Village

Batulayang Tourism Village, nestled in the renowned Puncak region of *Cisarua*, has risen as an outstanding tourist destination, celebrated for its bountiful natural and cultural riches. Originally an agricultural centre, the village astutely leveraged its resources, prompting a shift towards tourism. In 2002, the village adopted the *Dusun Wisata* or Tourism Village concept to fully harness its potential (Adira *et al.*, 2022). Recognizing Batulayang as a tourist village includes several stages, including submitting proposals by community groups to the village government, verification by regional authorities, and satisfying the stipulated criteria. Subsequently, in 2014, the village attained official recognition as a Tourism Village, as the Bogor Regency Tourism Office decreed under decree number 142.1/02/kpts/HUK.2014.

Through the collective endeavours of community groups, village authorities, and district officials, Batulayang Tourism Village has attained noteworthy achievements. In 2018, Batulayang Village Tourism received the first runner-up prize in the community-based tourism category as part of the Community Business Appreciation Award within the field of tourism. This Award is a village tourism national-level initiative by the Ministry of Tourism (Dwiyanto, 2018; Adira *et al.*, 2022).

Additional recognition came in 2019 when the village was honoured with an award for its climate village programme by the Ministry of Environment. The crowning achievement arrived in 2020, with the certification from the Indonesia Sustainable Tourism Council, Ministry of Tourism and Creative Economy, as a sustainable tourism village, following the criteria outlined in Regulation of the Minister of Tourism and Creative Economic Number 14 of 2016.

Academic, community, and media support played a pivotal role in propelling this development. Additionally, the invaluable contribution of the 'Ecovillage' community in preserving the village's natural environment has been instrumental in promoting sustainability and environmental awareness (Adira *et al.*, 2022). 'The Batulayang Ecovillage Community' is a community organization dedicated to the preservation and protection of the natural environment, operating in a non-governmental capacity. This group actively engages in activities that include maintaining the cleanliness of tourist destinations, managing waste through waste banks (emphasizing the principles of reducing, reusing, and recycling), and facilitating various other environmental conservation initiatives (Adira *et al.*, 2022; Rusmana, A., Bogor, 16–24 July 2023, personal communication).

Institutions and Key Stakeholders in Batulayang

Batulayang engages with several prominent stakeholders and institutions such as governments at all levels (national, regional, and village), *Pokdarwis*, *Perhutani*, Ecovillage Community, academia, and media (Latianingsih *et al.*, 2022; Rusmana, A., Bogor, 16–24 July 2023, personal communication). The *Pokdarwis* operate under the endorsement of the village and tourism office, actively fostering tourism awareness and implementing the 'sapta pesona' (seven charms: safe, orderly, clean, cool, beautiful, friendly, and memories) principles (Permenbudpar, 2008; Kepmenparpostel, n.d.). Sapta Pesona serves as a framework to promote tourism awareness, particularly emphasizing the community's role as hosts in fostering a favourable environment and ambience conducive to the growth and development of the tourism industry (Setiawati and Aji, 2020).

Moreover, *Perhutani* oversees forest resources within Batulayang, while the Ecovillage community is dedicated to safeguarding the natural environment. The government plays a pivotal role in shaping policies and managing the tourism village. At the same time, academia contributes to enhancing human resources, management, promotion, and innovation for Batulayang's ongoing development.

The Benefits of Batulayang Tourism Village

It is essential to recognize that the prevailing situation, exemplified by Batulayang Tourism Village in Bogor, underlines that the growth of emerging tourism villages often tends to benefit specific groups, particularly those directly involved. Consequently, noticeable disparities, particularly in terms of economic benefits, endure. These disparities have spawned inequalities within the local population, posing the potential for heightened social conflicts if government bodies and tourism village administrators neglect to address this pressing issue.

Nonetheless, the progress of Batulayang Tourism Village has yielded numerous benefits for the community and stakeholders in various ways. These include an enhancement of the general welfare, as seen through increased employment opportunities and augmented income, including earnings from homestay accommodations, foreign exchange gains, contributions to government revenues, and infrastructure development (Dewantara and Susanto, 2020). Furthermore, it has led to the preservation of the natural environment, encompassing the safeguarding of water systems and hydrology, soil fertility, erosion, and flood control, as well as the preservation of the area's beauty, uniqueness, and carbon absorption and storage (CNN Indonesia, 2021; Setiawan, 2023).

The Batulayang Tourism Village has also made strides in conserving its cultural heritage, encompassing Sundanese art, music, dance, customs, and culinary traditions. Additionally, this tourism village has played a pivotal role in enhancing community knowledge and overall quality of life, particularly in education and healthcare. Furthermore, it has stimulated the growth of small and medium-sized industries centred around local products within Batulayang village. Finally, it has contributed to the heightened promotion of the village and its locally produced goods to a broader market, achieved through social media networks and word-of-mouth marketing.

Through the comprehensive development of this tourist village, stakeholders can discern a noticeable improvement, particularly in the economy. Nonetheless, it cannot be denied that during this journey, the consequences of this development have been notably felt by specific stakeholders, namely *Perhutani*, the Ecovillage Community, *Pokdarwis*, the village government, and operators of micro, small, and medium-sized enterprises (MSMEs) encompassing homestays, handicrafts, glamping, tourist attractions, and culinary establishments. For *Perhutani*, it is evident that they derive benefits from preserving their forested area, resulting in reduced maintenance efforts and an enhancement in forest conservation education for both the local community and visitors. On the other hand, the village government benefits from promoting the tourist village as it elevates the village's image, contributes to the attainment of development objectives, and enhances the village's reputation. Meanwhile, for the local community, the direct advantages manifest as increased income, job opportunities, and a broad array of economic activities.

Supporting and Constraining Factors

The achievement of tourism village development, as exemplified in Batulayang, hinges upon many factors. These encompass community and regional income, investment levels, community proficiency, a harmonious and conflict-free environment, active community support and engagement, preservation of traditional values and local wisdom, land usage patterns, geographical conditions, pollution levels, technology adoption, community awareness, and willingness. Additionally, they include economic and environmental potential, the availability of tourist attractions, government commitment, management adherence, accountability, and transparency.

The success of Batulayang primarily derives from the active participation and consciousness of the community in managing the village (Latianingsih *et al.*, 2022). The community initiated the establishment of the village, and the Bogor Regency Tourism Village Association consistently lends its support to its development.

Adira *et al.* (2022) shed light on the pivotal factors underpinning the success of Batulayang Tourism Village. These encompass accessibility, appeal, central amenities, ancillary facilities, infrastructure, management of service systems, institutional components, and contributions to environmental quality. Nonetheless, to elevate the quality of services, the management should prioritize optimizing various elements, such as the condition of road networks, cleanliness of restrooms, availability of litter bins, the quality of telecommunications networks, and the standards upheld by accommodation service providers.

Additional critical factors that contribute to the prosperity of the tourism village encompass the aesthetic allure of the locale, showcasing attractions such as the *Kembar* waterfall, *Asmarandana* waterfall, and the *Cimandala* River. Competent tour guides, maintenance of cleanliness and security, the affability of residents and guides, access to hand washing facilities, active environmental conservation initiatives, and education on waste treatment also play crucial roles in sustaining the village's appeal (Adira *et al.*, 2022).

Tourism villages face significant challenges for successful development. Inadequate infrastructure, including transportation and basic facilities (Risdawati *et al.*, 2020), hinders progress and discourages tourists. A lack of knowledge, community engagement, and local perception, combined with a low level of community attitude (characterized by low responsiveness, reluctance to participate, and refusal), conflicts, policy and legislative discrepancies, low product/service quality, and insufficient collaboration among stakeholders (Tamir, 2020), represent substantial barriers. Moreover, Ngonya (2015) contends that insufficient government and community support, insufficient investment in education and training, infrastructure deficiencies, ineffective marketing strategies, security concerns, and suboptimal customer service further exacerbate these challenges.

In the context of Batulayang Tourism Village, several enduring challenges persist. These encompass issues related to accessibility, attractiveness, core amenities, auxiliary facilities, infrastructure, and the management of service systems, including the cleanliness of sanitation facilities, the provision of adequate waste disposal facilities, the quality of telecommunications networks, and the performance of accommodation service providers (Adira *et al.*, 2022). Moreover, these challenges extend to institutional components and their contributions towards preserving and enhancing environmental quality (Adira *et al.*, 2022).

Critical challenges for tourism villages include uncoordinated management and a need for more understanding of sustainable tourism practices. Environmental degradation, cultural identity erosion, and exploitation of local communities are prevalent issues. To address these challenges, collaboration among the government, local communities, *Pokdarwis*, entrepreneurs, and academia is vital, focusing on environmental sustainability, effective resource management, and community involvement. Inhibiting factors include the need to optimize cooperation and coordination among stakeholders and maximize the role of the private sector (Latianingsih *et al.*, 2022).

Stakeholders Involvement and Strategy

The success of Batulayang Tourism Village involves various stakeholders: the community, village government, *Cisarua* Tourism Activist Group, Tourism Village Management Group, Village-owned enterprises or *BUMDesa*, Bogor Regency Tourism Office, *Podomoro* University and *Politeknik Negeri Jakarta*, and the Batulayang Ecovillage Community (Latianingsih *et al.*, 2022; Adira *et al.*, 2022). They all play a vital role in its development and management.

Academics from *Podomoro* University have been collaborating with the Tourism Village Manager at Batulayang from its inception. Annually, they organize mentoring programmes focused on community development, boosting tourism awareness, and offering training for small and medium-sized enterprises (SMEs). These initiatives elevate the quality of service and entrepreneurship within Batulayang Tourism Village. Similarly, *Politeknik Negeri*

Jakarta has been involved in management consultancy and enhancing human resources.

Furthermore, *Perhutani* and *PT Jasa Puri* partake in profit-sharing arrangements due to their contributions to the tourist attractions in the area, specifically the *Kembar* waterfall and the *Ujung Rimba* Camp glamping site.

The Cisarua District Tourism Activist Group has played an active role in extending support to Batulayang Tourism Village, particularly emphasizing community development and promoting its tourism potential. The Village Government is pivotal in forming the village's management framework. In tandem, the Tourism Village Manager collaborates with online media, including CNN, to efficiently disseminate information and oversee tourism activities in Batulayang. Meanwhile, *BUMDesa* function similarly to businesses, with directors appointed by the village government utilizing village assets for various ventures, including tourism (Rusmana, A., Bogor, 16–24 July 2023, personal communication).

Effective stakeholder communication and collaboration are vital for successful tourism development and management in Batulayang (Adira *et al.*, 2022). In rural tourism, such collaboration is the key to addressing unique challenges that can impede success (McComb *et al.*, 2016). Therefore, to create an environmentally conscious village, Batulayang Tourism Village management and the local government hold quarterly deliberation sessions to discuss and resolve environmental issues, fostering mutual consensus, understanding, and trust among all parties involved (Adira *et al.*, 2022).

The local government acts as a facilitator, overseeing tourism operations and collaborating with other stakeholders to offer services and address issues (Liu *et al.*, 2020). In Batulayang, the village government assumes the role of a facilitator and mediator, acting as an intermediary between the community, the Tourism Village manager, and the environmental group (Rusmana, A., Bogor, 16–24 July 2023, personal communication). In this capacity, it ensures that the concerns of all parties are acknowledged and dealt with, thereby promoting efficient responses to environmental issues. The village government, the tourism village management, and the ecovillage group are responsible for tackling environmental concerns (Rusmana, A., Bogor, 16–24 July 2023, personal communication). Their collaborative efforts are centred on crafting a culturally aware village while safeguarding natural resources and cultural sustainability. The triumph in the development of Batulayang Tourism Village largely depends on their harmonious endeavours to establish an environmentally conscious community (Adira *et al.*, 2022).

In addition to their role as a facilitator and mediator, the village government, in partnership with the Bogor Regency Tourism Office, actively promotes Batulayang Tourism Village to attract more tourists and enhance regional tourism. The political interests behind this are centred on enhancing the village's image and promoting it to gain the attention of relevant parties for the advancement and development of the village in the future (Adira *et al.*, 2022).

Tourism Village Managers involve the local community in roles like SMEs, tour guides, and supporting businesses, boosting the village's economy. The ecovillage group contributes by organizing regular cleaning activities at tourist spots like *Asmarandana* waterfall and *Cimandala* River every 2–3 months. They also manage waste through a *Bank Sampah* (Waste Bank), correctly processing and separating organic and non-organic waste (Adira *et al.*, 2022; Rusmana, A., Bogor, 16–24 July 2023, personal communication).

Sustainable Tourism Certification of Batulayang Tourism Village

Sustainable tourism certification is essential for Batulayang tourism village. Certification carries substantial significance as it serves to heighten awareness regarding the adoption of sustainable tourism practices, bolster the preservation of the planet and environmentally conscious travel, acknowledge adherence to exacting quality standards, and promote responsible tourism (ISTC, n.d.). The certification criteria are based on globally recognized standards established by the Global Sustainable Tourism Council (GSTC) and endorsed by the United Nations World Tourism Organization (UNWTO) as adopted in the Permenparekraf (2016). It includes four primary categories, further divided into 10 subcategories, 38 criteria, and 174 indicators (Gambino, 2020). Certification is awarded through assessments by the Indonesian Sustainable Tourism Council (IST-Council), a government-accredited body (ISTC, n.d.).

In Indonesia, Law No. 10/2009 (Law, 2009) mandates the standardization of tourism services and businesses through Article 54, subsection (2), which outlines business certification for establishing and implementing standards. Sustainable tourism village certification aims to boost tourist satisfaction and stakeholder loyalty, fostering the development and enhancement of the tourism village (Habibie *et al.*, 2021). Certification implicitly protects managers, securing stakeholder support for village management (Laksmi, 2023). Moreover, sustainable tourism certification can boost a destination's image through branding (Bowman, 2010), where the destination brand shapes perceptions (Leisen, 2001).

Batulayang Village fulfils the certification criteria stipulated in Permenparekraf (2016). It is the sole village in West Java chosen by the Ministry of Tourism as a certified sustainable tourism village in November 2020, valid until November 2023. Sustainable tourism certification benefits Batulayang village by raising awareness, promoting eco-friendly travel and accommodation, recognizing high-standard businesses, and improving marketing and partnerships for sustainable tourism in Indonesia.

Stakeholders' Satisfaction

The performance of sustainable tourism in Batulayang tourism village has varying impacts on stakeholder satisfaction. Positive effects lead to higher satisfaction, while harmful consequences, like adverse impacts on the environment and social aspects, can result in disappointment over the short and long term. Stakeholder satisfaction becomes evident through increased participation and involvement in the development activities of the tourism village. The community's economic improvement and enhanced quality of life contribute to stakeholder satisfaction (Rusmana, A., Bogor, 16–24 July 2023, personal communication).

Various factors influence tourists' satisfaction, including the attraction of objects, services, facilities, and safety in Batulayang tourist village (Adira *et al.*, 2022). On the other hand, the government's satisfaction is determined by the village's ability to improve the local community's welfare and quality of life.

Conclusion

In conclusion, the Indonesian government has implemented policies to support tourism recovery after the COVID-19 pandemic. These policies encompass the National Economic Recovery Program (NERP), the Proud Tourism Movement, the empowerment of Micro, Small, and Medium Enterprises (MSMEs), technological innovation in e-tourism, and the CHSE Certification (Cleanliness, Health, Safety & Environment Sustainability). The development of tourism villages in Bogor Regency, West Java, has emerged as a crucial component of Indonesia's strategic agenda for tourism sector growth, underpinned by various governmental policies and regulatory frameworks.

Tourism villages align with global trends, favouring nature and local culture for unique travel experiences and offering sustainability, community well-being, climate action, and local engagement. Indonesia's focus on village tourism seeks broader welfare improvements, urban-rural equity, and central government support.

Successful tourism village development hinges on collaboration, community engagement, and adaptive policies. Batulayang Tourism Village overcame challenges with stakeholder support, sustainable practices, and certification. Indonesian tourism villages offer growth, welfare enhancement, and environmental benefits. Batulayang exemplifies these principles, showcasing their potential. Stakeholder satisfaction is apparent as they actively participate in the tourism village's development, benefiting from economic growth and an enhanced quality of life.

References

Adira, P., Herlambang, S. and Wipranata, B.I. (2022) Studi keberhasilan pengelolaan pada desa wisata berbasis masyarakat (objek studi: desa wisata Batulayang, Kecamatan Cisarua, Kabupaten Bogor). *Jurnal Sains, Teknologi, Urban, Perancangan, Arsitektur, (Stupa)* 4(2), 3127–3140. DOI: 10.24912/stupa.v4i2.22368.

Anon (2022) Profil Desa Wisata Batulayang Bogor, Keindahan Tersembunyi di Tanah Pasundan. Available at: https://www.perhutani.co.id/profil-desa-wisata-Batulayang-bogor-keindahan-tersembunyi-di-tanah-pasundan/ (accessed 4 July 2023).

Anon (2023) Company Profile. Available at: https://www.perhutani.co.id/en/tentang-kami/profil-perusahaan/ (accessed 30 October 2023).

Anon (n.d) Desa Wisata Batulayang. Available at: https://jadesta.kemenparekraf.go.id/desa/batulayang (accessed 2 July 2023).

Arora, S. and Sharma, A. (2021) *Covid-19 Impact on Overtourism: Diversion from Mass Tourism to Alternative Tourism*. Emerald Publishing Limited eBooks, Leeds, UK, pp. 275–283. DOI: 10.1108/978-1-83909-706-520211018.

Bahri, A.S., Basalamah, A., Abdillah, F. and Rahmat, T.A. (2023) Penerapan kriteria desa wisata pada desa wisata Batulayang, Bogor, Jawa Barat. *KONTAN: Jurnal Ekonomi, Manajemen dan Bisnis* 1(5), 8–17. Available at: https://jurnal.penerbitwidina.com/index.php/KONTAN/article/view/428/306

Bakti (2020) *Membangun Indonesia dari Desa (Building Indonesia from the Village)*. Bakti.or.id. Available at: https://bakti.or.id/berita/membangun-indonesia-dari-desa (accessed 26 October 2023).

Bowman, K.S. (2010) Sustainable tourism certification and state capacity: Keep it local, simple, and fuzzy. *International Journal of Culture, Tourism and Hospitality Research* 5(3), 269–281. DOI: 10.1108/17506181111156961.

Bramwell, B. and Lane, B. (1993) Sustainable tourism: An evolving global approach. *Journal of Sustainable Tourism* 1(1), 1–5. DOI: 10.1080/09669589309450696.

CNN Indonesia (2021) *Wisata edukasi ramah lingkungan di Desa Batulayang #DiIndonesiaAja*. [Youtube]. Available at: https://www.youtube.com/watch?v=Np0BENZlkh0 (accessed 3 July 2023).

Dewantara, Y.F. and Susanto, P. (2020) Analisis Dampak Positif terhadap Kesejahteraan dan Pertumbuhan Ekonomi dalam Penerapan Konsep Desa Wisata di Desa Batulayang, Kabupaten Bogor. *Destinesia: Jurnal Hospitaliti Dan Pariwisata* 2(1), 9–19. DOI: 10.31334/jd.v2i1.1067.

Dinas PUPR Kota Depok (Dinas Pekerjaan Umum dan Perumahan Rakyat Kota Depok; Public Works and Housing Office of Depok City) (2023) PUPR Spatial Data. Available at: https://puprspatialdata.depok.go.id/peta/dasar (accessed 29 September 2023).

Diskominfo Kabupaten Bogor (2021) Kemenparekraf RI berikan penghargaan Desa Wisata Batulayang Kabupaten Bogor sebagai Desa Wisata Berkelanjutan – Kabupaten Bogor. Available at: https://bogorkab.go.id/post/detail/kemenparekraf-ri-berikan-penghargaan-desa-wisata-batu-layang-kabupaten-bogor-sebagai-desa-wisata-berkelanjutan (accessed 2 July 2023).

Dwiyanto, M.F. (2018) *Disbudpar Bogor gelar Festival Desa Wisata*. AntaraNews Megapolitan. Available at: https://megapolitan.antaranews.com/berita/46853/disbudpar-bogor-gelar-festival-desa-wisata (accessed 30 October 2023).

Gambino, V. (2020) Sustainable Tourism for Destinations: Insights from the GSTC Destination Criteria 2.0 for Sustainable Tourism. Available at: https://www.gstcouncil.org/insights-from-the-gstc-destination-criteria-part-1/#:~:text=The%20GSTC%20criteria%20are%20organized%20in%20four%20sections%3A,to%20assess%20their%20progress%20on%20achieving%20the%20criteria (accessed 12 July 2023).

Government Regulation (2011) Government Regulation (PP) No.50 of 2011 concerning Rencana Induk Pengembangan Pariwisata Nasional/RIPPARNAS (the National Tourism Development Master Plan) 2010–2025. Available at: https://jdih.kemenparekraf.go.id/katalog-319-produk-hukum (accessed 8 July 2023).

Graci, S. (2013) Collaboration and partnership development for sustainable tourism. *Tourism Geographies* 15(1), 25–42. DOI: 10.1080/14616688.2012.675513.

Habibie, F.H., Arafah, W., Mustika, A. and Ratnaningtyas, H. (2021) Pelatihan dan Penyuluhan Sertifikasi Desa Wisata di Desa Cibuntu (Training and Counselling on Tourism village certification in Cibuntu Village). *Jurnal Pemberdayaan Pariwisata* 3(2), 78–83. Available at: https://garuda.kemdikbud.go.id/documents/detail/2581244 (accessed 31 October 2023).

Humaedi. D. (2023) in Widyanti, N.N.W (2023) Target Kunjungan Wisatawan ke Kabupaten Bogor Naik Jadi 10 Juta, 3 February 2023. Available at: https://travel.kompas.com/read/2023/02/03/111931927/target-kunjungan-wisatawan-ke-kabupaten-bogor-naik-jadi-10-juta (accessed 3 July 2023).

Intan, P. (2021) *Ini Desa Batulayang, Desa Wisata Berkelanjutan Pertama di Bogor*. detikTravel. Available at: https://travel.detik.com/domestic-destination/d-5489122/ini-desa-Batulayang-desa-wisata-berkelanjutan-pertama-di-bogor (accessed 29 June 2023).

ISTC (Indonesia Sustainable Tourism Council) (n.d.) Why Certification. Available at: https://ist-council.kemenparekraf.go.id/ISTC_certification_overview (accessed 31 October 2023).

Kepmenparpostel (n.d.) Kepmenparpostel (the Decree of the Minister of Tourism, Post and Telecommunications) No.KM.5/UM.209/MPPT-89 concerning Sapta Pesona Implementation Guidelines. Available at: https://jdih-dev.kemenparekraf.go.id/assets/8affcef7-a925-4fa2-8b5b-cf2e248c6d18 (accessed 31 October 2023).

Kohlbacher, F. (2005) The use of qualitative content analysis in case study research [89 paragraphs]. *Forum Qualitative Sozialforschung/Forum: Qualitative Social Research* 7(1), Art. 21. Available at: http://www.qualitative-research.net/fqstexte/1-06/06-1-21-e.htm (accessed 27 September 2023).

Krippendorff, K. (2004) *Content Analysis: An Introduction to Its Methodology*, 2nd edn. Sage, Thousand Oaks, CA.

Laksmi, S.G.W. (2023) *Perlindungan Pengelolaan Desa Wisata Melalui Sertifikasi Desa Wisata Berkelanjutan Ditinjau dari Undang-Undang No 10 Tahun 2009 tentang Kepariwisataan*. Undergraduate thesis, Universitas Kristen Indonesia. Available at: http://repository.uki.ac.id/12270/ (accessed 31 October 2023).

Latianingsih, N., Rudatin, C.L. and Sofia, M. (2022) Analisis Stakeholder pada Pengembangan Desa Wisata Batulayang Kabupaten Bogor. *Seminar Nasional Riset Terapan (SNRT) XI Desember 2022* 11(1), 353–360. Available at: https://prosiding.pnj.ac.id/index.php/snrtb/article/view/198 (accessed 17 July 2023).

Law (2009) Law No.10/2009 concerning Tourism. Available at: https://peraturan.bpk.go.id/Details/38598/uu-no-10-tahun-2009 (accessed 31 October 2023).

Law (2016) Law No.6/2016 Concerning Village. Available at: https://peraturan.bpk.go.id/Details/38582/uu-no-6-tahun-2014 (accessed 26 October 2023).

Leisen, B. (2001) Image segmentation: The case of a tourism destination. *Journal of Services Marketing* 15(1), 49–66. DOI: 10.1108/08876040110381517.

Liu, C., Dou, X., Li, J. and Cai, L.A. (2020) Analysing government role in rural tourism development: An empirical investigation from China. *Journal of Rural Studies* 79, 177–188. DOI: 10.1016/j.jrurstud.2020.08.046.

Liu, Y.L., Chiang, J.T. and Ko, P.F. (2023) The benefits of tourism for rural community development. *Humanities and Social Sciences Communications* 10(137), 1–12. DOI: 10.1057/s41599-023-01610-4.

Maryati, N. (2005) Partisipasi masyarakat dalam pengembangan desa Wisata: Penelitian di Padukuhan Bobung Desa Putat Kecamatan Patuk Kabupaten Gunungkidul. Available at: http://etd.repository.ugm.ac.id/home/detail_pencarian/28019 (accessed 4 August 2023).

McComb, E.J., Boyd, S. and Boluk, K. (2016) Stakeholder collaboration: A means to the success of rural tourism destinations? A critical evaluation of the existence of stakeholder collaboration within the Mournes, Northern Ireland. *Tourism and Hospitality Research* 17(3), 286–297. DOI: 10.1177/1467358415583738.

Nagarjuna, G. (2015) Local community involvement in tourism: A content analysis of websites of wildlife resorts. *Atna – Journal of Tourism Studies* 10(1), 13–21. DOI: 10.12727/ajts.13.2.

Ngonya, J.E. (2015) *Challenges facing Community-Based Tourism in Tanzania: A case study of Arumeru District in Arusha Region*. Master Thesis. Tourism Planning and Management, The Open University of Tanzania. Available at: http://repository.out.ac.tz/1267/1/Dissertation_-__Ngonya_ELisa_Jero_Final__-2.pdf (accessed 20 September 2023).

Ngo, T.H. and Creutz, S. (2022) Assessing the sustainability of community-based tourism: A case study in rural areas of Hoi An. *Vietnam. Cogent Social Sciences* 8(1), 1–21. DOI: 10.1080/23311886.2022.2116812.

Pamela, D.A. (2022) *Profil Desa Wisata Batulayang Bogor, Keindahan Tersembunyi di Tanah Pasundan*. liputan6.com. Available at: https://www.liputan6.com/lifestyle/read/5159713/profil-desa-wisata-Batulayang-bogor-keindahan-tersembunyi-di-tanah-pasundan?page=2 (accessed 10 July 2023).

Permenbudpar (2008) Permenbudpar (Regulation of the Minister of Tourism, Post and Telecommunication) No.PM.04/UM.001/MKP/2008 concerning Sadar Wisata (Tourism Awareness). Available at: https://

jdih-dev.kemenparekraf.go.id/assets/ceabd0ce-1f89-4182-acf8-140cfc602520 (accessed 31 October 2023).

Permendesa PDTT (2022) Permendesa PDTT (Regulation of the Minister of Villages, Development of Disadvantaged Regions, and Transmigration) No.8 Tahun 2022. Prioritas Penggunaan Dana Desa Tahun 2023 (Priority for the Use of Village Funds in 2023). Available at: https://peraturan.bpk.go.id/Home/Details/240997/permendesa-pdtt-no-8-tahun-2022 (accessed 26 June 2023).

Permenparekraf (2016) Permenparekraf (Regulation of the Minister of Tourism and Creative Economic) No.14 of 2016. Pedoman Destinasi Pariwisata Berkelanjutan. Available at: https://peraturan.bpk.go.id/Home/Details/171159/permenpar-no-14-tahun-2016#:~:text=Peraturan%20Menteri%20Pariwisata%20Dan%20Ekonomi%20Kreatif%20Nomor,14%20Tahun%202016%20tentang%20Pedoman%20Destinasi%20Pariwisata%20Berkelanjutan (accessed 25 June 2023).

Permenparekraf (2021) Permenparekraf (Regulation of the Minister of Tourism and Creative Economy) No.9 of 2021. Guidelines for Sustainable Tourism Destination. Available at: https://jdih.kemenparekraf.go.id/en/peraturan-english (accessed 27 June 2023).

Permenparekraf (2022) Permenparekraf (Regulation of the Minister of Tourism and Creative Economic) No.11 of 2022. Rencana strategis kementerian pariwisata dan ekonomi kreatif/ badan pariwisata dan ekonomi kreatif tahun 2020–2024 (The strategic plan of the ministry of tourism and creative economy/tourism and creative economy agency for 2020–2024). Available at: https://jdih.maritim.go.id/cfind/source/files/permenparekraf/2022/permen-parekraf-no.-11-tahun-2022.pdf (accessed 26 June 2023).

Rama, H. (2023) *Kawasan Puncak jadi primadona, 10 juta wisatawan desarguesian kunjungi Kabupaten Bogor tahun 2023*. Wartakotalive.com. Available at: https://wartakota.tribunnews.com/2023/02/06/kawasan-puncak-jadi-primadona-10-juta-wisatawan-ditargetkan-kunjungi-kabupaten-bogor-tahun-2023 (accessed 7 August 2023).

RBG, A. (2023) *Profil Desa Wisata Batulayang Bogor Yang Masuk Delapan Besar Tingkat Dunia*. RBG.id. Available at: https://www.rbg.id/daerah/9449751840/profil-desa-wisata-Batulayang-bogor-yang-masuk-delapan-besar-tingkat-dunia (accessed 10 September 2023).

Risdawati, A.A.P., Imron, D.K. and Pertiwi, C. (2020) Tourism village: Challenges and opportunities in new normal. In: Strielkowski, W., Black, J.M., Butterfield, S.A., Chang, C.C., Cheng, J. *et al.* (eds) *Advances in Social Science, Education and Humanities Research. Proceedings of the 6th International Conference on Social and Political Sciences (ICOSAPS 2020)*, pp. 510, 540–544. DOI: 10.2991/assehr.k.201219.082.

RPJMN (2020–2024) RPJMN (Rencana Pembangunan Jangka Menengah Nasional; the national medium-term development plan) 2020–2024. Appendix Presidential Regulation No.18 of 2020, Sekertariat Kabinet Republik Indonesia. Available at: https://jdih.bappenas.go.id/peraturan/detailperaturan/1037#:~:text=NOMOR%2018%20TAHUN%202020%20tentang%20Rencana%20Pembangunan%20Jangka,-%20NASIONAL%20-%20RPJMN%20-%20RPJMN%202020-2024%20- (accessed 25 June 2023).

Setiawan, M.F. (2023) *Desa Wisata Batulayang Bogor wakili Indonesia di Best Tourism Village*. Antara News. Available at: https://www.antaranews.com/berita/3673464/desa-wisata-Batulayang-bogor-wakili-indonesia-di-best-tourism-village (accessed 20 August 2023).

Setiawati, R. and Aji, P.S.T. (2020) Implementasi Sapta Pesona sebagai upaya dalam memberikan pelayanan prima pada wisatawan di Desa Wisata Pentingsari. *Jurnal Administrasi Bisnis Terapan (JABT)* 2(2), 128–142. DOI: 10.7454/jabt.v2i2.1071.

Sharma, G.D., Thomas, A. and Paul, J. (2021) Reviving tourism industry post-COVID-19: A resilience-based framework. *Tourism Management Perspectives* 37, 1–11. DOI: 10.1016/j.tmp.2020.100786.

Sianturi, J. (2007) *Sikap dan Partisipasi Masyarakat Lokal Terhadap Pengembangan Wanawisata Curug Kembar Batulayang (Studi Kasus di Desa Batulayang, Kecamatan Cisarua, Kabupaten Bogor, Propinsi Jawa Barat)*. Undergraduate Thesis. Fakultas Kehutanan, Institute Pertanian Bogor. Available at: https://123dok.com/document/myjnng2q-partisipasi-masyarakat-terhadap-pengembangan-kecamatan-cisarua-kabupaten-propinsi.html (accessed 4 July 2023).

Stone, M.T. (2015) Community-based ecotourism: A collaborative partnerships perspective. *Journal of Ecotourism* 14(2–3), 166–184. DOI: 10.1080/14724049.2015.1023309.

Tamir, M. (2020) Challenges and opportunities of community based tourism development in awi zone: A case study in Guagusa and Banja Woredas, Ethiopia. *Journal of Tourism, Hospitality and Sports* 11, 50–78. Available at: https://www.iiste.org/Journals/index.php/JTHS/article/view/24694

The Guidelines of Tourism Awareness Group or Pedoman Kelompok Sadar Wisata (2012) Buku Pedoman Kelompok sadar Wisata. Available at: https://asidewi.id/download/buku-pedoman-pokdarwis/ (accessed 29 November 2023).

Vitasurya, V.R., Hardiman, G. and Sari, S.R. (2018) Geographical conditions and cultural tradition as determinants in sustaining tourism village program case study tourism villages in Yogyakarta. *ARCHive-SR* 2(2), 1–9. DOI: 10.21625/archive.v2i2.240.

Widyanti, N.N.W. (2022) *Desa Wisata Semakin Menjamur di Bogor, Total Ada 55 Desa Tahun 2022*. Kompas.com. Available at: https://travel.kompas.com/read/2022/11/27/121825627/desa-wisata-semakin-menjamur-di-bogor-total-ada-55-desa-tahun-2022 (accessed 3 July 2023).

3 Overtourism and Tourism Sustainable Management in the Komodo National Park, Indonesia

Reiza Miftah Wirakusuma[1,2], Sarah Gardiner[1] and Alexandra Coghlan[2]
[1]Institute of Tourism, Griffith; Department of Tourism, Sport and Hotel Management, Griffith University, Gold Coast, Australia; [2]Universitas Pendidikan Indonesia, Bandung, Indonesia

Summary
Tourism is a central driver of economic growth and prosperity in Indonesia with Bali as arguably its best known and most successful destination. To replicate the success of Bali, disperse visitors and deliver economic outcomes to other regions in Indonesia, the Indonesian Government has identified five super-priority destinations to be the future 'New Balis'. Labuan Bajo is identified as one such destination. It is the gateway to the Komodo National Park which is famous for its Komodo Dragons. It is also popular among scuba divers due to its high marine biodiversity, including over 1,000 fish species and 260 coral species. Labuan Bajo also hosted the G20 side event and ASEAN Summit in 2023, attracting further global attention to the region. In many ways, Labuan Bajo is seen as another tourism success story.

Introduction

Yet, the large number of visitors is an emerging problem causing undesirable effects on the local environment and community. Increasing visitor numbers competing for the limited existing infrastructure are creating negative impacts (Westoby *et al.*, 2021). There have also been increasing negative media stories about tourism in recent years. The wild savannah fire caused by irresponsible tourists on 1 August 2018 (Mariani, 2018), safety issues on overpopulated dive boats, and congestion at a popular spot in Komodo National Park to have a scenic picture are examples of these impacts. It is, therefore, apparent that this destination might be suffering from overtourism, defined as excessive numbers of tourists at a specific destination either temporarily or during seasonal peaks (Milano *et al.*, 2019). Overtourism can have negative impacts on the host community, including the destruction of cultural attractions, as well as the physical, social and economic fabric of the community (Dodds and Butler, 2019; Mihalic, 2020). As a result, the long-term sustainability of tourism in the region is being questioned by local stakeholders.

This case study highlights three prominent issues for sustainable tourism in developing regions such as Labuan Bajo, Indonesia. The first issue is the vertical and horizontal governance complexity of managing a rapidly developing tourism destination across different levels of government (vertical management) and across multiple ministries/departments and across multiple stakeholders in a destination (horizontal management). Second, there are regulatory challenges of managing the

 Tourism Case Studies Asia-Pacific Region (eds Taufik Abdullah and Reni Polus)

impacts of overtourism, especially balancing commercial outcomes with sustainability. For example, visitor safety on the liveaboard diving boats and managing visitation to the Komodo National Park by increasing the entry price to visit the park are examined in this case study. Third, there are complex issues in managing the relationship between the local community's livelihoods and sustainable tourism. This case study highlights these three key issues in regions seeking to develop their visitor economy and manage overtourism to achieve sustainable tourism outcomes.

The Value and Interest of the Case Study

Through the viewpoint of a destination management organization, this case study explores the complex challenges of managing overtourism in developing destinations, with a specific focus on fragile ecosystems (both island-based national parks and marine parks) with complex governance structures and many varied stakeholders (Rahmafitria *et al.*, 2023).

Background

This case study investigates the issue of overtourism at a vulnerable island-based, protected area, and marine destination, identifying three key issues critical to understanding the current situation and, ultimately, developing strategies and policies to achieve a sustainable tourism future in developing countries. Managing overtourism is a critical issue for many destinations, including Labuan Bajo and Komodo National Park. This destination is both a World Heritage site and proclaimed as one of the new Seven Wonders of nature (Melita, 2012). It is known for its natural beauty and unique ecosystems as well as being an island destination. Therefore, it is more sensitive to disturbances than mainland destinations, as well as being one of Indonesia's new super-priority destinations. The region has numerous management challenges, many of which can have severe impacts on visitor satisfaction and safety, including capsizing boats and missing divers (Dwinanda, 2018). These challenges are attributed to the issue of overtourism. Consequently, many stakeholders question the sustainability of Komodo National Park. Despite the increasing management challenges, the Komodo National Park office and Komodo Survival Program have been able to curb the negative impacts and successfully maintain a stable dragon population (Hance, 2015; Global Conservation, 2023). A brief background of the national park is presented below to situate the case study.

The Komodo National Park is located in East Nusa Tenggara Province, in the subregion called Manggarai Barat Regency (Fig. 3.1). There are three main islands regularly visited by tourists in this national park, namely Komodo Island, Padar Island, and Rinca Island, while Nusa Kode and Gili Motang Islands are rarely visited by tourists. These islands, which are designated as being of Unique Outstanding Value (UOV), are the only Komodo dragon natural habitats in the world that draw tourists from abroad. Tourists must use private or shared boats as the main transportation to enter the national park from the islands' gateway city, Labuan Bajo, located at the western end of Flores Island.

Labuan Bajo itself is a focal super-priority destination in the tourism marketing of Indonesia. In terms of tourism infrastructure, it has a domestic airport and a major inter-island harbour. Through marketing campaigns, increased awareness of the region has led to rapid visitation growth and demand for popular tourist sites (Fig. 3.2). Pre-COVID-19 pandemic, visitation peaked at over 221,000 visitors but dropped to 51,618 people in 2020. Following the pandemic, visitation and tourism industries rapidly accelerated again and as of September 2023, the industry is back to pre-pandemic levels, with 219,693 visitors.

Tourism development is sprawling throughout Labuan Bajo, with accommodation, restaurants, diving operators, travel agencies and other supporting businesses highly visible across the city. The wide range of accommodations makes a significant contribution to Labuan Bajo's revenue. Table 3.1 shows the accommodation data from 2017 to 2021, according to hotel classification, total units, total rooms, total beds, and total guests in Labuan Bajo. As shown in Table 3.1, the growth of room occupancy between 2017 and 2019 was approximately

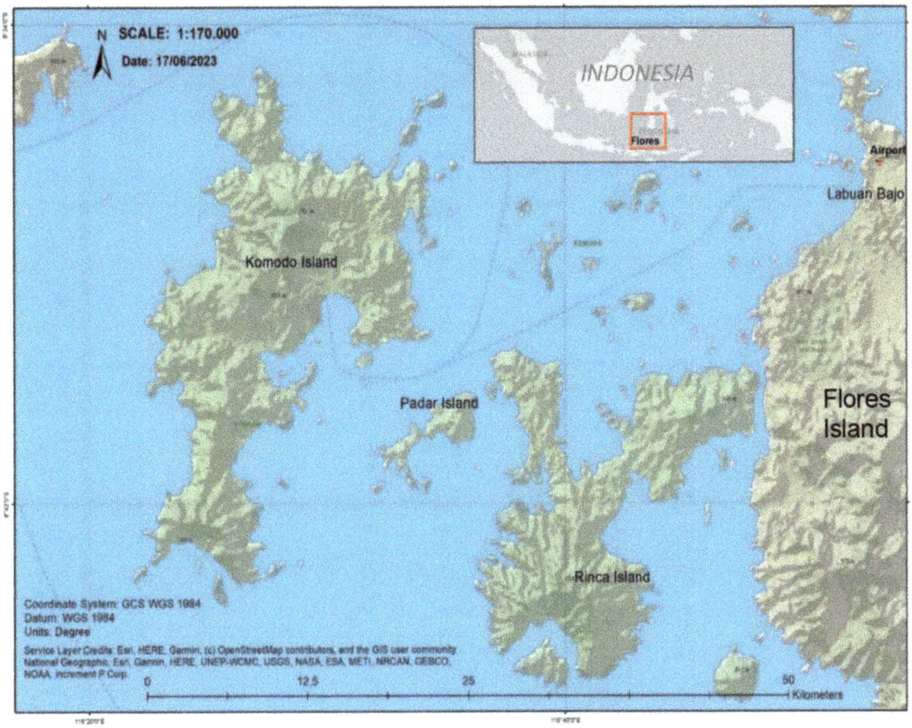

Fig. 3.1. Map of Labuan Bajo and Komodo National Park.

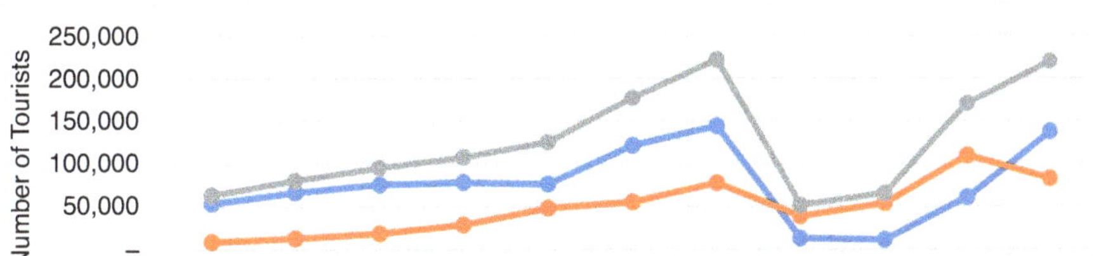

	2013	2014	2015	2016	2017	2018	2019	2020	2021	2022	2023 (Sep)
Foreign	54,147	67,089	76,195	78,617	76,612	121,411	144,068	13,089	11,538	60,770	137,239
Domestic	9,654	13,537	19,215	29,094	48,457	55,423	77,635	38,529	53,824	109,307	82,454
Total	63,801	80,626	95,410	107,711	125,069	176,834	221,703	51,618	65,362	170,077	219,693

Fig. 3.2. Tourist statistics in Komodo National Park in 2013–2023, Office of Tourism, Creative Economy and Cultural in 2023 (Badan Pusat Statistik, 2021; Lasso and Dahles, 2021).

2–3% annually, with a concomitant drop during the pandemic period.

The accommodation profile of Labuan Bajo in 2016 (the latest available data, according to a Swiss Contact Report by Remmer, 2017) is presented in Fig. 3.3. Budget hotels (28%), star-rated hotels (15%) and beach resorts (8%) represent the largest share of accommodation offerings. Homestays (13%), hostels (9%) and guest houses (4%) are also an important part of the accommodation offering, representing approximately one-quarter of the sector in 2016.

Table 3.1. Hotels statistics in Labuan Bajo from 2017 to 2021.

	Star hotels		Non-star hotels					
Year	Units	Room occupancy (%)	Units	Room occupancy (%)	Total accommodation	Total rooms	Total beds	Total guests
2017	7	53.22	69	22.70	76	1191	2262	142,905
2018	7	57.44	91	25.91	98	1462	3029	179,081
2019	8	58.19	93	29.50	101	1798	3939	220,125
2020	12	49.78	99	32.14	111	1691	3488	62,448
2021	10	24.43	71	14.61	81	1966	4047	139,112

Statistic Buerau of Manggarai Barat Regency (2021).

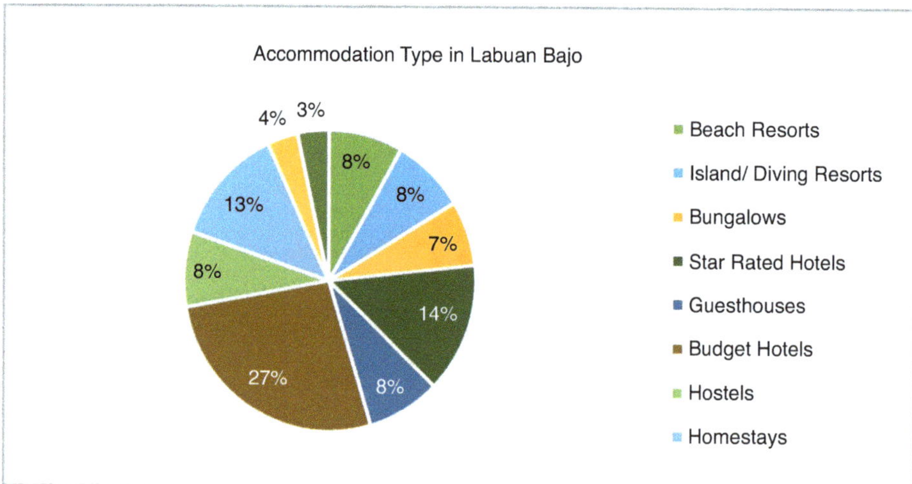

Fig. 3.3. Accommodation profile in Labuan Bajo (Remmer, 2017).

In addition to accommodation providers, another highly visible part of Labuan Bajo's tourism is the vessels used to move travellers around the marine-based destination to go trekking and diving, and accommodate them in what is known as liveaboards. These boats are locally owned (30%) as well as owned by outsiders (40% Indonesian, and 27% foreign-owned, Fig. 3.4). The fleet is diverse, ranging from luxury vessels to more modest and traditional boats, generally holding 16–24 passengers. Each boat has multiple type of cabins and is equipped with a kitchen, dining area, showers, toilets and dedicated snorkelling and scuba diving area with the relevant equipment. As the boats are self-sufficient, their itinerary is adjustable to accommodate weather and environmental conditions (Wirakusuma *et al.*, 2021). As the popularity of this activity has increased, the number of boats offering this experience has expanded. As a result, the harbour in Labuan Bajo often exceeds the capacity to service the tourist vessels and overcrowding occurs, particularly during the peak visitor seasons of approximately May–September and November–December.

It should be noted that the absence of current data on key management statistics, such as the accommodation and business ownership profile of the region, is another challenge for the management of Labuan Bajo and Komodo National Park.

With the rapid growth of accommodation providers and tours at Labuan Bajo, the region (including islands) is arguably in the development phase in Butler's Tourism Area Life Cycle Model (Butler, 2006). Finding a sustainable

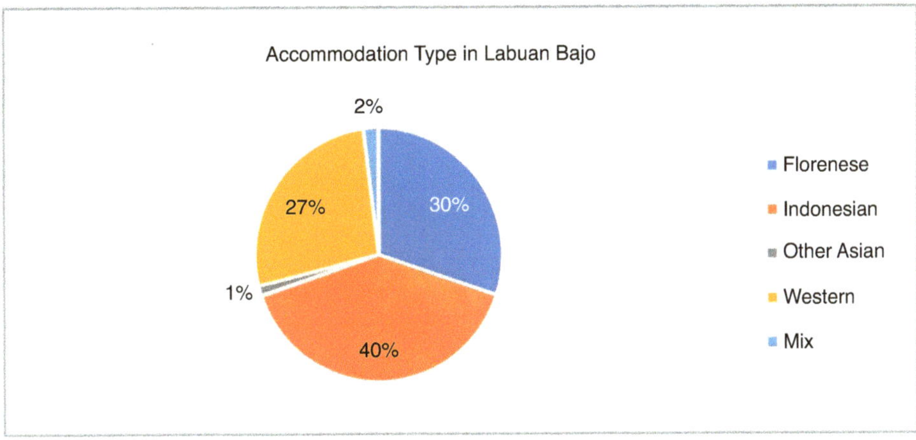

Fig. 3.4. Tourism business ownership profile in Labuan Bajo (Remmer, 2017).

management model to manage this development and consolidate outcomes that achieve a more sustainable future for the destination is, therefore, needed. As highlighted by Westoby *et al.* (2021), this is a challenge for all super-priority developing destinations in Indonesia and, more broadly, all developing countries. In particular, managing complex governance structures and multiple and varied stakeholders to balance diverse needs is a key component of achieving sustainable outcomes. Furthermore, the involvement of the local community is essential to avoid tensions from top-down, elite-led development from ministries to provinces. Planned tourism infrastructure development and a strong policy framework between stakeholders to manage tourism in developing destinations is also essential to manage impacts (Damanik, 2016).

This is proving to be a challenge within this case study, as depicted in the images in Fig. 3.5, taken at the centre of the region's national parks. Thousands of tourists visit this site on Pader Island to look at its aesthetic beauty and to take desirable photos. As a result, the site has become overcrowded, particularly on weekends, which is the peak visitation time. The tour guides manage the flow of the tourists according to the boats' arrival time. There is limited infrastructure to manage the high visitation to the site. Some hard infrastructure does exist, such as fencing along the path and a rocky staircase to keep visitors to the track but as shown in the photos, the path is narrow and visitors often walk outside of the defined walkway. In terms of soft infrastructure, which is the services provided by national park staff, visitors are able to go sightseeing under the guidance and direction of the national park employees at the beginning trekking point. However, due to limited human resources, the staff are quite often overwhelmed by the tourists.

This case study explores the issue of overtourism at a fragile island-based, protected area and marine destination, and identifies three key issues that are central to understanding the current situation and, ultimately, developing strategies and policies to achieve a sustainable tourism future in developing countries. Each of these issues is presented in their own section below.

Vertical and Horizontal Governance Complexity of Managing the Destination

Of the many complex management issues in Labuan Bajo, the relationship between the various levels of stakeholders (vertical linkages) and across stakeholders in the destination (horizontal linkages) is central to its sustainability. Understanding the power dynamics and who has influence on decision making among the stakeholders and which axis (i.e. vertical or horizontal governance) has the most influence

Fig. 3.5. Tourist flooded Padar Island (image courtesy of @ardhymotret on Instagram).

is critical to determining the dynamics of the tourism system as they relate to the issue of overtourism.

Vertical governance complexity of destination management

The relationship between national, provincial, and local stakeholders in government and industry is known as the vertical governance of the destination. In Indonesia, there are six ministries at the national level that regulate tourism destination development. They include the Ministry of Tourism and Creative Economy, Ministry of Environment and Forestry, Ministry of Transportation, Coordinating Ministry of Maritime and Investment, Ministry of Finances, and Ministry of Public Works and Human Settlement.

Managing tourism demand and planning tourism policy is the responsibility of the Ministry of Tourism and Creative Economy, through the Tourism National Plan known as RIPPARNAS. It is RIPPARNAS that identifies Labuan Bajo as one of the five new super-priority destinations. As a result, the region's funding for domestic and international marketing comes from four central government ministries. This funding increases annually and is allocated to Labuan Bajo and Flores Tourism Authority programs.

At a district level, West Manggarai Regency and the Statistics Bureau of West Manggarai have their own representative offices. These offices are called 'Dinas' and they are on the local level. These Dinas are responsible for collecting revenue at a local level on behalf of the government at the national level. Meanwhile, the national level government has policies to manage Labuan Bajo and Komodo National

Park such as regulating operating standards and issuing permits for tour operators and activity providers to deliver tourism services and experiences in the destination. Examples of such standards are provided below. They also manage the policy on the admission prices to access key tourist sites, including guide services for trekking.

To operate in any of the three main islands (all World Heritage listed), tourism businesses are required to comply with the high environmental awareness requirements set by the Ministry of Environment and Forestry. This ministry also selects and assesses business tenants in a 5-year lease agreement. In addition, the ministry formulates the admission price for guiding services, tourist entry permits, and documentation permits. The ministry also has a direct extension in the form of the Komodo National Park office to issue the legal aspects of permits, operational standards, and patrol the national park. Some examples of their purview include setting standards for engine emissions, waste management systems, preventing wildlife poaching and local guide recruitment.

A key component of managing visitation and avoiding the detrimental impacts of tourism is having adequate infrastructure. Investment in infrastructure is critically linked to regional development. This portfolio is the responsibility of the Ministry of Maritime Coordination and Investment, the Ministry for Public Works and Human Settlement and West Manggarai Regency. The Ministry of Transportation is responsible for sea accessibility and harbour operations. The Regent/Mayor is also supported by the Statistics Bureau of West Manggarai which provides data and statistics to assist in formulating civic policies.

As can be seen in Fig. 3.6, vertical governance provides a direct chain of command from the ministry to the Komodo National Park Office

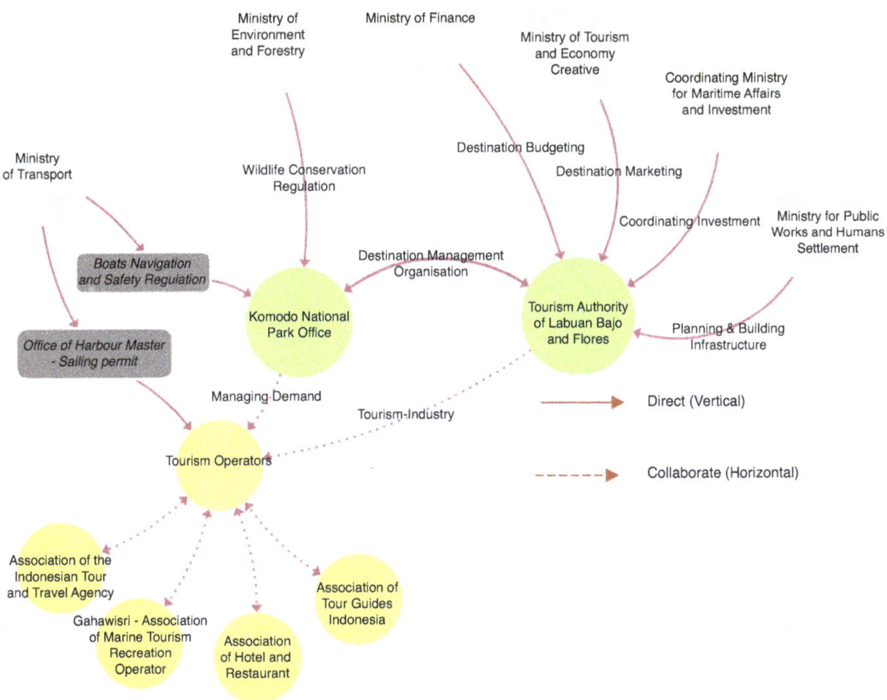

Fig. 3.6. Diagram of government stakeholders, destination management organizations and tourism operators.

and Labuan Bajo Flores Tourism Authority, and also active supervision and guidelines of the field practice. The program of activities to manage this site is evaluated by the Komodo National Park Office annually to assess its effectiveness. However, the bureaucracy between offices and ministries tends to be rigid. There are many programs in each organization and offices and, therefore, collaboration between entities is required to achieve a collectively positive outcome. Accordingly, communication between ministries, offices and local associations plays an essential role.

Horizontal collaboration of destination management organizations

Fig. 3.6 highlights the types of vertical collaboration within the region, which are also needed to achieve sustainable tourism, ensure conservation outcomes are met and manage supply and demand at key visitor sites. Here, two key destination management organizations (DMOs) exist side by side: Komodo National Park Office and Labuan Bajo and Flores Tourism Authority.

Komodo National Park Office is a direct extension of the Ministry of Forestry and Environment in the region's local government. This office has three main functions: (1) managing biodiversity conservation and education; (2) managing the flow of tourists to and around the region; and (3) managing the relationship between the government DMOs and the local community. This office also collaborates with the harbourmaster, police and rescue teams to maintain safety in the vast territory of Labuan Bajo. The national park is vulnerable to environmental degradation, deer poachers, boat and tourist accidents. Therefore, one of their most important programs is a frequent patrol throughout the national park, both during the day and at night.

Labuan Bajo and Flores Tourism Authority collaborates with tourism operators and associations to attract visitors. Key associations include the Tour and Travel Agent Association (ASITA), Hotel and Restaurant Association (PHRI), Marine Tourism Recreation Operators (GAHAWISRI), and Association of Indonesian Tour Guide (HPI). Other associations focusing on tourism boats, such as ASKAWI and JANGKAR, and the dive guides community, such as DOCK and P3KOM also play roles in creating a healthy business environment and minimizing conflicts between competing businesses in the region. The tourism operators include premium liveaboard and deluxe liveaboard tour operators, scuba diving operators and local travel agents.

As can be seen in Fig. 3.6, this case study reveals a particularly high level of both vertical and horizontal complexity, which is likely contributing to the problem of overtourism in the national park. Further research on the complexity of this system and its effects on managing overtourism and conservation is underway using interviews, observations and focus groups with these key governance agencies and other stakeholders.

Regulatory Challenges of Managing (Over)tourism in Komodo National Park

The rapid growth of tourist visitation to Komodo National Park has resulted in a number of illegal boat operators, putting pressure on the regulation of operators within the destination. To explore this further, we use Butler and Dodds (2022) conceptual framework of failure action to address the overtourism summarized in Table 3.2. Their study considers the shifting responsibilities of tourism development and marketing organizations as well as how public sector organizations have failed to manage and control tourism. Here we argue that a similar phenomenon is happening in Labuan Bajo and Komodo National Park. Hence we apply their framework to the destination (Table 3.2).

The Relationship Between the Local Community's Livelihoods and Sustainable Tourism

The third and final issue that we explore in this case study is the relationship between the local community's livelihoods and sustainable tourism.

Table 3.2. Applying Butler and Dodds (2022) framework to the present situation in Labuan Bajo.

No.	Frameworks from Butler and Dodds	Present situation in Labuan Bajo
1	Multiple difficulties in integrating tourism policy with other sectors and levels of government. This phenomenon occurred when formulating the regulations to maintain tourism demand.	As a result, there was an inevitable dispute on implementing national park entrance fee and massive protest from tourism operators in 2022.
2	Tourism stakeholders are partially involved in the new regulations. Communication and feedback from the government, DMO and tourism operators should be of the utmost importance.	Communication among six ministries still partially occurred. Aspirations, opinions, and suggestions from tourism operators submitted to the associations (PHRI, ASITA and GAHAWISRI) in Fig. 3.6, suffer from a delayed response.
3	Stakeholders support in managing the destination. Individual stakeholders need to have a collective responsibility.	Tourism operators were helping each other during the COVID-19 pandemic, especially managing low occupancy. As a result, the level of business collapse was minimized.
4	There is ambiguity for the new regulation to be implemented.	The regulations such as admission price, natural guide services and tour package arrangements are volatile and a lack of communication happened.
5	Efficiency in using resources is minimal. Bigger proportion of the annual budget on marketing rather than sustainability.	Tourism authority office has a tremendous budget for the marketing campaign and creating new programs as instructed by the central government of Indonesia.
6	Focus on votes which refers to the political will that influences the regulations.	National elections occur once every 5 years. Therefore, many regulations are amended by different political leaders.
7	Economic aspects are the priority and rely on numbers rather than yield. There is a focus on the short term rather than the future.	Managing demand for this destination is more challenging than organizing the supply side. According to Fig. 3.2, the number of visitations in 2023 will exceed those of 2022.

Labuan Bajo and Komodo National Park are rich in natural resources. The livelihoods in this destination are heavily reliant on those natural resources. It is, therefore, imperative that communities have access, control and decision making capabilities to guide the development of tourism in this region (Westoby *et al.*, 2021). There should be evident trade-offs for individuals, communities, and the environment as a by-product of tourism expansion. Yet, arguably the local community was sidelined and experienced marginalization and power imbalances in the development of tourism.

The Komodo Village showcases this occurring in practice. Since this village is a part of the conservation area, the Komodo National Park authority rules and regulations have a direct impact on the locals' ability to support themselves. For example, the wood crafters of Komodo statues are no longer able to use Kanawa logs as this tree is now protected by law. This outcome was evident in Lasso and Dahles (2018) study of community livelihood in Komodo National Park. They found that the development of tourism first prompted a diversification of the region's fishing-based economy. However, once it had become established, tourism took the place of the prior means of livelihood, and the community returned to being economically dependent on a single source of income, from tourism. A return to fishing has become impossible as locals have sold their boats and given up their knowledge. The community fears this occurring in the super-priority destination.

Conclusions

Overtourism management is a serious issue for many places, including Labuan Bajo and Komodo National Park, which has been designated as a UNESCO World Heritage Site and recognized as one of the new Seven Wonders of Nature. This place is well-known for its natural beauty and distinct ecosystems, as well as the fact that it is an island destination (and hence more vulnerable to disruptions than mainland destinations). This all leads to several issues that make managing tourism in this region particularly complex. On the one hand, the government and destination marketers want to attract visitors to drive economic outcomes. One lesson from this case study is that there are still many tourism destinations across the globe that are arguably over-reliant on increasing the number of visitors as their tourism strategy. Yet, sudden increase in the number of tourists make it difficult for government stakeholders to implement policies due to complex management systems and vulnerability to conflict creation. As a result, managing overtourism may lead to sustainable or destructive outcomes, as originally posited in Butler's TALC model once a visitation threshold is crossed. Without actions that disrupt the status quo, overtourism will only become more serious and a deteriorating outcome is envisaged. A shift in perspective, strategic planning and the implementation of policies and regulations co-designed by government, industry and local people is needed to change this trajectory and achieve a more sustainable outcome.

Acknowledgement

The first author would like to convey gratitude to the Indonesia Endowment Fund for Education (LPDP) as the provider of scholarships and also granting research funds.

References

Badan Pusat Statistik (2021) Kabupaten Manggarai Barat dalam Angka 2021. *BPS Kabupaten Manggarai Barat*. DOI: 10.1016/j.encep.2012.03.001.

Butler, R. (ed.) (2006) *The Tourism Area Life Cycle: Application and Modifications*, Vol. 1. Channel View Publications, Bristol, UK.

Butler, R. and Dodds, R. (2022) Overcoming overtourism: A review of failure. *Tourism Review* 77(1), 35–53. DOI: 10.1108/TR-04-2021-0215.

Damanik, J. (2016) Lack of stakeholder partnerships in destination management: Lessons learned from Labuan Bajo, Eastern Indonesia. *Asian Journal of Tourism Research* 1(2), 167–191. DOI: 10.12982/AJTR.2016.0019.

Dodds, R. and Butler, R. (2019) The phenomena of overtourism: A review. *International Journal of Tourism Cities* 5(4), 519–528. DOI: 10.1108/IJTC-06-2019-0090.

Dwinanda, R. (2018) Swedish Tourist Dies after Diving in Sea Off Komodo Island. Available at: https://republika.co.id/berita/p1z0gx414/swedish-tourist-dies-after-diving-in-sea-off-komodo-island (accessed 10 October 2023).

Global Conservation (2023) Komodo National Park Progress Report 2022–2023. Available at: https://globalconservation.org/news/komodo-national-park-progress-report-2022-2023/ (accessed 9 September 2023).

Hance, J. (2015) Komodo Dragon: One of Indonesia's Rare Conservation Success Stories. Available at: https://news.mongabay.com/2015/12/komodo-dragon-one-of-indonesias-rare-conservation-success-stories/ (accessed 12 September 2023).

Lasso, A. and Dahles, H. (2018) Are tourism livelihoods sustainable? Tourism development and economic transformation on Komodo Island, Indonesia. *Asia Pacific Journal of Tourism Research* 23(5), Article 5. DOI: 10.1080/10941665.2018.1467939.

Lasso, A. and Dahles, H. (2021) A community perspective on local ecotourism development: Lessons from Komodo National Park. *Tourism Geographies* 0, 1–21. DOI: 10.1080/14616688.2021.1953123.

Mariani, E. (2018) Police Hunt Cigarette Tosser Who Allegedly Caused Fire in Komodo National Park. Available at: https://www.thejakartapost.com/news/2018/08/05/police-hunt-cigarette-tosser-who-allegedly-caused-fire-in-komodo-national-park.html (accessed 16 January 2024).

Melita (2012) Komodo Island Officially Confirmed as One of the new7wonders of Nature. Available at: https://about.new7wonders.com/2012/05/16/komodo-island-officially-confirmed-as-one-of-the-new7wonders-of-nature/#:~:text=Komodo-,Komodo%20Island%20officially%20confirmed%20as%20one%20of%20the%20New7Wonders%20of,of%20the%20New7Wonders%20of%20Nature (accessed 1 September 2023).

Mihalic, T. (2020) Concpetualising overtourism: A sustainability approach. *Annals of Tourism Research* 84(July), 103025. DOI: 10.1016/j.annals.2020.103025.

Milano, C., Novelli, M. and Cheer, J.M. (2019) Overtourism and tourismphobia: A journey through four decades of tourism development, planning and local concerns. *Tourism Planning & Development* 16(4), 353–357. DOI: 10.1080/21568316.2019.1599604.

Rahmafitria, F., Dirgahayani, P., Putro, H.P.H., Rosyidie, A. and Hudalah, D. (2023) Tourism accessibility in protected islands: The case of the Komodo National Park, Indonesia. *Tourism Review* 78(3), 966–985. DOI: 10.1108/TR-03-2022-0110.

Remmer, S. (2017) *Tourism Impacts in Labuan Bajo*. Swiss Contact. Available at: https://tourism4sdgs.org/wp-content/uploads/2018/08/Labuan_Bajo_Impact_Assessment.pdf (accessed 3 September 2023).

Statistic Buerau of Manggarai Barat Regency (2021) Hotels Statistic of Manggarai Barat Regency. Available at: https://manggaraibaratkab.bps.go.id/publication/2023/01/03/ce84416fc967d10b192d2804/statistik-perhotelan-kabupaten-manggarai-barat-2021.html (accessed 29 August 2023).

Westoby, R., Gardiner, S., Carter, R.W. and Scott, N. (2021) Sustainable livelihoods from tourism in the '10 New Balis' in Indonesia. *Asia Pacific Journal of Tourism Research* 26(6), 702–716. DOI: 10.1080/10941665.2021.1908386.

Wirakusuma, R.M., Luck, M., Schanzel, H., Widiawaty, M.A., Pramulatsih, G.P, *et al.* (2021) How risky is liveaboard diving in Indonesia? An empirical investigation on the divers' perceived risk and oceanic geomorphology. In: *Promoting Creative Tourism: Current Issues in Tourism Research*. Routledge, Oxfordshire, UK, pp. 297–304. DOI: 10.1201/9781003095484-43.

4 Adopting Sustainable Practices at an Island Ecotourism Destination in Sabah Through a Sea Turtle Conservation Project

Jennifer Kim Lian Chan
Centre for Economic Development and Policy, Faculty of Business, Economics and Accountancy, Universiti Malaysia Sabah, Kota Kinabalu, Sabah, Malaysia

Summary
This case centres on the sustainable ecotourism practices carried out on Libaran Island, Malaysia, through a sea turtle conservation project called the Walai Penyu Conservation Park. This widely recognized and award-winning project was launched and led by an ecotourism practitioner in collaboration with the island community, supported by governmental and non-governmental organizations. The project's sustainable approaches have had significant positive impacts on the islands' economy, environment, society, and culture. Among these impacts are the transformation of Libaran Island into an ecotourism destination, the preservation and promotion of local culture, and the generation of income for islanders through tourism job opportunities, village tour services, and handicraft sales. The project also enriches the pristine island environment by keeping its beaches clean, promoting recycling, and conserving endangered marine species (i.e., green turtles and hawksbill turtles). In summary, the project embodies the concept of sustainability and underscores the strong, interdependent relationships among the tourism industry, island communities, government agencies and NGOs.

The Value and Interest of the Case Study

The Walai Penyu Conservation Park on Libaran Island in the north-east of Sandakan, Sabah, has garnered multiple awards. It showcases a successful sea turtle conservation project and the sustainable practices it has implemented on the island. With support from both government agencies and non-governmental organizations (NGOs), the project was initiated and remains led by an ecotourism practitioner in close collaboration with the island community.

By implementing sustainable practices on Libaran Island, the project has contributed substantially to realizing the Reduce, Reuse, and Recycle (3Rs) principle, as well as achieving multiple Unit Nations Sustainable Development Goals (SDGs), including numbers 1, 8, 12 and 17. In doing so, it has brought significant value to turtle conservation while simultaneously generating extensive positive impacts on the local economy, environmental and societal well-being, and islanders' quality of life. Notably, the project has successfully transformed Libaran Island into an attractive and sustainable marine ecotourism destination that focuses on the conservation of endangered marine species (i.e. green turtles and hawkbill turtles). Thus, the project exemplifies the practical application of sustainability concepts and emphasises the strong interconnectedness and collaboration among the tourism industry, island communities, government agencies, and NGOs.

Background and Context

According to the International Sea Turtle Society, sea turtle populations around the planet are facing serious threats from human activities (Pilcher, 2022). However, conservation efforts towards this species are limited. Libaran Island, situated on the north-eastern coast of Sabah, a Borneo state in Malaysia, has played a pivotal role in addressing this challenge. From November 2011 to December 2015, a total of 21,672 baby turtles (14,851 green turtles and 6821 hawksbill turtles) were released into the ocean from Taman Hadiah on Libaran Island (Vanar, 2016). As more sea turtles continue to land and nest on the island with each passing year, Libaran Island has emerged as a conservation success story, attracting island seekers and wildlife enthusiasts to visit the island to witness the complete life cycle of sea turtles, from their landing and nesting to the hatching and release of baby turtles into the sea.

A brief background of Libaran Island

Libaran Island covers an area of approximately 480 ha (4.8 sq km). Nonetheless, it is the largest of the islands in the Turtle Islands National Park, which are encircled in Fig. 4.1 (Sabah Parks Board of Trustees, 2023). The island is home to an estimated population of 800 people and is surrounded by the Sulu Sea. It is situated approximately 40 min away, by boat, from Sandakan, the main city on the east coast of Sabah. The core economic activity on the island is fishing, supplemented by basic agriculture, with an estimated average monthly income of US$33 per capita (UPM *et al.*, 1996).

Libaran Island boasts approximately 3 km of beaches, which serve as vital landing and nesting sites for turtles. Remnants of early settlements have also been found within the dense

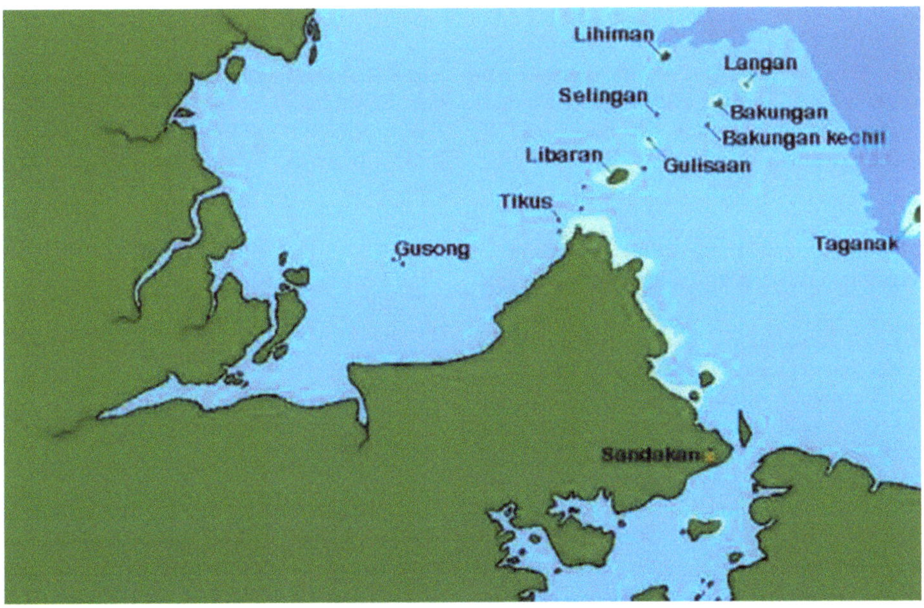

Fig. 4.1. Location of Libaran Island within the Turtle Island Parks (Alin and Yee, 2022).

Fig. 4.2. Locations of Walai Penyu Conservation Park, Taman Hadiah Hatchery, and Walai Penyu Resort Land (shaded in maroon) on the Land and Survey Department's land title of Libaran Island (Alin and Yee, 2022).

mangrove forest in the north-eastern region of the island. However, there has been limited documentation regarding the island and its socio-economic situation.

Walai Penyu Conservation Park, Sman Hadiah Hatchery, and Walai Penyu Resort

Fig. 4.2 indicates the precise location of the Walai Penyu Conservation Park, Taman Hadiah Hatchery, and Walai Penyu Resort Land on Libaran Island. The Walai Penyu Conservation Park (Fig. 4.3) was established in 2013 alongside the Taman Hadiah Turtle Hatchery (Fig. 4.4) to ensure the safeguarding of the sea turtle eggs and their nesting sites.

The objective of the Walai Penyu Conservation Park is to protect and preserve the sea turtle population in Sabah. The green turtle (*Chelonia mydas*) is fully protected under Schedule 1, Part 1, of the Wildlife Conservation Enactment, 1997. Notably, the conservation park has earned recognition through three prestigious tourism awards at various levels:

1. ASEAN Sustainable Tourism Standard Award 2022–2023, rural category, regional level.
2. Malaysia Tourism Council Gold Award for a Conservation Destination 2022, national level.
3. Sabah Tourism Board's Best Tourism Product 2019, state level.

The Taman Hadiah Sea Turtle Hatchery is a main tourist attraction and serves as an alternative economic activity on the island. Throughout the year, varying numbers of sea turtles arrive at the beach each night to nest (Fig. 4.5). The island's

Fig. 4.3. Aerial view of Libaran Island (photo courtesy of Deechin). The Walai Penyu Conservation Park is located on the sandy beach (encircled in pink) in the foreground (Alin and Yee, 2022).

Fig. 4.4. Taman Hadiah Hatchery, a 6-acre (2.4 ha) turtle hatchery on Libaran Island (Photo courtesy of Alex, 2022).

Fig. 4.5. Sea turtle landing on shore for nesting (photo courtesy of Alex, 2021).

Fig. 4.6. Female sea turtle and her nest (photo courtesy of Alex, 2022).

Fig. 4.7. Fenced hatchery site to protect the turtle eggs. A plastic mesh is placed around each nest for protection (photo courtesy of Alex, 2022).

location in the open sea facilitates their ease of landing. Observing female turtles nest during the night has become the island's primary attraction for visitors (Fig. 4.6). Indeed, the presence of sea turtle nesting sites on the island's long sandy beaches represents a unique tourism resource.

Sea turtles return annually for nesting, landing at night to lay around 150 eggs, which they cover with sand. Rangers carefully dig up, mark, date, and bury these eggs in the Taman Hadiah Hatchery, a securely fenced area (Fig. 4.7). Once hatched, baby turtles wait to be released into the sea, either by visitors or by rangers (Figs. 4.8 and 4.9).

Visitors can enjoy their stay at the nearby Walai Penyu Resort, which offers clean and comfortable tented campsites with stunning sea views.

Main tourism activities on Libaran Island

While its core attractions are sea turtles and coral reefs, Libaran Island's main tourist activities are swimming, snorkelling, and scuba diving. Other activities and programmes carried out on the island include the following:

1. Hatchling release programme, the core attraction and highlight.
2. Education and awareness of sea turtle programmes.
3. Village tour, including traditional mat-weaving and local cooking activities.
4. Sea turtle conservation research activities.

To further strengthen the implementation of park and conservation activities, a Memorandum

Fig. 4.8. Baby turtles in the hatchery, waiting to be released (photo courtesy of Jensen Chua).

Fig. 4.9. Release of baby turtles into the sea (photo courtesy of Alex, 2022).

Fig. 4.10. Alex, known as the 'turtle man' by the local village and the leader of the conservation park (photo courtesy of Alex, 2022).

of Understanding was signed between Trekkers Lodge Sendirian Berhad (TLSB), the tour company that promotes sea turtles at Libaran Island, and the Sabah Wildlife Department in 2013, aiming to enhance local awareness by:

1. Promoting sea turtle conservation through an approved hatchery on the island.
2. Creating awareness programmes for locals and visitors on the conservation value of sea turtles.
3. Conducting research to better understand the life cycle of sea turtles.
4. Communicating and sharing research findings with students and the public.

Method of Enquiry

This case study was written based on the review of secondary published sources related to Libaran Island and personal interviews. The secondary sources included the review of books, local newspapers, and government documents. A comprehensive overview of the case background was provided by these sources. The primary data collection entailed conducting personal interviews with both members of the local community and the initiator of marine turtle conservation at Walai Penyu Conservation Park, Alex, pictured in Fig. 4.10. He expressed that his motivation for setting up the Walai Penyu Conservation Park was to retain, protect, and conserve the endangered sea turtle species on the island (Omar, 2016). The data from the interviews provided invaluable insights into the project.

Seeing sea turtles landing and nesting every year motivated Alex to establish the conservation project known as the Walai Penyu Conservation Park. In the local language, the term 'Walai' ' means 'home', and 'penyu' means 'turtle'. For funding, a conservation levy was introduced on island tour packages sold in both domestic and international markets. The setup further required strong support and collaboration from the 382 residents of Kg Pulau Libaran, the only village on Libaran Island. In Alex's words, 'I wanted to do something, i.e. to develop a specific

Fig. 4.11. The polluted beachfront (photo courtesy of Alex, 2019).

site for nesting and opportunities to release baby turtles later ... I feel this is an opportunity for me and the local community to care about the turtles' landing and nesting'.

Findings of the Case Study

Prior issues faced on Libaran Island

Despite its captivating appearance, Libaran Island has faced several challenges related to environmental degradation, including the accumulation of plastic waste and other rubbish along the shoreline and in the sea (Fig. 4.11). Such waste erodes the beachfront, affects the pristine environment, and harms marine life, making it difficult for sea turtles to land and nest.

Moreover, historically there was a lack of conservation efforts by the islanders to ensure the hatching and preservation of turtle eggs. The island community had limited knowledge and skills in the conservation of turtle eggs. Instead, they would typically harvest turtle eggs for economic gain and personal consumption. A village chief, Sarief Nasidip Uyung, said, 'Before 2013, villagers here would consume turtle eggs regularly as the eggs could be easily found on the beach. It had never crossed our minds that we were endangering the turtles. It was just like chicken eggs, only that this came for free'. In fact, selling turtle eggs was an alternative form of livelihood for the island community. Meanwhile, the authorities made limited efforts to prevent the sale and consumption of turtle eggs on the island.

Essential basic facilities for promoting island tourism, such as clean water and electricity, also used to be scarce on the island. Limited basic amenities and infrastructure made it challenging for the island's local community and ecotourism to thrive.

In short, there were serious socio-economic disparities and limited economic opportunities for the island community before the establishment of the sea turtle conservation project. These had restricted the potential of developing the island as a tourist destination.

Fig. 4.12. The launching of the park and hatchery site, supported by Tourism Malaysia (photo courtesy of Alex, 2022).

Establishment of a marine turtle conservation and hatchery site

In consideration of these issues, Alex worked closely with the island community and the wildlife department to establish a hatchery centre called the Walai Penyu Conservation Park in 2013 (Fig. 4.12). As part of this project, the local community's responsibility is to protect the turtle eggs, and they are prohibited from selling or consuming them.

Additionally, the community has received training to enhance their skills in sea turtle management, conservation practices, waste management, and recycling. Training programmes are conducted by the NGO called Friends of Sea Turtles Education and Research (FOSTER), the wildlife department, and tour operators. Notably, two islanders became certified wildlife wardens after receiving training from the wildlife department. Ecotourism operators have also supported and collaborated with the community to promote the development and sustainable growth of the destination.

Today, the park has evolved into a tourist attraction for both domestic and international visitors. It has yielded significant economic benefits and employment opportunities, ensured the sustainability of marine resources, and improved the livelihood and well-being of the local community.

Tourism facilities provided through the Walai Penyu Conservation Park

Next to the hatchery, the Walai Penyu Resort offers eight dome-shaped luxury tents with a maximum capacity of 16 guests, enabling visitors to stay overnight on the island. Basic tourist amenities, including public shower facilities and dining areas, have been constructed with a focus on minimizing environmental impacts (Fig. 4.13).

Additionally, lighting is kept to a minimum to prevent disturbance to nesting turtles. A designated nesting site allows tourists to observe sea turtles and the release of baby turtles on the shore. Notably, access to clean water and solar energy has been made available on the island to meet tourists' needs.

Evidence of sustainable practices on Libaran Island

Both the private sector and the island community have taken proactive actions to reduce

(a) (b)

Fig. 4.13. (a) external, and (b) internal views of Walai Penyu Resort's public shower facilities (photo courtesy of Alex, 2022).

Fig. 4.14. Collaborative weekly beach-cleaning by islanders and the Walai Penyu Resort, sponsored by a private tour operator (photo courtesy of Alex).

environmental pollution and food waste on Libaran Island, including beach-cleaning collaborations (Figs. 4.14 and 4.15). The following are other sustainable practices undertaken by various stakeholders.

Plastic bottles are recycled into building materials and decorative items for village houses. Fig. 4.16 depicts a unique 300 sq ft house built with 3500 used plastic bottles.

A few villagers have also learned how to repurpose used plastic bottles by painting and hanging them as house decorations and artificial flowers (Figs. 4.17 and 4.18).

The empirical evidence from this case study demonstrates a clear alignment between Walai Penyu Conservation Park's sustainable practices and the United Nations' SDGs. Specifically, SDG 1 (no poverty) is supported

Fig. 4.15. Collaborative beach-cleaning efforts among FOSTER, local villagers, and Alex (photo courtesy of Alex).

Fig. 4.16. A unique 300 sq ft house built by local villagers with 3,500 used plastic bottles (photo courtesy of Alex).

by the generation of job opportunities and income sources for the community. SDG 8 (economic growth) is supported by increased tourism. SDG 12 (responsible consumption) is supported by the sustainable practices and activities of the park. SDG 17 (forging partnerships) is supported by the partnerships among private practitioners, the local community, the wildlife department, and the Sabah Tourism Promotion Board. In short, the collaborative efforts to conserve marine turtles through these partnerships have metamorphosed the island into a sustainable marine ecotourism destination. Such stakeholder partnerships represent a vital and integral component of effective ecotourism destination management for sustainability.

Positive impacts of sustainable practices on Libaran Island

Alex states that 'through conservation initiatives, both villagers and visitors benefit

Fig. 4.17. Painted used plastic bottles hung as decorative items (photo courtesy of Alex).

Fig. 4.18. Artificial flowers made from plastic bottles, decorating a house in Kg Pulau (photo courtesy of Alex).

economically, culturally, and socially'. As a result of the Walai Penyu Conservation Park, local villagers actively participate in turtle conservation, beach clean-ups, recycling programmes, souvenir mat weaving, and bird-watching tours (Lee, 2023). By engaging villagers in these activities, the park has fostered their understanding and realization of the potential for sustainable economic development and tourism on the island.

Locals themselves acknowledge that conservation efforts have played a significant role in maintaining the cleanliness of their village. The village chief stated that, 'We now never throw rubbish into the sea anymore. We clear rubbish that is washed ashore from other places to ensure that it doesn't stay on our beach'. A female villager added, 'We bury and burn our rubbish. Whatever we do, we ensure that the rubbish will not go into the sea'.

The park's efforts have also created numerous job opportunities, including roles such as boatmen, wildlife wardens, beach cleaners, gardeners, and waiters/waitresses. In addition, seven households are being employed as village guides to conduct village walking tours as well, which include visiting village homes, interacting with locals, learning about local food and culture, and observing traditional mat weaving using rattan – a nearly extinct material on the island. This initiative generates additional income for the locals, which is crucial as many islanders lack a fixed income to support their families. Additionally, visitors to the villages purchase food, beverages, and handicrafts from the villagers, further strengthening the local economy.

Community engagement and collaboration towards the sustainability of the Walai Penyu Conservation Park

According to the United Nations World Tourism Organization (n.d.), or UNWTO, strong commitment and collaborative efforts among tourism practitioners, local communities, and government agencies are imperative for the achievement of sustainable development and destination management. Correspondingly, in the present context, the Walai Penyu Conservation Park's successful sea turtle conservation and positive environmental impacts have been made possible by the strong commitment of various stakeholders, as follows:

1. FOSTER, an NGO that supports turtle conservation through volunteering services and advice, sponsors the community beach-cleaning project and operates the turtle hatchery on Libaran Island in collaboration with the Sabah Wildlife Department.
2. Private tour companies and a hotel in Sabah sponsor beach-cleaning activities and promote the Walai Penyu Conservation Park as a tour package.
3. Gamuda Land (a property developer) adopted a 100m plot of beach land as part of corporate social responsibility to ensure the beach is cleaned.
4. The Federal and State Tourism Promotion Boards offer support, feedback, and sponsorship to the Walai Penyu Conservation Park for the conservation of sea turtles and as an ecotourism destination to attract tourists.
5. The honorary wildlife wardens of the Sabah Wildlife Department provide advice, technical support, and monitoring. Villagers will enhance their skills and knowledge in conservation practices and protection of the sea turtles.
6. The Tanjung Aru Marine Ecosystem Centre conducts workshops and study sessions to elevate villages as tourist attractions. This will enhance the quality of service and experiences of Libaran Island as a marine ecotourism destination.

Sustainable Marine Ecotourism Framework

Based on the findings, this case study highlights the significance of a comprehensive framework for a sustainable marine ecotourism destination, comprised of three critical elements: sustainable practices and pillars, destination resources, and the involvement of stakeholders from the local community, tourism sector, and government (Fig. 4.19). The framework underscores the interdependence of these three elements, emphasizing the need for effective collaboration and partnerships among the private sector, local community, and relevant government agencies.

For the long-term sustainability and growth of Libaran Island, the conservation of destination resources must be a priority. Simultaneously, active engagement from various groups (i.e. the local community, tourism stakeholders, government agencies, and NGOs) is essential to promote sustainable practices. By fostering

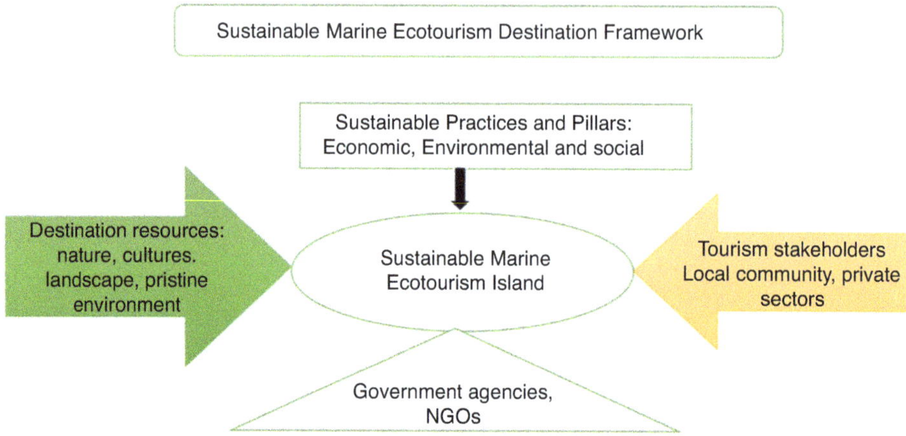

Fig. 4.19. Proposed sustainable marine ecotourism framework.

effective collaboration among these groups, the island can evolve into a sustainable and thriving ecotourism destination.

Conclusion

The concept of sustainability is multifaceted and contextual, encompassing economic, environmental, and socio-cultural aspects. Notably, strong commitment and collaboration among various stakeholders is vital for creating sustainable tourism destinations (Ruggerio, 2021). From this perspective, Libaran Island has demonstrated significant potential as a marine ecotourism destination that embodies the three pillars of sustainability: the economy, the environment, and society. It serves as a tangible, positive example of successfully restoring an island ecosystem. The case points to the importance of continuous support from stakeholders and government agencies in protecting sea turtles and their eggs for long-term sustainability.

In conclusion, this case study illustrates the successful development of a marine ecotourism destination through conservation efforts and sustainable practices. Thus, it offers valuable insights as a learning case and contributes to understanding the theoretical and practical implications of sustainable destination management for marine ecotourism sites.

References

Alin, J. and Yee, A. (2022) *A Tale of 30 Thousand Turtles: Walai Penyu Conservation Park*. FOSTER (Friends of Sea Turtles Education and Research), Libaran Island, Sabah.

Lee, S. (2023) Sandakan conservation park involves villagers in activities for win-win partnership. *The Star*. Available at: https://www.thestar.com.my/news/nation/2023/02/02/sandakan-conservation-park-involves-villagers-in-activities-for-win-win-partnership (accessed 10 July 2023).

Omar, A.A. (2016) Turtle landings in Libaran bring cheer to locals. *New Straits Times*. Available at: https://www.nst.com.my/news/2016/05/147846/turtle-landings-libaran-bring-cheer-locals (accessed 10 July 2023).

Pilcher, N. (2022) Foreword. In: Alin, J. and Yee, A. (eds) *A Tale of 30 Thousand Turtles: Walai Penyu Conservation Park*. FOSTER (Friends of Sea Turtles Education and Research), Libaran Island, Sabah, p. ii.

Ruggerio, C.A. (2021) Sustainability and sustainable development: A review of principles and definitions. *Science of The Total Environment* 786, 147481.

Sabah Parks Board of Trustees (2023) Turtle Islands Park. Available at: https://www.sabahparks.org.my/turtle-islands-park/turtle-conservation (accessed 10 July 2023).

United Nation World Tourism Organization (UNWTO) (n.d.) Sustainable Development. Available at: https://www.unwto.org/sustainable-development (accessed 5 October 2023).

UPM, UMS and Perhilitan, J. (1996) *Development and Management Plan: Turtle Islands Park.* Sabah Parks, Sabah, Malaysia.

Vanar, M. (2016) Malaysia: Turtle Conservation Awareness Bearing Results. Available at: https://wildsingaporenews.blogspot.com/2016/03/malaysia-turtle-conservation-awareness.html (accessed 10 July 2023).

Section 2: Cultural Preservation and Identity Through Tourism

In this section the focus shifts to cultural sustainability, exploring cases from Indonesia, Singapore, Cambodia, Japan and China. These case studies demonstrate how cultural sustainability is achieved through the preservation of heritage sites, traditional arts, and cultural practices, while also emphasizing the role of stakeholders in shaping destination identity.

5 Improving Cultural Sustainable Tourism: Evidence of Saba Budaya Baduy from the Baduy Tribe, Indonesia

I. Putu Gede Eka Praptika, Amanda Natasanti, Della Ade Saputri, Muhammad Ibnu Azzulfa, Brigita Talentiana Mahen and Fahmi Prihantoro
Universitas Gadjah Mada, Yogyakarta, Indonesia

Summary
Baduy Cultural Tourism is the activity of visiting the Baduy Tribes in Indonesia, which potentially threatens the resilience of the Baduy's culture. Responding to this, the Baduy Customary Institution changed the nomenclature of Baduy Cultural Tourism to *Saba Budaya Baduy*. This case study explores the value and specialty of *Saba Budaya Baduy* as the identity of Baduy Tribes, to analyze the potential of threats and the cultural resilience from the usage of Baduy Cultural Tourism and *Saba Budaya Baduy* nomenclatures. It explores the values of *Saba Budaya Baduy* as the effort of cultural resilience to realize cultural sustainable tourism. This qualitative research is based on tourism anthropology and social psychology perspectives in line with the interpretative phenomenological analysis (IPA) technique. The initial findings indicate that Saba Budaya Baduy is guided by the principle of fostering a strong relationship between hosts and guests, rooted in the cultural values of the Baduy Tribe. This approach aims to enhance the representation of the Baduy Tribe and improve the overall experience for visitors engaging in activities with the tribe. The results also highlight how psychological resilience, in responding to external threats, is reflected in the Baduy indigenous people, who are obedient in maintaining and implementing their ancestral traditions. Finally, *Saba Budaya Baduy* is oriented towards the social, cultural, tourism economy, and environmental resilience dimension of visiting activities within the Baduy Tribe. Therefore, *Saba Budaya Baduy* can realize and improve cultural sustainable tourism which is oriented towards cultural resilience and sustainable tourism activities.

Background

Cultural practices can attract the tourism sector, especially cultural tourism, improving the community's economy (Arcana and Wiweka, 2016; Yeniasir and Gökbulut, 2018). However, the existence of tourism activities not only has a positive impact but can also threaten cultural and environmental resilience. This risk can arise from interactions between local people and tourists from different cultures. These two communities can influence each other, with the possibility of domination of one party over the other. This occurs in Indonesia, famous for its ethnic and cultural diversity. The Baduy tribe is one of the indigenous communities in Indonesia that live in Kanekes Village, Lebak Regency, Banten Province (Budiaman *et al.*, 2018). The Baduy indigenous people have experienced various impacts from tourism.

Tourism slowly influences the environmental conditions and culture of the Baduy Tribe. Kanekes Village, where the Baduy indigenous people live, was designated a tourism destination by the Lebak Regency Government in 1992 (Solikhah, 2020). However, the nomenclature of Baduy Cultural Tourism as an identity often contained in tour packages or instructions for accessing tourist destinations of the Baduy is allegedly threatening cultural resilience. According to Sunaryo (2013), cultural tourism is an activity associated with observing tourist objects based on human creations, both in the form of objects and cultural values that have remained constant until now. Based on this understanding, Baduy Cultural Tourism seems to position the Baduy Tribe as merely an exciting object to visit and watch. There was a surge in tourists visiting the Baduy to over one hundred thousand in 2019 (Rifa'i, 2020). This massive visitation was accompanied by environmental pollution issues and the incompatibility of tourist behavior with the Baduy Tribe's culture, thus threatening the existence of the Baduy Tribe.

In principle, the Baduy indigenous people uphold customs, customary law, natural preservation, and sustainable life (Kartika and Edison, 2019). The Baduy indigenous people live based on customs and traditions passed down from generation to generation. The Baduy Tribe's culture requires the division of groups based on adherence to custom, namely the *Baduy Luar* (*Panamping*) and the *Baduy Dalam* (*Tangtu*). *Baduy Luar* refers to the Baduy indigenous people group residing in the outer regions, which are open to interactions with the outside world, while *Baduy Dalam* denotes the Baduy indigenous people living in inner areas, adhering to a more traditional way of life and maintaining strict limitations on contact with outsiders. However, these two groups have the same objective to maintain the cultural and environmental nobility of the Baduy Tribe. This objective is reflected in the persistence of the community in complying with customary provisions called *Pikukuh* (Budiaman *et al.*, 2018).

The realization of obedience to *Pikukuh* is reflected in traditional farming life and simple house structures without electricity. However, research by Kartika and Edison (2019) shows that there has been a shift in behavior patterns among the *Baduy Luar* who have started using cell phones and batteries for their energy needs. One of the causes of this is the high intensity of interaction between the Baduy indigenous people, especially the *Baduy Luar* and tourists who have inharmonious behavior patterns with the Baduy Tribe's culture (Firdaus and Herminasari, 2020). Various changes in the community behavior and environmental conditions in the Baduy area and circumstances due to tourism activities that are inharmonious with the Baduy Tribe's culture reflect the negative impact of tourism. This reality reinforces the indication that tourism and the Baduy Tribe's culture are not yet harmonious. This has the potential to threaten cultural resilience.

Cultural resilience is characterized by a desire to maintain traditions and cultural values as a form of respect for ancestors (Rugebregt *et al.*, 2020). Similarly, the Baduy indigenous people strive to preserve their cultural values amidst challenges posed by tourist behavior, such as the use of smartphones, inappropriate attire, chemical goods, and incongruous lifestyles. This endeavor reflects their dedication to cultural resilience and the imperative of conserving their cultural heritage. Therefore, responding to various threats to cultural resilience, the Baduy Customary Institution (*Lembaga Adat Baduy*) seeks to emphasize the need to change the terminology of 'Baduy Cultural Tourism' to become '*Saba Budaya Baduy*'.

The existence of *Saba Budaya Baduy* nomenclature has been regulated by the Baduy Customary Institution. Since 2007, the term 'Baduy Cultural Tourism', referring to community visits to the Baduy Tribe area, has been changed to '*Saba Budaya Baduy*'. The new nomenclature was ratified in Village Regulation (*Peraturan Desa*) Number 1 of 2007 concerning *Saba Budaya* and the Protection of Baduy indigenous people. This endeavor was undertaken to sustain tourism activities within a framework aligned with the cultural values of the Baduy indigenous people. Obviously, the positive impact of tourism that contributes to the Baduy indigenous people's economy still needs to be acknowledged by the alignment between local cultural interests and tourism.

This case study analyzes the value of *Saba Budaya Baduy* as an identity that is harmonious with tourism. Specifically, this research aims to analyze the potential of the *Saba Budaya Baduy* nomenclature, which is more suitable than Baduy Cultural Tourism in strengthening the resilience of the Baduy Tribe's culture.

This case study also explores the value of *Saba Budaya Baduy* as an effort to create cultural resilience to strengthen cultural sustainable tourism while maintaining the Baduy Tribe's culture without eliminating tourism activities.

Primary data was collected through purposive sampling, employing in-depth interviews with informants selected based on specific criteria aligned with the research objectives: individuals involved in formulating the term *'Saba Budaya Baduy'* and those engaged in tourism and research activities within the Baduy Tribe. Seven informants were identified: three key informants from the *Baduy Luar*, one representative from the Baduy Customary Institution, called *Jaro Pamarentah*, one from the Banten Province Tourism Office, an activist specializing in observing Baduy customs, and an expert lecturer focusing on the Baduy indigenous people. Data analysis employed a descriptive-analytical method rooted in a phenomenological approach, complemented by tourism anthropology and social psychology perspectives, guided by the National Resilience Assessment Scale (NRAS).

Values and Characteristics of *Saba Budaya Baduy*

The Head of Kanekes Village is customarily known as *Jaro Pamarentah*. *Jaro Pamarentah* is one of the positions within the Baduy indigenous people responsible for traditional and legal affairs. The person holding this position serves as the head of customary or traditional leadership. Their role includes overseeing and ensuring the implementation of the Baduy Tribe's culture, as well as mediating disputes within their community. *Jaro Pamarentah* stated that visits from outside communities to the Baduy area had occurred since the 1980s. On the one hand, the enduring customs preserved by the Baduy indigenous people across generations have been tourist attractions, attracting visitors from the outside to visit the Baduy Tribe. For instance, as depicted in Fig. 5.1, Baduy indigenous people are observed carrying their harvest from the fields barefoot. On the other hand, by adhering to a cultural pattern that shields itself from the influence of new cultures and external lifestyles,

Fig. 5.1. Baduy indigenous people are carrying their harvest from the fields barefoot. Source: radarjabar.disway.id.

the Baduy indigenous people express their reverence for ancestors, aiming to maintain the balance and harmony of the universe (Saragih *et al.*, 2023).

Arcana and Wiweka (2016) classify the types of tourist attraction in Baduy into natural and sociocultural. Natural attractions include forests with various flora and fauna, natural lakes, and bridges formed from tree root vines. Sociocultural tourism attractions are associated with traditional Baduy house architecture, traditions, and religious rites. These attractions include the religious system of the Baduy indigenous people, namely *Sunda Wiwitan*. *Sunda Wiwitan* is a traditional Sundanese faith or belief system in Indonesia. It is a form of indigenous religion practiced by some of the Sundanese people, of which one group of them is the Baduy indigenous people. Based on *Sunda Wiwitan*, the Baduy Tribe carries out customary rules called *Pikukuh*, which are determined by the ancestor spirits, who are supernatural figures (Herandy, 2022).

The combination of natural beauty and sociocultural richness has established the Baduy Tribe's region as a destination for cultural tourism within Lebak Regency, known as Baduy Cultural Tourism. According to insights from the Baduy Customary Institution, the term 'tourism' within the context of Baduy Cultural Tourism has taken on an unfamiliar connotation, allowing outsiders visiting the Baduy area considerable freedom to pursue satisfaction as tourists. Nonetheless, the Baduy indigenous people are committed to upholding their local customs, which visitors are obliged to observe. Therefore, 'Baduy Cultural Tourism' differs from the term *Saba Budaya Baduy*, *Saba* – meaning *'silaturahmi'* (friendship-visitation) – derived from the language used by the Baduy indigenous people in their daily communication, namely Sundanese. Then, the term *Budaya Baduy* shows that the foundation in *Saba* for the Baduy indigenous people is their culture and customs.

The difference between *Saba Budaya Baduy* and Baduy Cultural Tourism lies in the definition and responsibilities of visitors from outside the Baduy Tribe. Within the framework of cultural tourism, which serves as the foundation of Baduy Cultural Tourism, visitors are typically positioned as tourists who come to observe and enjoy cultural-based tourism objects and experiences (Sunaryo, 2013). As a result, the Baduy indigenous people have the potential only to become tourist objects and managers of tourist attractions whose job is to provide services to visitors. Further, Waluya *et al.* (2022) argue that the existence of tourism can pose threats to social change. This can include the phenomenon of cultural shock which can be understood as the sensation of confusion and being out of place that an individual undergoes when they are abruptly exposed to an unfamiliar culture presented by tourists. Second, it can cause impact events from social inequality. However, in the concept of *Saba Budaya Baduy*, visitors must adjust to the Baduy Tribe's culture and comply with all the prescribed rules.

Visitors are positioned as guests who aim to stay in touch or establish friendships with the Baduy indigenous people. When arriving at Baduy, visitors are required to show respect to the residents, adhere to local customs, and uphold traditional practices. As depicted in Fig. 5.2, the visitor (a man on the right wearing eyeglasses) is seen wearing local attire while visiting a Baduy indigenous people's house. The other residents on the left side of the visitor, who are Baduy indigenous people themselves, warmly welcomed him and even offered a welcome drink. The Baduy indigenous people, as hosts, are responsible for providing advice or warnings to guests who violate customary rules. This action can prevent the Baduy Tribe from the potential of cultural exploitation by outsiders who seem to position the Baduy Tribe as mere objects for spectacle. This issue was conveyed by an informant from the *Baduy Luar* during an interview:

> It has happened several times, in Baduy, we truly are the residents, like a spectacle, when visitors arrive, sometimes we are ignored.

The implementation of *Saba Budaya Baduy* contains values of social life such as kinship, decency, and order. The value of kinship is manifested in the friendly attitude of the Baduy indigenous people in welcoming visitors and receiving them as family members returning home. The results show that the Baduy indigenous people enjoy being friendly with visitors. The value of decency or politeness is reflected in the reciprocal attitude of visitors who must respect the culture and people. For instance, visitors are

Fig. 5.2. Tourist visitation based on *Saba Budaya Baduy*.

expected to behave and speak respectfully, which includes adhering to cultural norms, avoiding improper behavior, and expressing positive ideas and words. The value of order requires visitors to behave according to the culture and customary rules of the Baduy Tribe to maintain public peace and environmental sustainability. *Jaro Pamarentah* emphasized the importance of adhering to customary rules to prevent adverse consequences such as pollution or accidents. As he stated below:

> You cannot use soap in the river, you cannot scream, you cannot take pictures. This is what the Baduy indigenous people should know. If they meet visitors, they must be informed. If visitors are not notified, those who are in danger will be the Baduy indigenous people. But if you have been informed but still do not follow the rules, visitors will be in danger.

The values contained in the *Saba Budaya Baduy* must be implemented and maintained by the whole of society, from the Baduy indigenous people to visitors. Changes in society can occur and cannot be avoided, but that does not mean that changes cause a society to abandon their values. Society has the right to choose and determine the values that will be abandoned, maintained, and used. The life of the indigenous Baduy people experiences a similar thing. Both the *Baduy Dalam* and *Baduy Luar* people are willing to accept visitors from outside the Baduy with certain conditions. The main requirement is that visitors must comply with customary rules in the Baduy in order to maintain the cultural order that the indigenous Baduy people uphold.

Cultural Resilience: The Implementation of *Saba Budaya Baduy*

The use of the *Saba Budaya Baduy* as a nomenclature reflects identity and is expected to strengthen the cultural resilience of the Baduy indigenous people. From the perspective of social psychology, it is known as collective resilience. It relates to how people face the challenges and pressures of life and strive for sustainability

(Kirmayer *et al.*, 2009). Thus, a desire arises to stay together for a better life (Nashori and Saputro, 2021). Collective resilience is built on forming a cultural strategy to deal with change and how people adapt themselves to existing normative values and act as creative agents. *Saba Budaya Baduy* follows the principles of cultural resilience, which includes language, religion, art, family, and social systems (Makmur and Biantoro, 2014). Through the principle of cultural resilience, showing cultural forms that are different from other cultures can also become the identity of a culture.

The National Resilience Assessment Scale (NRAS) method is used in this case study to analyze the resilience of community culture (Kimhi *et al.*, in Nashori and Saputro, 2021). This method is a valuable tool utilized to meticulously analyze and gauge the resilience of community culture. This comprehensive method delves into four fundamental aspects of resilience, which serve as its foundational factors, including identifying with the state, solidarity and social justice, trust in national institutions, and trust in public justice.

In the first aspect, identifying with the state includes confidence in the ability of leaders and the government to make the right policies for the local community. In this aspect, there are two reference indicators: the Baduy indigenous people believe in the Baduy Customary Institution, and the Baduy indigenous people love their culture. All elements of the Baduy indigenous people obey the customs they adhere to (Ngamanken, 2013). This obedience is stated in the *Pikukuh*, which also serves as a guideline for the behavior of the Baduy indigenous people to maintain cultural values amidst the progress of civilization (Suparmini *et al.*, 2013). Using *Saba Budaya Baduy* in place of the term Baduy Cultural Tourism has been agreed upon by the Baduy Customary Institution and respected by the Baduy indigenous people. There are also threats of punishment for anyone who violates the stipulated provisions, starting from being *ditegor*/reprimanded, *dipapatahan*/admonished, to being expelled from the Baduy area (Fathurokhman, 2010).

The second aspect is solidarity and social justice. The behavior of the Baduy indigenous people who adhere to ancestral traditions can become a solid foundation of cultural resilience to limit threats from outside cultural influences. This is supported by Rutter's statement (in Fletcher and Sarkar, 2013) that psychological resilience contributes positively to individual differences in society in response to adversity and threats. This fosters a sense of caring and mutual need for one another to maintain the harmonization of customs (Weiss, 2017). The Baduy indigenous people believe in one ideology, namely, 'humans born into this world are entrusted with protecting nature by their ancestors, namely *Batara Tunggal* (God Almighty) and *Batara Cikal* (the first person born into the world).'

Tourism has positively impacted the Baduy indigenous people economically in various ways. Tourism activities can provide income-generating opportunities through the sale of local products such as souvenirs and handicrafts, natural resources such as local sugar and honey, and even local guided tours. Obviously, tourism has created job opportunities for the Baduy indigenous people, especially in the *Baduy Luar*, from tour guides to artisans, providing a source of livelihood. In terms of education and quality of human resources, tourism among the Baduy Tribe can lead to increased education and awareness among the Baduy indigenous people regarding the importance of their culture and natural environment.

The rise in tourism activities and tourist visits has caused the Baduy indigenous people to worry about environmental and cultural sustainability. The influence of modern culture, which is synonymous with the use of telecommunications equipment, clothing styles, and visitor behavior patterns, has the potential to trigger the degradation of people's adherence to customs. This influence is proven by the community's desire to own objects commonly used by visitors but prohibited by customs. In addition, tourism activities in the Baduy have made it seem that indigenous people are mere spectacles or objects, not subjects that play a central role in tourism activities. This was explicitly conveyed by informants from the Baduy Customary Institution regarding the negative impact of tourism activities on the Baduy. These findings reflect the threat that the indigenous people of Baduy feel from Baduy Cultural Tourism activities. Therefore, the nomenclature of *Saba Budaya Baduy* is considered appropriate

Fig. 5.3. The Baduy Customary Institution, together with the local government, discussed the *Saba Budaya Baduy*.

to represent the Baduy Tribe and visit activities to the Baduy.

To implement *Saba Budaya Baduy*, this concept was agreed upon by the Baduy Customary Institution, Baduy Indigenous Activist, and the Tourism Office of Lebak Regency. Fig. 5.3 shows that the Baduy Customary Institution collaborated with the local government in meaningful discussions concerning *Saba Budaya Baduy*. This significant deliberation took place in a traditional house in *Baduy Luar*, hosted by the Baduy Customary Institution, with participants attired in traditional uniforms, as depicted in Fig. 5.3. The outcomes of this discussion culminated in a pivotal agreement involving the Baduy Customary Institution, Baduy Indigenous Activist, and the Tourism Office of Lebak Regency. This agreement led to the issuance of Village Regulation (*peraturan desa*) Number 1 of 2007 concerning *Saba Budaya Baduy* and the Protection of the Baduy indigenous people. This agreement is important when referring to the third aspect of the NRAS, namely trust in national institutions. It has been reflected in the Baduy indigenous people, who believe in a system that is regulated jointly with the government to maintain cultural integrity. This trust is related to one of the tasks of the Baduy indigenous people, namely *Ngasuh Ratu Ngayat Mena*, which means recognizing the existence of a legitimate government. This is reflected in the *Seba Baduy Ceremony*, a form of gratitude for establishing a friendship with the Government of Lebak Regency and Banten Province. The *Seba ceremony* contains various tips, advice, and mandates, among which a leader must be a role model, not arbitrary, and protect his people (Rusnandar, 2013). Related to this, the Baduy indigenous people also need legal protection and government recognition of their customs.

The fourth aspect of the NRAS emphasizes trust in public justice. This aspect describes two indicators, namely the evaluation of the Baduy indigenous people towards the news in the media and the role of the media in reporting on the Baduy indigenous people. The presence of the media has a strategic role in acquainting the public with the traditional lifestyle of the Baduy indigenous people centered around the enduring principles of *Pikukuh*, which is still upheld. One is news related to hospitality towards visitors who come to the Baduy area. The media also emphasizes that the Baduy indigenous people are not an isolated community, but they deliberately isolate themselves and have a concept of life that is at one with nature (Nashori and Saputro, 2021). This perspective reveals that while the Baduy indigenous people are susceptible to potential threats and changes, they selectively accommodate changes while steadfastly adhering to their established customs.

Table 5.1. Method analysis *National Resilience Assessment Scale* (NRAS).

Aspect	Indicator	Findings
Identifying with the state	Believe in Baduy Customary Institution	The Baduy indigenous people exhibit compliance and adherence to the directives issued by the Baduy Customary Institution.
	Have a love for Baduy culture	The Baduy indigenous people have a love for culture and are obedient to *Pikukuh* as a guideline of behavior to maintain cultural values amid the progress of civilization.
Solidarity and Social Justice	Feel safe in the Baduy area	The Baduy indigenous people are assigned to live a simple life based on honesty, as it is, not changing behavior and glorifying the universe where they live in harmony.
	Have a high sense of concern	The principle of life respects fellow human beings and glorifies the life of the universe and its contents.
	Show a sense of mutual need for one another	Preservation of ancestral traditions based on kinship.
Trust in national institutions	Believe in a system regulated by the government regarding the community	The government has a central role in facilitating and assisting management of tourism activities with an orientation towards customary preservation.
Trust in public justice	The Baduy Tribe's assessment of the news in the media	The news about the Baduy indigenous people asking to be removed from the Baduy tourist destination in July 2020 was not confirmed by the Baduy Customary Institution through *Jaro Pamarentah*.
	The role of the media in reporting about the Baduy indigenous people	Reporting on real conditions in the Baduy Tribe has a strategic role to provide education for visiting outsiders and to be a blessing for the Baduy indigenous people.

The explanation regarding the cultural resilience of the Baduy Tribe, which was examined through the NRAS method, is strengthened by several findings shown in Table 5.1.

Based on Table 5.1, the nomenclature change from Baduy Cultural Tourism to *Saba Budaya Baduy* is the right step to strengthen the resilience of the Baduy Tribe's culture. Cultural resilience, as suggested by Hall *et al.* (2017), can enhance the community's collective response to crises. This is reflected in the analysis using the NRAS method, which identified the psychological resilience of the Baduy indigenous people in maintaining the Baduy Tribe's culture. Then, *Saba Budaya Baduy* is suitable as a reference for optimizing the positive impact of tourism activities. Thus, implementing *Saba Budaya Baduy* can maintain the nobility of the Baduy Tribe's culture amidst the intensity of tourism activity. This follows one of the philosophies of life adopted by the Baduy indigenous people: *lojor teu meunang dipotong, pendek teu meunang disambung* (long cannot be cut, short cannot be spliced). This philosophy means that the Baduy indigenous people must live with honesty and simplicity, not changing behavior, not destroying nature, and glorifying life, both with humans and the universe and everything in it. Hopefully, the nomenclature of *Saba Budaya Baduy* can become a guideline for the Baduy indigenous people, enabling long-term cultural resilience during the increasing presence of foreign cultures through tourist presence, tourism development, and attachment between communities.

Fig. 5.4. The tourist learned the typical weaving of the Baduy Tribe.

Cultural Sustainable Tourism: The Harmonization Between Tourism and the Baduy Tribe's Culture

The internalization of *Saba Budaya Baduy* aims to strengthen cultural resilience without compromising tourism activities. This orientation aligns with the concept of cultural sustainability, which promotes the harmony of sustainable development with cultural values and natural systems to sustain the lives of indigenous people (Finlayson, 2015; Handayani *et al.*, 2018). Cultural heritage and vitality constitute the fourth pillar of sustainability that actively encourages the welfare of society, parallel to the other three pillars of sustainability, namely ecological, social, and economic (Hawkes, 2001). In cultural sustainability, cultural aspects also play a significant role in achieving economic values that positively impact social and ecological development (Soini and Birkeland, 2014). *Saba Budaya Baduy* represents a concrete step that is taken by the Baduy indigenous people to protect and preserve their culture, prioritizing it above tourism interests. For instance, Fig. 5.4 strengthens this point, depicting a visitor in the center holding traditional woven fabric while being introduced to Baduy's traditional weaving by the Baduy indigenous people standing beside him. It is hoped that visitors from outside communities will extend their support to the Baduy Tribe's endeavor to safeguard the universe through simplicity and coexistence (Septiana, 2018). The informants' statements further reinforce this perspective:

> For the sake of goodness for the future, Baduy can open Saba Budaya with a note that it must be orderly. It must follow the cultures that exist here. It is best to come here to interact with the community and take care of the environment together. Because the Baduy people live here, there is a reason, not our own volition. It is just that we can't do it arbitrarily because we remember that we have to maintain the trust of our ancestors.

Baduy indigenous people's efforts to preserve their culture while also deriving positive benefits from tourism activities through *Saba Budaya Baduy* align with the concept of sustainable tourism. In their study, Falatoonitoosi *et al.*

Fig. 5.5. Tourists buy typical weaving works of the Baduy Tribe.

(2022) stated that sustainable tourism can have a positive impact on environmental quality and prosperity. Sustainable tourism is oriented towards achieving harmony in natural social and cultural dimensions while minimizing negative impacts and maximizing economic benefits for local communities (Moswete and Thapa, 2015). This orientation is based on three main aspects. First, preserving the natural dimension is reflected in the behavior patterns of the Baduy indigenous people, who are committed to protecting their nature. Second, the life of the Baduy indigenous people, who strictly adhere to their ancestral traditions to resist outside cultural influences, aligns with sociocultural sustainability. Finally, optimizing economic gains is seen through the utilization of tourism activities for the introduction and selling of locally distinctive products inherent to the Baduy Tribe's culture. As depicted in Fig. 5.5, tourists are encouraged to purchase traditional Baduy woven items, bags, and hats to maximize economic benefits for locals.

Saba Budaya Baduy, rooted in the nobility of the Baduy Tribe's culture, is closely related to the concepts of cultural sustainability and sustainable tourism. It is a potential indicator for realizing cultural sustainable tourism, which comprises three main aspects, sociocultural dimensions, cultural tourism economics, and environmental resilience. These aspects are measured by 12 indicators, providing a benchmark for internalizing cultural sustainable tourism within each aspect (Durovic and Lovrentjev, 2014). Yeniasir and Gökbulut (2018) define cultural sustainable tourism as an activity aimed at preserving natural, cultural, and historical values for the benefit of future generations, guided by the principle of usefulness. Based on research findings, the correlation between *Saba Budaya Baduy* and the concept of cultural sustainable tourism is presented in Table 5.2.

Based on the findings presented in Table 5.2, *Saba Budaya Baduy* holds significant value as it represents the Baduy indigenous people's commitment to preserving their culture amidst tourism activities. Additionally, the Baduy indigenous people capitalize economically on their cultural attractiveness by offering crafts and processed products unique to the Baduy. Moreover, the implementation of *Saba Budaya Baduy* showcases the Baduy Tribe's harmonious coexistence with nature and reflects the embodiment of environmental resilience.

The concept of *Saba Budaya Baduy*, which is aligned with cultural sustainable tourism, is oriented towards realizing the resilience of the Baduy Tribe's culture without compromising

Table 5.2. *Saba Budaya Baduy* values based on cultural sustainable tourism indicators.

Aspect	Indicator	Findings
Sociocultural dimension	Protection of cultural heritage	*Saba Budaya Baduy* obliges every tourist to always respect the traditions of the Baduy indigenous people. Protection of the Baduy Tribe through the implementation of the *Saba Budaya Baduy* is represented by Perdes No. 1 of 2007 concerning *Saba Budaya Baduy* and the Protection of the Baduy Indigenous People.
	Social carrying capacity	Baduy indigenous people have tolerance towards their community and outsiders.
	Authentic presentation of cultural heritage	Saba Budaya Baduy represents the identity of the Baduy indigenous people, who maintain traditions and culture.
	The level of local community satisfaction in general	The use of *Saba Budaya Baduy* is in line with the wishes of the Baduy indigenous people to create harmony between tourism and the life of local cultural traditions.
Cultural tourism economics	Improving the economy of local communities	*Saba Budaya Baduy* activities can still encourage the economic progress of indigenous people from the sale of products and services offered by communities in the *Baduy Luar* area.
	Human resource training	There is training from the local government regarding the manufacture and sale of local products as souvenirs.
	The volume of tourist demand	*Saba Budaya Baduy* activities are still able to attract tourists who want to experience life directly in the Baduy Tribe. Market opportunities for selling handicrafts and processed food products resulting from the creativity of the community remain open.
Environmental resilience	Protection of natural ecosystems	*Saba Budaya Baduy* has an impact on maintaining the natural ecosystem of the Baduy region according to local traditions.
	Energy management	Electronic use is permitted in the Baduy Luar area. However, it is prohibited for use in Baduy Dalam.
	Water management	Baduy indigenous people maintain the environment by not using soap, toothpaste, and other chemicals so as not to pollute the river.
	Pollution management	There is a ban on the construction of roads for motorized vehicles to support tourism activities.
	Environmental management	*Saba Budaya Baduy* supports the efforts of indigenous people in preserving their culture and environment.

tourism advantages. Consequently, tourists who visit play an important role in implementing *Saba Budaya Baduy*. This concept has universal values to maintain cultural and environmental sustainability amid tourist activities. As highlighted in Fig. 5.6, the Baduy Tribe has embraced tourism and visitation activities, maintaining the traditional appearance of elements like houses and pathways. Thus, *Saba Budaya Baduy* can be an inspiration for reference for sustainable

Fig. 5.6. Activities of the Outer Baduy tribe and tourists. Source: suara.com, 2022.

management of cultural tourism destinations throughout Indonesia and worldwide.

Baduy indigenous people demonstrate through *Saba Budaya Baduy* the urgency of safeguarding their cultural heritage against the detrimental impacts of tourism activities. According to Waluya *et al.* (2022), *Saba Budaya Baduy* serves to revitalize tourism in the Baduy Tribe, emphasizing sustainable development anchored in the values of the indigenous community. This approach aligns with community-based tourism (CBT), which empowers local communities to own and manage tourism while upholding sociocultural principles (Waluya *et al.*, 2022). This perspective is deeply ingrained in the Baduy indigenous people, as evidenced by *Saba Budaya Baduy's* foundation in the ancestral tradition of *Pikukuh*. When all stakeholders – the Baduy indigenous people, the government, and tourists – collaborate to preserve the Baduy Tribe's culture, it fosters an environment where future generations can still experience the essence of cultural sustainable tourism. Considering that all tourism destinations and typologies must now adopt responsible practices across sociocultural, environmental, and economic dimensions for long-term sustainability (Mondéjar-Jiménez and Ferrari, 2022), the cultural sustainability of the Baduy Tribe paves the way for tourism sustainability. Both cultural sustainability and tourism sustainability can flourish together, always prioritizing mutual respect and care.

Conclusion

Saba Budaya Baduy is a proper nomenclature representing the cultural identity of the Baduy indigenous people rather than cultural tourism. The presence of tourists is based on the principle of hospitality in the values of *Saba Budaya Baduy* – kinship, decency, and order – reflecting the harmony between the Baduy indigenous people and the outside community. This philosophy is contrary to Baduy Cultural Tourism, which places the Baduy Tribe as mere spectacle objects that can threaten cultural resilience. The change in the use of the Baduy Cultural Tourism nomenclature to the *Saba Budaya Baduy* is one of the solutions to increasing the cultural resilience of the Baduy indigenous people. This nomenclature is supported by social psychological resilience in responding to threats reflected in the obedience of the Baduy indigenous people to maintain ancestral traditions. Moreover, *Pikukuh* becomes a moral and customary bond that strengthens the psychology of the Baduy indigenous people in preserving their culture.

In addition, *Saba Budaya Baduy*, accompanied by a determination to maintain culture while taking positive benefits from tourism activities, is a reference for realizing cultural harmony and tourism. This harmony represents cultural sustainable tourism which focuses on cultural resilience, natural preservation, and historical understanding to ensure continued benefits for the next generation. Internalization of *Saba Budaya Baduy* is expected to be able to bind visitors who visit to refer to the Baduy Tribe's cultural order. As a result, *Saba Budaya Baduy* has the potential to strengthen the resilience of Baduy culture while still taking positive benefits from sustainable tourism. The response from the tourist side to the existence of the *Saba Budaya Baduy* nomenclature as a substitute for the term Baduy Cultural Tourism has the potential to be studied further to enrich academic and broader perspectives and discourse regarding sustainable tourism and culture.

References

Arcana, K.T.P. and Wiweka, K. (2016) A study of indigenous tourism product development case study: The baduy, South Banten, West Java. *Jurnal Ilmiah Hospitality Management* 7(1), 65–74.

Budiaman, M., Hidayat, R. and Mukhtar, S. (2018) The model of Baduy's community system to anticipate the social environment change. *Journal of Organizational Culture, Communications and Conflict* 22(2), 1–11.

Durovic, M. and Lovrentjev, S. (2014) Indicators of sustainability in cultural tourism. *The Macrotheme Review* 3(7), 180–189.

Falatoonitoosi, E., Schaffer, V. and Kerr, D. (2022) Does sustainable tourism development enhance destination prosperity? *Journal of Hospitality & Tourism Research* 46(5), 1056–1082.

Fathurokhman, F. (2010) Hukum pidana adat baduy dan relevansinya dalam pembaharuan hukum pidana. *Jurnal Law Reform* 5(1), 1–38.

Finlayson, M. (2015) Cultural sustainability of African Canadian heritage: Engaging students in learning, the past, the present and the future. *Improving Schools* 18(2), 142–156.

Firdaus, M.N. and Herminasari, N.S. (2020) Peran Komunitas Pramuwisata Baduy Luar Dalam Mempertahankan Kearifan Lokal. *Jurnal Edukasi IPS* 4(1), 11–20.

Fletcher, D. and Sarkar, M. (2013) Psychological resilience: A review and critique of definitions, concepts, and theory. *European Psychologist* 18(1), 12–13.

Hall, C.M., Prayag, G. and Amore, A. (2017) *Tourism and Resilience: Individual, Organisational and Destination Perspectives.* Channel View Publications, Clevedon, UK.

Handayani, R.D., Wilujeng, I. and Prasetyo, Z.K. (2018) Elaborating indigenous knowledge in the science curriculum for the cultural sustainability. *Journal of Teacher Education for Sustainability* 20(2), 74–88.

Hawkes, J. (2001) *The Fourth Pillar of Sustainability: Culture's Essential Role in Public Planning*. Common Ground, Melbourne.

Herandy, C. (2022) Pariwisata Baduy: Antara Perubahan dan Kontinuitas. *Lembaran Antropologi* 1(2), 204–216.

Kartika, T. and Edison, E. (2019) Masyarakat Baduy dalam Mempertahankan Adat Istiadat di Era Digital. *ISBI* 1(1), 56–62.

Kirmayer, L.J., Sehdev, M., Whitley, R., Dandeneau, S.F. and Issac, C. (2009) Community resilience: Models, metaphors and measures. *Journal of Aboriginal Health* 5(1), 62–117.

Makmur, A. and Biantoro, S. (eds) (2014) Ketahanan budaya: pemikiran dan wacana. Pusat Penelitian dan Pengembangan Kebudayaan, Badan Penelitian dan Pengembangan, Kementerian Pendidikan dan Kebudayaan.

Mondéjar-Jiménez, J. and Ferrari, G. (2022) Sustainable tourism destinations. *Journal of Hospitality & Tourism Research* 46(7), 1239–1240.

Moswete, N. and Thapa, B. (2015) Factors that influence residents' support for community-based ecotourism development in Kgalagadi Transfrontier Park, Botswana. *Journal of Ecotourism* 12(2–3), 243–263.

Nashori, F. and Saputro, I. (2021) *Psikologi Resiliensi*. Universitas Islam Indonesia, Yogyakarta.

Ngamanken, S. (2013) Fenomena Keterlantaran Suku Baduy di Pulau Jawa. *Humaniora* 4(2), 1064–1073.

Rifa'i, B. (2020) Banyak Turis Bandel dan Nyampah, Baduy Dikunjungi 100 Ribu Orang/Tahun. Available at: https://travel.detik.com/travel-news/d-5085141/banyak-turis-bandel-dan-nyampah-baduy-dikunjungi-100-ribu-orangtahun (accessed 2 September 2021).

Rugebregt, J., Prapunoto, S. and Kristijanto, A. (2020) Mendedah Ketangguhan dan Resiliensi Perempuan Nuaulu dalam Tradisi Adat Nuhune di Kabupaten Maluku Tengah. *Jurnal Humaniora Yayasan Bina Darma* 7(1), 87–102.

Rusnandar, N. (2013) *Seba*: The culmination of Baduy's religious ritual in kabupaten(regency) lebak, the province of banten. *Balai Pelestarian Nilai Budaya Bandung* 5(1), 83–100.

Saragih, F.L., Turnip, P.K. and Simbolon, B. (2023) Indonesia's abundant wealth from the Baduy culture. *Indonesia Journal of Social Science Research* 4(1), 69–73.

Septiana, Y. (2018) Peranan suku Baduy di Desa Kanekes Kabupaten Lebak dalam melestarikan budaya tradisional dan kearifan lokal 2014–2017. Dissertation. Universitas Islam Negeri Sunan Gunung Djati, Bandung.

Soini, K. and Birkeland, I. (2014) Exploring the scientific discourse on cultural sustainability. *Geoforum* 51, 213–223.

Solikhah, N. (2020) Ethnic tourism and sustainable of vernacular settlement in Cibeo village, Baduy Dalam. *IOP Conference Series: Earth and Environmental Science* 452(1), 012021.

Sunaryo, B. (2013) *Kebijakan Pembangunan Destinasi Pariwisata Konsep dan Aplikasinya di Indonesia*. Edisi ke-1, Gava Media, Yogyakarta.

Suparmini, S., Setyawati, S. and Sumunar, D.R.S. (2013) Pelestarian lingkungan masyarakat Baduy berbasis kearifan lokal. *Jurnal Penelitian Humaniora* 18(1), 8–22.

Waluya, B., Malihah, E., Ruhimat, M. and Wiyanarti, E. (2022) Tourism revitalization based on local wisdom in 'Saba Budaya'of Baduy. *Jurnal Geografi Gea* 22(1), 1–13.

Weiss, M.L. (2017) Resistance and resilience: Coping with/against the state. *Journal of Social Issues in Southeast Asia* 32(2), 374–404. DOI: 10.1355/sj32-2e.

Yeniasir, M. and Gökbulut, B. (2018) Perception and attitudes of local people on sustainable cultural tourism on the islands: The case of Nicosia. *Sustainability* 10(6), 1892.

6 Hawker Centres, Cultural Heritage and Tourism – A Case Study of Lau Pa Sat, Singapore

Lan Wang
School of Art, Design and Media, Nanyang Technological University, Singapore

Summary
In the dynamic matrix of urban development, the conservation of heritage sites and the imperative of modernization often find itself at a challenging nexus. This case study delves into the intricate interplay between urban heritage conservation and contemporary socio-economic demands, using the transformative narrative of Lau Pa Sat, Singapore, as a focal case. Drawing upon the dual paradigms of 'use value' and 'exchange value', the research underscores the potential of heritage sites to both conserve cultural memory and catalyse modern economic revitalization. Lau Pa Sat's adaptive repurposing from a historical edifice to a dynamic hawker hub exemplifies a successful synthesis of architectural preservation and entrepreneurial initiative. Such an approach ensures the site's functional sustainability while amplifying its cultural resonance. The research further highlights the significance of multi-stakeholder participation in rejuvenating urban memory through visual representation, communal spaces, and cultural paradigms. The findings underscore that, with strategic planning and community engagement, heritage sites can transcend historical confines, emerging as vibrant spaces that harmonize cultural conservation with contemporary relevance. This study provides invaluable insights for urban planners, policy makers, and heritage management professionals, advocating for a holistic approach that balances cultural stewardship with economic imperatives in urban heritage contexts.

The Value and Interest of the Case Study

The case study of Lau Pa Sat underscores the multifaceted value of iconic historic buildings in shaping urban landscapes. Central to this narrative is the building's image, a testament to architectural grandeur and cultural significance. Its iconicity contributes not only to the aesthetic fabric of the city but also serves as a potent draw for tourism. Local and international tourists are invariably attracted to such historical edifices, seeking authentic experiences that meld architectural splendour with cultural narratives. Furthermore, the study delves deep into the symbiotic relationship between tourism and the preservation of communal identity. In its revitalization, Lau Pa Sat is a bastion for Singapore's cherished hawker culture, reinforcing and perpetuating a salient facet of communal identity. As a crucible for gastronomic traditions, it offers a nostalgic touchstone for the local populace, serving as a locus of shared memories and collective experiences. The adaptive repurposing of this venerable structure underscores the potential of such edifices to transcend their historical confines. These monuments can metamorphose into dynamic, utilitarian spaces that fortify communal identity and respond adeptly to contemporary imperatives, thereby contributing meaningfully to a region's socio-economic dynamism. This case study elucidates the delicate equilibrium between conservation and contemporaneity, proffering invaluable insights for urban planners, heritage conservationists, and community stakeholders.

Background – Hawker Culture as an Intangible Cultural Heritage in Singapore

In Singapore, hawker culture represents a social practice of gastronomy embedded in intangible heritage. The cultural, social, and economic importance of hawker food centres in Singapore has caught the attention of many researchers. Historically, hawker stalls, predominantly offering affordable and quick meals, were a common sight in less developed countries of Southeast Asia. These stalls, primarily operated by the grassroots class, were strategically situated near bustling residential zones or construction sites. Their food often bore testament to the region's mixed-culture, embodying inclusivity and diversity (Kowalczyk, 2014). More than just addressing hunger, hawker food is crucial in celebrating culinary diversity, strengthening community ties, and representing culinary traditions across borders. In Singapore, the emergence of hawker stalls can be dated back to the 1930s, as per Henderson's (2000a) findings. While hawker food provided significant convenience to residents, the unregulated and makeshift nature of these stalls posed potential threats to public health and urban order. As reported by Tung (2023), the Hawkers Inquiry Commission, spearheaded by Governor F. Gimson in 1950, was instituted to examine the multifarious challenges pertaining to hawking, emphasizing its socio-economic and public health ramifications. Under the aegis of Professor T.H. Silcock, the commission engaged in extensive consultations with key stakeholders, encompassing the Hawkers Association, municipal health authorities, and the broader public, to amass a holistic perspective. The insights gleaned from the Hawkers Inquiry Report culminated in the formulation of a strategic policy, facilitating the methodical relocation of hawkers to delineated zones. This enhanced oversight and governance over subsequent years. By 1986, all registered hawkers were relocated to these centres, and street vending became illegal without specific permits. These centres function as community dining rooms, bringing together people from diverse backgrounds to share dining experiences (UNESCO, 2020; Roots, 2021). Not only do they cater to locals, but they also serve as a vibrant attraction for tourists, offering an authentic insight into Singapore's culinary and cultural practices (Henderson, 2000b).

Henderson *et al.* (2012) highlight the role of cooked food hawkers as a sometimes overlooked component of the tourism and hospitality industries. They provide affordable food options that also serve as a window into the history and heritage of the destination (Bessiere and Tibere, 2013). Tam (2017) expands on this, depicting hawker food as a reflection of Singapore's multicultural city-state, with its origins deeply rooted in colonial and immigrant influences. Hawking is a prominent feature in Southeast Asia, particularly in Singapore, characterized by its small unit size and ease of market entry due to low start-up and operating costs. The presence of hawkers brings life and colour to locations, attracting visitors as observers and potential customers. While the regulation of Singaporean hawkers has generally been viewed positively, there are voices expressing regret over the disappearance of more mobile and unregulated hawkers (Native Food and Wine, 2010). Despite this, hawker centres continue to foster an appreciation of Singapore's history and heritage, acting as unique community spaces for social interaction (Ghani, 2011). Residents frequently visit them, with some survey data suggesting that half of the respondents ate hawker food at least six times a week (Euromonitor, 2010, as cited in Henderson *et al.*, 2012).

Tam (2017) also discusses the role of hawker centres as symbols of Singapore's multiculturalism. Once considered an administrative problem, these centres have been transformed into symbols of Singapore's heritage. Tam warns that the selectivity by which cultures are represented in hawker centres could inadvertently reinforce racial hierarchies. Radomskaya and Bhati (2022) focus on the importance of hawker centres in the production of social space, stating that these centres balance the need to maximize space usage in land-scarce Singapore with the need to provide a comfortable spot for hungry residents. Beyond their functionality, hawker centres serve as spaces for dining and mingling, contributing to greater social cohesion.

Efforts to safeguard hawker culture are being made at various levels, including community organizations, NGOs, educational institutions, and the public and private sectors (Imai, 2019; Roots, 2021). The importance of preserving this multicultural culinary heritage, which is fundamental to the Singaporean diasporic identity, is emphasized (Imai, 2019). Despite their significance, concerns have yet to be raised about the future of hawker centres. The rapid urban development in Singapore may affect the architecture of existing hawker centres, altering how patrons interact and perceive these spaces. There is also a risk of losing their intrinsic character due to an overemphasis on tourism markets (Henderson *et al.*, 2012).

In conclusion, while hawker centres in Singapore encapsulate the country's multicultural identity, serve as social spaces for interaction, and contribute to the economy, the challenges posed by urban development and modernization necessitate further research and conservation efforts to preserve this intangible cultural heritage. Within this context, Lau Pa Sat stands out due to its rich historical narrative and architectural significance, symbolizing the essence of Singapore's hawker culture. Transitioning from its origins as a wet market during the colonial period to its current status as a thriving hawker centre, Lau Pa Sat exemplifies Singapore's broader trajectory of modernization while maintaining its intrinsic cultural values. As urban planners, conservationists, and policy makers confront the complexities of preserving heritage in contemporary urban landscapes, Lau Pa Sat offers a compelling case study. It underscores the nuanced interplay between historical preservation, cultural identity, economic considerations, and urban development.

Methodology

In contemporary urban studies, the adaptive reuse of historic buildings has emerged as a potent tool for safeguarding local culture and identity, especially when examined through the prism of heritage tourism. This research is anchored in the theoretical foundations provided by Kelvin Lynch's theory of city image[1] and Leslie Sklair's conceptualization of iconic buildings.[2] Within this theoretical framework, the study critically examines the rejuvenated image of Lau Pa Sat, integrating insights from both community participation and the tourism sector. The analysis focuses on two main dimensions: the nuances of district imagery and the evolving functional significance of spaces with historical import contextualized within the trajectory of modern urban development. A systematic review of secondary sources, spanning historical documents, reports, maps, and documentaries, forms the bedrock of this study, enabling a comprehensive understanding of the metamorphosis of Lau Pa Sat.

Methodologically, the research is rooted in a qualitative paradigm, harnessing both observational techniques and historical participation methods. This qualitative lens offers an intricate exploration of the inherent values and cultural significance of Lau Pa Sat. For a detailed representation, the study employs visual data in figures and diagrams, spotlighting Lau Pa Sat's pivotal role at the nexus of tourism, evolving societal dynamics, and cultural progression.

An onsite field study was undertaken to augment the empirical depth of the research in 2023. This endeavour provided direct insights into the lived experiences and ongoing transformations within Lau Pa Sat. The fieldwork comprised both semi-structured and unstructured interviews, targeting diverse respondents, including general operators (stakeholders), hawkers, and diners, encompassing tourists. This methodological choice captures varied perspectives and gauges the intensity of stakeholder engagement in the adaptive reuse of spaces process. Furthermore, the decision making framework surrounding Lau Pa Sat's renewal was critically appraised using Arnstein's ladder[3] of citizen participation, shedding light

Fig. 6.1. Lau Pa Sat. © Lau Pa Sat.

on the extent of decentralized decision making involved.

Analysing Lau Pa Sat

History of Lau Pa Sat

Lau Pa Sat, derived from the Hokkien dialect meaning 'the old market,' is a venerable hawker centre of great historical significance in Singapore (Fig. 6.1). Located centrally, this octagonal marvel, embellished with Victorian architectural motifs, silently chronicles Singapore's metamorphosis from a colonial port city to a contemporary global hub.

Positioned strategically within Singapore's financial epicentre, Lau Pa Sat's unique octagonal configuration, coupled with its decorative columns, was initially conceptualized by George Coleman (Visiting Singapore, 2023). Its historical roots can be traced back to 1825, when it was inaugurated as the city's premier wet market, catering to the sale of fresh perishables and meats (Tam, 2017). As part of the urban redevelopment initiatives under the Raffles Town Plan, Sir Stamford Raffles directed the fish market relocation to Telok Ayer (Fig. 6.2). After Chinatown expanded and land reclamation moved the shoreline south, in 1894 the market was moved from its waterfront location and quadrupled in size (Expat Living, 2022). George Coleman also designed numerous other colonial edifices in Singapore, notably the Old Parliament House (Visiting Singapore, 2023). However, the construction was fraught with challenges. The timber foundations on which the structure rested were subpar, necessitating swift replacements post-construction (National Heritage Board, 2021). Over the years, the building's vulnerability to monsoons and proximity to the sea compromised its stability. Moreover, its original tap roof flouted building codes, leading to its replacement with tile roofing. Despite multiple renovations, the infrastructure was eventually deemed 'extremely unsafe' and a comprehensive rebuild was mandated (Liu, 2001, p. 48).

Nonetheless, the essence of Lau Pa Sat transcends its architectural significance. Its vibrancy emanates from the daily activities within, vendors engaged in animated negotiations, the whiff of fresh produce, and, eventually, the mouth-watering aromas from a myriad of culinary delights. As the surrounding financial district flourished, Lau Pa Sat transitioned into a hawker centre in 1972, epitomizing Singapore's diverse gastronomic heritage (Savage and Yeoh, 2013). In the same year, its declaration as a national monument by the National Heritage Board underscored its historical and cultural

Fig. 6.2. The old Telok Ayer Market painting by John Turnbull Thomson in 1847.

import. Subsequent decades have ushered in more transformations, with Lau Pa Sat evolving into a bustling festival market reminiscent of the evocative street food culture from days gone by. Despite its many incarnations, its core ethos remains unchanged: Lau Pa Sat is an epicentre of community, culture, and cuisine. After its recent refurbishment, Lau Pa Sat was redeveloped as Singapore's first combined food and beverage and retail venue (Quek, 2020a), which reopened in 2021. Today, Lau Pa Sat boasts an eclectic mix of nearly 80 food outlets interspersed with retail stalls offering many distinctive products. Operating round-the-clock, it beckons patrons with its rich culinary offerings, all set against a backdrop echoing the heritage and traditions of historic Singapore (Lau Pa Sat, 2020).

Lau Pa Sat is emblematic of more than a mere structure. It encapsulates the rich tapestry of Singapore's historical trajectory. Its continual evolution mirrors the city's indomitable spirit and flexibility. For locals and tourists, Lau Pa Sat is not just a gastronomic haven but a portal into the very essence of Singapore, paying homage to its illustrious past while embracing its forward-looking aspirations.

Physical image of Lau Pa Sat: The architecture

According to Sklair (2005), architecture serves as a symbolic representation of cultural ideology. This perspective emphasizes the intricate nexus between multidisciplinary practices and the generation of contemporary and prospective commercial opportunities. One salient aspect of architecture is its iconicity, which is prominently evident in the visual appeal that emphasizes the aesthetic facade of a structure. The notion of architectural iconicity encapsulates both tangible and intangible elements. Primarily, the iconic status of a building is often anchored in its distinct appearance within a specific locale. This conceptualization resonates with Kevin Lynch's urban 'imageability' theory (Lynch, 1964, p. 9), which underscores the demarcation of recognizable boundaries within urban communities.

Fig. 6.3. Cast iron structures. © photographed by author.

Historical precincts, exemplified by clusters of shophouses, epitomize this concept. Their uniform facades and external walls form a cohesive boundary, demarcating the heritage zone amidst contemporary architectural designs.

From an architectural standpoint, Lau Pa Sat stands as a distinguished icon, embodying Lacaton and Vassal's philosophy of 'Never demolish, never remove or replace, always add, transform, and reuse!' (Huber, 2016). Its distinctive octagonal design endures not only historically but is also emblematic of themes such as regeneration, rebirth, and renewal. These themes resonate with the broader narrative of Singapore's independence (lioncitysites, 2012). As a hawker centre, Lau Pa Sat's expansive layout facilitates a communal ambience, fostering a dining culture quintessential to Singapore. Furthermore, the building's internal structure artfully melds Victorian aesthetics with functional engineering. The retained cast-iron structures (Fig. 6.3), painted in their original green hue, epitomize the era's preference for neat and respectable housing (Lioncitysites, 2012; Smith, 2018). Concurrently, design elements like the soaring arches, delicate fretted eaves, and slender columns have a pragmatic purpose, ensuring adequate air circulation and optimal ventilation (Expat Living, 2022; Roots, 2022; Visiting Singapore, 2023), reflecting a contemporary commitment to sustainable practices aimed at minimizing carbon footprints. These adaptive reused structures align with Steemers (2021) theory of enhancing quality of life through holistic architectural approaches.

Regarding its scale, Lau Pa Sat is an expansive octagonal pavilion, its sloping roof sheltering an area of 52,000 square feet. The space encompasses eight hawker zones, housing over 70 stalls and is strategically situated at the confluence of four major roads in the downtown Central Business District (CBD). Notably, the building's design contrasts starkly with its neighbouring structures. Lau Pa Sat's unique design exudes an aura of isolation, making it a distinct entity within the CBD. The vast space it occupies appears to challenge the prevailing concept of the 'modern skyscraper,' which is often seen as a testament to the triumph of capitalist ideologies in urban land use. Instead, Lau Pa Sat's presence underscores the value placed on cultural and traditional preservation in contemporary Singapore. When juxtaposed against the modern glass and steel skyscrapers surrounding it, Lau Pa Sat's distinct shape and design demarcate a clear boundary, preserving its historical ambience amidst the modernity of the adjacent office buildings. This stark contrast

Fig. 6.4. Top view of Lau Pa Sat. © Google map.

accentuates its iconic stature, especially when viewed from the towering edifices that envelop it (Fig. 6.4).

Mental image: Lau Pa Sat as a cultural resource

Beyond its architectural significance, Lau Pa Sat serves as a beacon of Singapore's hawker culture. Its structure's meticulous conservation and adaptive reuse underscore its pivotal role as a national intangible cultural heritage repository. Posited by Kamel-Ahmed (2015), the absence of tangible or physical manifestations that reinforce collective and spatial memories could render intangible cultural expressions vulnerable to erratic transformations. As a historical landmark, Lau Pa Sat is a testament to Singapore's architectural conservation ethos, harmonizing aesthetic value with preserving historical authenticity. Its image offers a symbolic representation that transcends physicality. This monument amplifies its iconic significance through its adaptive reuse, serving as a built form and a memory carrier. It interweaves the present with historical reminiscences, thereby crystallizing the hawker culture – a quintessential element of Singapore's intangible cultural heritage. Drawing inspiration from Lynch (1964) seminal work on urban spaces, a profound nexus emerges between urban environments and individual perceptions. Lynch (1964) asserted that individuals cultivate intricate affiliations with distinct urban settings based on their experiences. Deeply anchored in a tapestry of memories and intrinsic values, these affiliations transform tangible urban environs into resonant cultural and personal narrative repositories. Consequently, Lau Pa Sat's significance transcends mere symbolic or functional attributes, emerging as a linchpin for cultural preservation amidst urban evolution.

Lau Pa Sat's allure is not confined to its architectural prowess or prime location. It manifests as a sensory mosaic, where each element contributes to a collective hawker culture memory. While honouring its historical essence, this adaptive reuse has been instrumental in preserving its physical form and etching a vivid mental imprint. The myriad sensory experiences it offers, akin to Sklair's (2005) insights on the nexus between vision and memory, serve as potent mnemonic anchors. Every aroma wafting from the stalls, every dish that speaks of Singapore's multicultural tapestry, and every corner that resonates with stories of yesteryears conjures vivid mental images. Intricately woven with individual and collective recollections, these images construct a rich mosaic of memories. Thus, Lau Pa Sat transcends its physical realm, carving a significant niche within the collective psyche, and symbolizing Singapore's rich heritage and vibrant life.

In the modern context, characterized by the rise of tourism consumerism, Lau Pa Sat epitomizes its role as a cultural linchpin, highlighting its integral contribution to Singapore's urban narrative. Its evolution from a historical landmark to a mediated monument[4] underscores a transformation that surpasses mere commercialization. It is evident in crafting the city's outward image, which aligns with its functional evolution and realization while enlightening residents about the advantages of urban revitalization (Elhagla *et al.*, 2020). As a historic space, Lau Pa Sat evidences Singapore's urban development and regeneration. After being redefined as a corporate space, this building accommodates the diverse inclinations of indigenous inhabitants and global visitors. In this light, the contemporary Lau Pa Sat leans towards being a commercial architectural entity, especially in its public prominence and aesthetic appeal. Conveniently located in the heart of the Central Business District, it integrates into the daily lives of myriad professionals, serving as a gastronomic sanctuary from morning to night. In doing so, it elevates the living standards of hawkers and the community and offers a unique insight into the hawker culture amidst the urban landscape. Lau Pa Sat's image resonates powerfully at the crossroads of city branding and the distinct Singaporean identity.

A special hawker centre

Nestled in the heart of Singapore's Central Business District, Lau Pa Sat is more than just an ordinary food centre. Its strategic position among the city's skyscrapers and bustling streets has become a cultural postcard, offering a glimpse into Singapore's diverse, mixed-culture society through its vast culinary offerings.

Lau Pa Sat is a culinary time capsule, bearing witness to immigration influences on Singapore from the colonial era to the present day. According to recent data, over 50% of its stalls serve traditional and local food, paying homage to Singapore's rich history and catering predominantly to the local diners (Lau Pa Sat, 2020). These authentic dishes, brimming with flavours and traditions, transport diners back in time, allowing them to savour the culinary legacy of Singapore's early inhabitants. Meanwhile, the remaining stalls introduce international cuisines, from American fast foods to Mexican. This cosmopolitan array reflects the city's global nature and is tailored to cater to the varied tastes of locals, tourists and expatriates. The centre's location in the CBD also makes it a popular spot for local and western business professionals, who can easily step out from their nearby offices for a taste of local delights.

Unlike most hawker centres located in residential areas, Lau Pa Sat is situated amidst the commercial structures of Singapore's Central Business District. The location suggests that a significant number of hawkers operating there are not residents of the immediate vicinity, deviating from the standard model where local inhabitants manage family-owned stalls. Influenced by the corporate environment, these vendors have opted to conduct their business in this historic building, merging contemporary practices with traditional values. A remarkable point is that, even in western food stalls, most of the hawkers are Singaporean Chinese. Over time, the role of Lau Pa Sat has been subtly transforming. While it remains a haven for delectable and affordable meals, its image is slowly shifting. No longer just a traditional 'hawker' centre in the eyes of many, it is emerging as a tourist must-visit destination. Visitors, eager to immerse themselves in an authentic Singaporean experience, flock to Lau Pa Sat, seeking both the flavours and ambience of local life.

Lau Pa Sat is much more than a venue to satiate one's culinary cravings. It is a pivotal community space, potentially evolving into a social enterprise or even a cosmopolitan hub for novel culinary experiences (Yong, 2021). Beyond its role as a marketplace, Lau Pa Sat significantly contributes to the life quality of a diverse set of stakeholders. For local hawkers, it provides an avenue to earn a livelihood, thereby supporting economic stability and job satisfaction at a grass-root level. It also preserves cultural heritage by enabling these hawkers to continue their traditional cooking methods and share them with a broad audience.

For residents and tourists alike, Lau Pa Sat offers a unique dining experience beyond mere consumption, enhancing leisure and cultural education. It provides a space to interact socially, contributing to social belonging and wellbeing, and community cohesion. Furthermore, its cosmopolitan nature welcomes diversity, reflecting the multicultural fabric of the city and offering a wider variety of food choices that cater to different tastes and dietary requirements. Thus, Lau Pa Sat is a multifaceted institution that enhances the quality of life, offering gastronomic delights and economic, social, and cultural benefits to the community.

Tourism

When framed within the tourism narrative, historical architecture serves as cultural capital, amplifying a city's image. For tourists, these iconic historic structures epitomize a city's unique and authentic traits, forging a connection between them and the destination (Elhagla *et al.*, 2020). Studies indicate that the primary allure of a historical site lies in the quality of the place, intricately woven from elements of 'architecture,' 'people,' 'food,' 'culture,' and 'diversity' (Gross, 2006 as cited in Maitland and Newman, 2010). Through adaptive reuse, buildings are repurposed, fostering a deeper cultural understanding and ensuring cultural continuity for tourists. This enriches not only the tourist experience but also generates significant economic benefits. Consequently, adaptively reused historical structures hold immense value in cultural and heritage tourism.

Culinary practices serve as a direct manifestation of a city's authenticity, as posited by Tam (2017). Lau Pa Sat, a rejuvenated hawker centre, exemplifies this by intertwining historical culinary traditions with geographically influenced consumption patterns. Lau Pa Sat, emblematic of a diverse culinary hub, offers an array of cuisines from Singapore and other Asian nations, as well as further afield. Many stalls within Lau Pa Sat are dedicated to showcasing Singapore's local and traditional culinary delights. To accurately assess the gastronomic offerings of Lau Pa Sat for tourism potential, it becomes crucial to differentiate between local and traditional cuisines, given that both contribute to the multifaceted hawker culture yet originate from distinct sources. Traditional food is characterized by its profound historical and cultural roots, intricately tied to a specific community and occasionally limited by seasonal local produce (Sproesser *et al.*, 2019). Conversely, local food, while representing everyday dietary practices confined to a particular geographic area, might evolve over time to symbolize a unique regional 'identity', although it lacks the historical or cultural depth inherent to traditional food (Vargas *et al.*, 2021). Intriguingly, with the passage of time and consistent cultural integration, certain local foods have the potential to transition into the realm of traditional foods, particularly when they become deeply ingrained in community practices and social events over extended durations. Diving into Lau Pa Sat's hawker cuisines, diners can find traditional Singapore food (Fig. 6.5) and drinks such as Laksa (Peranakan rice noodles), Bak Kut Teh and Kopi (Nanyang coffee), and local cuisine (Fig. 6.6) like Satay (Indonesian kebab), Popiah (thin pancake from Taochew), and Roti Prata (Indian flatbread). Therefore, Lau Pa Sat's rich tapestry of culinary offerings underscores the profound impact of immigration and colonization on Singapore's gastronomic landscape. This hawker centre stands as a testament to the city's multicultural heritage, blending diverse culinary traditions and reflecting the historical integration of cultures over time.

Fig. 6.5. Singapore traditional cuisines (left to right Laksa, Bak Kut Teh, Kopi). © Lau Pa Sat.

Fig. 6.6. Singapore local cuisines (left to right Satay, Popiah, Roti Prata). © Lau Pa Sat.

Building upon the gastronomic appeal of Lau Pa Sat, its broader impact on the tourism sector of Singapore is evident. Drawing a substantial annual influx of tourists, predominantly English and Chinese speakers, the centre ensures a seamless and accessible experience for its visitors. Post its 2014 refurbishment, satisfaction rates from patrons have consistently exceeded 80%.[5] Most diners perceive Lau Pa Sat as a laid-back venue, ideal for savouring traditional local dishes and fostering familial connections. Beyond gastronomy, the centre has expanded its appeal by incorporating diverse touristic and recreational itineraries. Numerous tourism packages feature Lau Pa Sat, integrating innovative experiences such as cycling excursions, food-focused tours, and historical site explorations. Such initiatives have enhanced its global tourist appeal and substantially augmented local hawkers' revenue streams. Furthermore, 'Food Folk @ Lau Pa Sat' offers an invaluable platform for local vendors, showcasing distinctive products that captivate tourists. This surge in tourism, in turn, catalyses economic growth, reinforcing the centre's commitment to architectural conservation and the preservation of its rich culinary heritage.

Community participation

Historical landscapes embody a sophisticated interplay between architectural excellence and community dynamism. Within urban regeneration paradigms, adaptively reused edifices transcend mere structures, serving as vibrant community nexuses and revitalizing their intrinsic societal values. The 1987 Washington Charter, formally titled the Charter for the Conservation of Historic Towns and Urban Areas, emphasizes the paramount importance of resident engagement in successful conservation endeavours (Yung *et al.*, 2014). Moreover, proactive community participation accentuates sustainability, seamlessly interweaving environmental, social, and economic dimensions. Within such topographies, adaptively reused structures emerge as vital assets, enhancing the quality of life of hawkers and diners, offering adaptability, nurturing harmonious socio-environmental ecosystems, and optimizing resource utilization (Othman, 2007, as cited in Othman and Elsaay, 2018).

Steeped in historical significance, Lau Pa Sat exemplifies the harmonious union between individuals and their architectural milieu.

Celebrated as an emblematic hawker centre, it resonates with monumental public heritage value. The adaptive reuse and operational management of Lau Pa Sat have remained separate from the community as it is manipulated under the jurisdiction of the National Heritage Board due to its status as a national monument. This scenario reflects the broader heritage conservation situation in Singapore, where a top-down decision making approach is predominantly employed (Teo, 1992, as cited in Nguyen and Chen, 2018). In this framework, the Heritage Conservation Centre (HCC), a pivotal entity under the National Heritage Board, pursues three core objectives: establishing and upholding high professional standards in heritage conservation, enhancing public accessibility to the National Collection, and disseminating heritage care knowledge to the public (National Heritage Board, 2021).

In the specific context of Lau Pa Sat, decision making authority is chiefly vested in a select cohort of stakeholders, primarily comprising governmental and professional bodies. This centralized decision-making structure aligns with Arnstein (1969) delineation of the manipulation and therapy stage, where the public's role is substantially circumscribed. The available documentation underscores a focus on preserving the historical and architectural integrity of Lau Pa Sat, with lesser emphasis on fostering public engagement in adaptive reuse and operational strategies. Consequently, the publicly accessible information concerning community engagement or public consultations during the adaptive reuse and operational phases of Lau Pa Sat is scant.

Although the physical structure is subject to strict national oversight, its functional adaptive reuse facilitates a level of community engagement by promoting local culture and businesses. This dichotomy illustrates a controlled approach towards heritage conservation while allowing for localized economic activity and cultural expression within the framework established by national heritage policies. Under the aegis of the FairPrice Group and the National Trades Union Congress (NTUC), Lau Pa Sat has metamorphosed into a gastronomic and cultural nexus, championing indigenous brands. The national 'Made with Passion' programme, curating diverse local brands, harmoniously melds tradition and contemporaneity. The advent of 'Food Folks @ Lau Pa Sat' (Fig. 6.7) epitomizes this ethos, curating a space for over 120 brands, with a significant 70% bearing Singaporean origins. This initiative amplifies contemporary F&B innovations and venerates the legacy of seasoned hawkers, ensuring their culinary traditions flourish. Central to Lau Pa Sat's renaissance is

Fig. 6.7. Retail store of Food Folk. © photographed by author.

a steadfast commitment to fostering a cohesive community vision. This vision manifests in the support extended to nascent brands, exemplified by Edens and its acclaimed shiitake mushroom snacks (Quek, 2020b).

In addition to this tangible metamorphosis, strategic digital integrations have been orchestrated. A meticulously crafted website offers insight into Lau Pa Sat, while agile Instagram campaigns ensure sustained digital rapport. The overarching ambition of exceptional customer satisfaction underpins this initiative. These online platforms effectively promote Lau Pa Sat's cultural and commercial activities to the general public and collect feedback from end-users that contribute to further development and successfully catering to a variegated client. Furthermore, the strategic vision delineated by Mr Seah Kian Peng, of both NTUC Enterprise and FairPrice Group, underscores an unwavering commitment to the community (FairPrice Group, 2020). By proffering robust marketing support and abrogating initial establishment fees, especially amidst the COVID-19 pandemic's tumult, their dedication to fostering community vibrancy and fostering economic prosperity is palpably evident.

The above approaches align with Arnstein (1969) theory of 'tokenism', which encompasses the stages of informing, consultation, and placation. The 'Informing' stage is mirrored in disseminating information through digital channels about Lau Pa Sat, while the 'Consultation' stage is reflected in the initiatives promoting local brands, albeit within a predefined framework. The 'Placation' stage is subtly represented where the community is given a platform within controlled parameters defined by larger governing bodies and private stakeholders. These stages highlight a controlled engagement, where community input is sought, but decision making remains largely with authoritative entities, such as giant business groups and national institutions, resonating with the controlled adaptive reuse approach of Lau Pa Sat.

Impact of adaptive reused Lau Pa Sat

The impacts of Lau Pa Sat's regeneration refer to the follows:

Image building

The architectural image of Lau Pa Sat serves as an epitome of the intricate balance between preserving historical integrity and meeting contemporary urban demands. While visually arresting, this venerable structure extends beyond mere aesthetics. It is deeply entrenched within the collective psyche, representing a tangible nexus of Singapore's storied hawker heritage juxtaposed against a modern urban backdrop. The concept of 'use value' is exemplified in the adaptive repurposing of this historic space. The seamless integration of this structure into the contemporary urban landscape not only imbues it with a new socio-cultural significance but also ensures its ongoing functional relevance, preserving its visual character within an evolving cultural context. Thus, Lau Pa Sat transcends its physical confines, fostering a renewed sense of communal identity, evoking deep-seated nostalgic resonances, and reaffirming its integral role within Singapore's evolving socio-cultural milieu.

In examining the built environment's impact on Lau Pa Sat's image it is imperative to recognize Singapore's urban development trajectory. Characterized by a market-centric orientation, the nation's urban strategy is deeply rooted in a neoliberal development ethos (Liow, 2012) and is further propelled by a survivalist ideology from the formative years of its independence (Dugo, 2022). This development paradigm primarily seeks to magnetize global capital and investments, as exemplified by the Central Business District's evolution in the Raffles Place. This market-driven orientation sculpted the cityscape, frequently favouring commercial ventures and high-value land utilizations. Such a preference often translates to constructing towering skyscrapers, which cater to a broader consumer base, over preserving traditional shophouses. Consequently, a need emerges to recalibrate Lau Pa Sat's role, ensuring its alignment with its surrounding milieu's evolving commercial and upscale valuations. However, this developmental strategy has inadvertently catalysed gentrification, distancing original inhabitants from their native environment. Particularly poignant is the spatial dislocation of the hawker centre, an emblematic grassroots dining space nailed in a high-end commercial district. Compared with other community hawker centres, Lau Pa Sat's image alludes to the overarching influence of neoliberal urban governance, crafting a unique aesthetic or economic blueprint divergent from its counterparts.

Tourism

Lau Pa Sat's adaptive reuse serves as a significant model in leveraging historical architecture for cultural and heritage tourism, enhancing Singapore's image while offering tourists a tangible connection to the city's authentic traits through its diverse gastronomic offerings. Within the ambit of urban tourism, the iconic stature of Lau Pa Sat emerges not merely as an architectural marvel but as a potent outreach of Singapore's hawker heritage. As a public emblem, it meets not only the cultural aspirations of the populace but also facilitates commercial imperatives, thereby optimizing economic returns for its custodians. For the local connoisseur and the global tourists, its carefully curated image acts as a magnetic beacon, encapsulating and echoing the nuances of heritage dining traditions and cultural narratives. This strategic transformation of Lau Pa Sat's persona elevates it from a historical vestige to a pulsating epicentre of tourism.

Tourism in the case of Lau Pa Sat adaptive reuse serves both financial and grassroots interests. The adaptive reuse strategy amplifies Lau Pa Sat 's 'use value,' endowing it with renewed socio-cultural relevance and ensuring its operational and intangible heritage preservation. Simultaneously, this rejuvenation manifests a tangible 'exchange value.' The embedded cultural and heritage dimensions of Lau Pa Sat become commodified assets, harmonizing with the expansive aspirations of the contemporary tourism sector. Lau Pa Sat financially benefits the local economy and stakeholders involved in the project by attracting tourist spending, while also serving grassroots interests by showcasing local culture, heritage, and entrepreneurship. This symbiotic relationship augments economic traction via augmented visitor engagement, fostering the creation of novel tourism modalities and beckoning investment avenues. Given that commercialization often underpins the fiscal viability of heritage conservation, Lau Pa Sat exemplifies a confluence of traditional Singaporean ethos and modern consumer-driven strategies, laying the foundation for economically sustainable conservation endeavours. In essence, Lau Pa Sat encapsulates the delicate balance between preserving cultural heritage and navigating economic exigencies, highlighting the efficacy of intertwining 'use' and 'exchange' values in the quest for cultural preservation and economic vibrancy. Additionally, ensuring the sustainability of other infrastructure can be enhanced by integrating environmentally friendly technologies and practices in urban renewal projects, and aligning them with broader sustainability goals of the city-state, which in turn can create a more holistic and sustainable urban development model.

Community

Within the socio-cultural framework of Singapore, Lau Pa Sat elevates beyond an architectural landmark to a dynamic representation of the nation's valued hawker traditions. It instils a strong sense of belonging among the local population, serving as a vivid reminder of shared cultural narratives, thus fortifying local identity amidst Singapore's post-colonial discourse. The adaptive reuse of Lau Pa Sat diverges from the often static museum-like preservation associated with historic buildings. Instead, it adeptly incorporates Singapore's rich cultural heritage into the vibrant flow of contemporary urban life, exemplifying a form of cultural hybridity. This revitalized emblem of hawker culture enhances local cultural awareness, preserving the unique characteristics and essence that define the local milieu while resisting the erasure of historical and cultural memories. Beyond its symbolic significance, Lau Pa Sat holds a tangible role in the local economy. Its transformation has stimulated the growth of indigenous businesses, generating employment opportunities within its landscape and reflecting a form of economic self-determination. Consequently, the reciprocal bond between Lau Pa Sat and its neighbouring communities deepens, fostering a sense of collective pride and elevating the overall quality of life. Through its adaptive reuse, Lau Pa Sat illustrates how heritage conservation can concurrently serve as a beacon of cultural identity and a catalyst for socio-economic growth amidst the broader financialisaton of Singapore's urban landscape, showcasing a balanced approach to urban development that aligns financial interests with social and cultural values.

While Lau Pa Sat stands as an incubator for entrepreneurship, its management reflects a controlled participatory approach. The adaptive reuse narrative of Lau Pa Sat exhibits a nuanced interplay between authoritative governance and localized engagement, akin to the lower and middle echelons of the Ladder of Citizen Participation. The stringent national oversight epitomizes the 'Manipulation' and 'Therapy' rungs, while initiatives such as 'Food Folks @ Lau Pa Sat' resonate with the 'Informing,' 'Consultation,' and 'Placation' rungs, fostering a controlled conduit for local cultural and economic expression. The collaboration with entities like the FairPrice Group and NTUC potentially mirrors a modicum of 'Partnership,' albeit devoid of extending towards the zenith rungs of 'Delegated Power' or 'Citizen Control,' thus encapsulating a structured, albeit constrained, paradigm of community engagement within the prescribed ambit delineated by national heritage and economic stratagems.

Challenge and Opportunities for Future Lau Pa Sat

Over the years, Lau Pa Sat has evolved into a paragon of successful historic adaptive reuse, adeptly catering to the demands of both residents and tourists while simultaneously generating substantial economic benefits. Nestled within the gastronomy sector, it is not immune to the inherent challenges that the industry grapples with. As Singapore's hawker food culture serve as 'cultural ambassadors' (Henderson *et al.*, 2012), they also tread the delicate line between attraction and apprehension. To many tourists, exploring hawker fare is an enthralling culinary odyssey, motivated by the thrill of novelty. Nevertheless, this very novelty can sometimes breed unease and reservations.

A significant proportion of the criticism levelled at Lau Pa Sat pertains to perceived lapses in customer service, inflated pricing, and sanitation concerns. Illustratively, issues such as uncleared trays and exposed waste bins raise hygiene red flags. A dearth of cleaning personnel in many hawker centres, Lau Pa Sat included, exacerbates these challenges, leading to suboptimal table maintenance and tray return practices. This is echoed in the Public Cleanliness Satisfaction Survey, which reveals a rather tepid cleanliness satisfaction rate of 71.4% for hawker centres (Muhd and Shah, 2021). Further compounding these concerns are the sanitary conditions of restrooms, which, when compared to those in adjacent commercial establishments, may deter potential patrons. Such apprehensions, grounded in health and hygiene concerns, can inhibit footfall (Lues *et al.*, 2006). This reluctance undermines not only the gastronomic tourism potential but also impedes the propagation of the rich hawker culture, casting a shadow over the tourism economy and cultural heritage promotion.

Urban transformation in Singapore, illustrated by Lau Pa Sat's adaptive reuse, strives for a balance between modernization, heritage preservation, and the promotion of indigenous business and identity, in line with broader sustainability and liveability goals (URA, 2019). With its historic prestige and prime location, Lau Pa Sat stands poised as a veritable incubator for local brands and budding entrepreneurs. This iconic hawker centre, weaving together Singapore's rich culinary tapestry with its bustling urban milieu, offers an unparalleled platform for local enterprises to showcase their innovations and traditional offerings alike. In the current globalised market, where local brands often grapple to find their foothold amidst multinational conglomerates, Lau Pa Sat provides them with the visibility and credibility they yearn for. The centre's emphasis on championing home-grown talent is evident in initiatives like 'Food Folks @ Lau Pa Sat', which reflects a step towards nurturing indigenous businesses within a controlled yet supportive framework, aligning with the nation's urban renewal and economic strategies. By offering these brands a stage in the heart of Singapore's Central Business District, Lau Pa Sat nurtures not only their entrepreneurial spirit but also facilitates their interaction with diverse clientele, ranging from locals to global tourists. This symbiotic relationship accelerates not only the growth trajectories of these local brands but also enriches the gastronomic landscape of Lau Pa Sat, ensuring its continual evolution and relevance in a rapidly changing market dynamic. In essence, beyond being a culinary hub, Lau Pa Sat is evolving into a crucible of innovation and entrepreneurship, fostering the next generation of Singaporean brands.

The approach of adaptive reuse fosters a level of inclusivity through local engagement and entrepreneurship. Yet, its controlled participatory nature, as exhibited in Lau Pa Sat, could constrain inclusivity, with the degree of community involvement, the openness of the political framework, and the provision of meaningful opportunities for local stakeholders in decision making processes pivotal to enhancing inclusivity in urban renewal projects. To enhance inclusivity, adopting a more open regulatory framework that encourages broader public participation is advisable, ensuring that local stakeholders (especially hawkers) have meaningful opportunities to contribute to the decision making processes in Lau Pa Sat renewal projects and its further operation.

Conclusion

In the dynamic landscape of urban evolution, juxtaposing historical preservation and contemporary demands presents challenges and opportunities. Lau Pa Sat, a quintessential hawker centre in Singapore, is a compelling case study for the transformative potential of adaptive reuse in urban environments. Positioned at the crossroads of architectural adaptive reuse, heritage conservation, tourism, and community involvement, its transformation embodies a subtle equilibrium that manages intricate needs and goals. As a cultural monument, Lau Pa Sat is a testament to Singapore's commitment to preserving its built heritage, offering a dialogue between past and present.

Lau Pa Sat's role in tourism is twofold: it serves as a gastronomic destination and a cultural landmark, amplifying Singapore's image on the global stage. By offering a diverse array of culinary options that reflect local and global influences, Lau Pa Sat caters to a broad audience and serves as an authentic expression of the city-state's multicultural fabric. This symbiosis between economic interests and cultural preservation underscores the centre's contribution to the tourism sector and the local economy. In terms of community engagement, Lau Pa Sat presents a nuanced narrative. While its management and operational strategies are centrally controlled, the centre is a vital community space fostering local businesses and cultural expression. The controlled-supportive framework aligns with broader national strategies, offering a sustainable development stage for the evolution of local brands and gastronomic traditions. Despite this, there is scope for greater inclusivity and participation from local stakeholders, which could enrich the centre's social and cultural dimensions.

In summary, Lau Pa Sat epitomizes the intricate choreography of adaptive reuse. Its success story offers valuable insights into how cities can strategically repurpose historical landmarks to serve contemporary needs while honouring their past. It stands as a model for effectively integrating heritage conservation, economic development, and community engagement harmoniously and sustainably. As cities worldwide grapple with the complexities of urban regeneration, the Lau Pa Sat experience offers a roadmap for how to navigate these challenges thoughtfully and effectively.

Notes

[1] Lynch's (1964) city image theory posits that individuals interpret and navigate urban spaces based on their perceptions and interpretations of the physical environment, which he terms 'mental maps.' These mental maps are shaped by five key elements: paths, edges, districts, nodes, and landmarks. Lynch argues that a well-organized and distinct physical environment can lead to more apparent mental maps, enhancing a city's legibility and its inhabitants' sense of place. His theory has been instrumental in urban planning and design, guiding city developers in creating spaces that resonate with residents and visitors alike. In this case study, Lynch's city image theory is used to elucidate the isolated image of Lau Pa Sat, a distinctive landmark encircled by the towering skyscrapers of Singapore's CBD.

[2] Sklair's (2017) conceptualization of iconic buildings posits that such structures transcend mere architectural achievements, instead acting as symbols of global capitalism and consumer culture. Sklair emphasizes that these iconic edifices often serve the interests of transnational capitalist class elites by shaping urban skylines and representing economic and cultural dominance. Furthermore, these buildings play pivotal roles in constructing a city's global image, while reinforcing the ideologies of cultural consumerism. In this case study, Sklair's theory is adapted to explore the cultural consumerism ideology inherent in food tourism, using Lau Pa Sat as a prime example of a mediated monument anchoring tourists' imaginations with the image of the destination city.
[3] Arnstein's (1969) Ladder illustrates varying citizen involvement degrees in decision making processes. The Ladder's eight rungs start from 'Manipulation' and 'Therapy' at the bottom, representing non-participative engagement, ascending to 'Citizen Control' at the top, indicating full citizen power in planning, policy making, and management. Applying this to the Lau Pa Sat adaptive reuse project can help evaluate community engagement levels, assessing how decisions were made and the extent of public engagement in the functional transformation.
[4] From Sklair (2017), the term 'mediated monument' as coined by Vale (1999), alludes to historically significant structures amplified and reinforced in their iconic status through various forms of media and publicity. Historically, renowned global icons like the Egyptian Pyramids, the Taj Mahal, and the Acropolis have captured the collective imagination. While a structure's inherent architectural and historical merits contribute to its greatness, its fame is often augmented by continuous media exposure, as evidenced by travel guides and promotional materials. With the advent of tourist consumerism and the burgeoning realm of cyber-tourism, this pool of iconic monuments has been expanding, encompassing historical landmarks and modern edifices. While rooted in historical significance, many of these icons have their iconicity bolstered in the modern era through media representation. This phenomenon aligns with Riegl's distinction between 'age-value' and 'newness-value' (Gubser, 2005), suggesting a dynamic interplay between historical reverence and contemporary appeal in determining a monument's iconic status.
[5] Data extracted from Trip Advisor, covering from 2016 to 2023.

References

Arnstein, S.R. (1969) A ladder of citizen participation. *Journal of the American Institute of Planners* 35(4), 216–224. DOI: 10.1080/01944366908977225.

Bessiere, J. and Tibere, L. (2013) Traditional food and tourism: French tourist experience and food heritage in rural spaces: Traditional food and tourism in France. *Journal of the Science of Food and Agriculture* 93(14), 3420–3425. DOI: 10.1002/jsfa.6284.

Dugo, A. (2022) Neoliberal singapore: Nation-state and global city. In: Tan, K.P. (ed.) *Singapore's First Year of COVID-19: Public Health, Immigration, the Neoliberal State, and Authoritarian Populism*. Springer, Singapore, pp. 23–51. DOI: 10.1007/978-981-19-0368-7_2.

Elhagla, K., Nassar, D.M. and Ragheb, M.A. (2020) Iconic buildings' contribution toward urbanism. *Alexandria Engineering Journal* 59(2), 803–813. DOI: 10.1016/j.aej.2020.01.020.

Expat Living (2022) Do You Know the History of Lau Pa Sat Hawker Centre?. [WWW Document]. Lifestyle Guide To Moving To & Living in Singapore – Expat Living. Available at: https://expatliving.sg/history-lau-pa-sat-satay-hawker-centre-singapore/ (accessed 17 August 2023).

FairPrice Group (2020) *Annual and Sustainability Report: 2020 Everything Food Made Easy*. NTUC FairPrice Group, Singapore.

Ghani, A. (2011) A recipe for success: How Singapore hawker centres came to be. *Institute of Policy Studies* 3, 1–15.

Gubser, M. (2005) Time and history in Alois Riegl's theory of perception. *Journal of the History of Ideas* 66(3), 451–474.

Henderson, J. (2000a) Food hawkers and tourism in Singapore. *Hospitality Management* 19, 109–117.

Henderson, J. (2000b) Attracting tourists to Singapore's Chinatown: A case study in conservation and promotion. *Tourism Management* 21, 525–534. DOI: 10.1016/S0261-5177(99)00108-9.

Henderson, J.C., Yun, O.S., Poon, P. and Biwei, X. (2012) Hawker centres as tourist attractions: The case of Singapore. *International Journal of Hospitality Management* 31(3), 849–855. DOI: 10.1016/j.ijhm.2011.10.002.

Huber, D. (2016) Lacaton & Vassal Have Pioneered a Strategy for Saving France's Social Housing—Metropolis. [WWW Document]. Available at: https://metropolismag.com/profiles/lacaton-vassal-pioneered-strategy-saving-france-social-housing (accessed 9 October 2023).

Imai, A. (2019) Safeguarding street food: Hawker culture in singapore. In: *What Is the Essence of Conservation?* ICOFOM, Paris, pp. 76–83.

Kamel-Ahmed, E. (2015) What to conserve? Heritage, memory, and management of meanings. *ArchNet-IJAR* 9, 67. DOI: 10.26687/archnet-ijar.v9i1.469.

Kowalczyk, A. (2014) From street food to food districts–gastronomy services and culinary tourism in an urban space. *Turystyka Kuturowa* 9, 136–160.

Lau Pa Sat (2020) Lau Pa Sat – The Taste of Singapore. Available at: https://laupasat.sg/ (accessed 19 August 2023).

Lioncitysites (2012) Lau pa sat – The Architecture. *LionCitySites*. Available at: https://lioncitysites.wordpress.com/2012/11/05/lau-pa-sat-the-architecture/ (accessed 17 August 2023).

Liow, E.D. (ed.) (2012) The neoliberal-developmental state: Singapore as case study. *Critical Sociology* 38(2), 241–264. DOI: 10.1177/0896920511419900.

Liu, G. (2001) *Singapore: A Pictorial History, 1819–2000*. Psychology Press, Singapore.

Lues, J.F., Rasephei, M.R., Venter, P. and Theron, M.M. (2006) Assessing food safety and associated food handling practices in street food vending. *International Journal of Environmental Health Research* 16, 319–328.

Lynch, K. (1964) *The Image of the City*. MIT Press, Bosto, MA.

Maitland, R. and Newman, P. (2010) Visitor–host relationships: Conviviality between visitors and host communities. In: *City Spaces-Tourist Places*. Routledge, pp. 223–242.

Muhd, Z.I. and Shah, H.P. (2021) The Ultimate Guide to Cleaner Hawker Centres. [WWW Document]. Temasek Polytechnic[. Available at: https://www.tp.edu.sg/about-tp/media-centre/campus-news/stop-press-news/the-ultimate-guide-to-cleaner-hawker-centres.html (accessed 19 August 2023).

National Heritage Board (2021) Heritage Conservation Centre. [WWW Document]. Available at: https://www.nhb.gov.sg/what-we-do/our-work/preserve-our-stories-treasures-and-places/the-national-collection/heritage-conservation-centre (accessed 10 August 2023).

Native Food and Wine (2010) Singapore's Food Hawkers. [WWW Document]. Available at: http://www.nativefoodandwine.com/features-journal/singapores-food-hawkers.html (accessed 19 August 2023).

Nguyen, H. and Chen, J. (2018) Towards more participatory governance in Singapore. *Lee Kuan Yew Centre for Innovative Cities*. Available at: https://lkycic.sutd.edu.sg/blog/towards-participative-governance-singapore/ (accessed 10 August 2023).

Othman, A.A.E. and Elsaay, H. (2018) Adaptive reuse: An innovative approach for generating sustainable values for historic buildings in developing countries. *Organization, Technology and Management in Construction: An International Journal* 10, 1704–1718. DOI: 10.2478/otmcj-2018-0002.

Quek, E. (2020a) Meet the new folks at Lau Pa Sat. The Straits Times.

Quek, E. (2020b) FairPrice group opens food hall at Lau Pa Sat. The Straits Times.

Radomskaya, V. and Bhati, A.S. (2022) Hawker centres: A social space approach to promoting community wellbeing. *Urban Planning* 7(4), 167–178. DOI: 10.17645/up.v7i4.5658.

Roots (2021) Hawker Culture. [WWW Document]. ROOTS[. Available at: https://www.roots.gov.sg/en/ich-landing/ich/hawker-culture (accessed 31 March 2023).

Roots (2022) Former Telok Ayer Market (now known as Lau Pa Sat). [WWW Document]. Available at: https://www.roots.gov.sg/places/places-landing/Places/national-monuments/former-telok-ayer-market-now-known-as-lau-pa-sat (accessed 17 August 2023).

Savage, V.R. and Yeoh, B. (2013) *Singapore Street Names: A Study of Toponymics*. Marshall Cavendish International Asia Pte Ltd, Singapore.

Sklair, L. (2005) The transnational capitalist class and contemporary architecture in globalizing cities. *International Journal of Urban and Regional Research* 29, 485–500. DOI: 10.1111/j.1468-2427.2005.00601.x.

Sklair, L. (2017) *The Icon Project: Architecture, Cities, and Capitalist Globalization*. Oxford University Press, New York.

Smith, D. (2018) The bells at Lau Pa Sat. *Medium*. Available at: https://devinsmithwork.medium.com/the-bells-at-lau-pa-sat-30f5c471995e (accessed 18 August 2023).

Sproesser, G., Ruby, M.B., Arbit, N., Akotia, C.S., Alvarenga, M. *et al.* (2019) Understanding traditional and modern eating: The TEP10 framework. *BMC Public Health* 19(1), 1606. DOI: 10.1186/s12889-019-7844-4.

Steemers, K. (2021) *Architecture for Well-Being and Health.* [WWW Document] Velux Commercial. Available at: https://commercial.velux.com/blog/daylight/architecture-for-well-being-and-health (accessed 13 November 2023).

Tam, A. (2017) Singapore hawker centers: Origins, identity, authenticity, and distinction. *Gastronomica* 17, 44–55.

Tung, A.J. (2023) Travelling Hawkers. [WWW Document]. National Heritage Board[. Available at: https://www.nlb.gov.sg/main/article-detail?cmsuuid=ede5738c-465b-49ce-9b3f-4001ad6826e5 (accessed 5 October 2023).

UNESCO (2020) Hawker culture in Singapore Community Dining and Culinary Practices in a Multicultural Urban Context. [WWW Document]. Available at: https://ich.unesco.org/en/RL/hawker-culture-in-singapore-community-dining-and-culinary-practices-in-a-multicultural-urban-context-01568 (accessed 30 July 2023).

URA (2019) Highly Liveable and Sustainable City. [WWW Document]. Available at: https://www.ura.gov.sg/Corporate/Data/Resources/Publications/Annual-Reports/Web-AR/Web-AR-19/AnnualReport2018-2019/Theme-1-landing (accessed 9 October 2023).

Vale, L.J. (1999) Mediated monuments and national identity. *The Journal of Architecture* 4(4), 391–408.

Vargas, A.M., de Moura, A.P., Deliza, R. and Cunha, L.M. (2021) The role of local seasonal foods in enhancing sustainable food consumption: A systematic literature review. *Foods* 10, 2206. DOI: 10.3390/foods10092206.

Visiting Singapore (2023) Lau Pa Sat [WWW Document]. Available at: https://www.visitsingapore.com/see-do-singapore/architecture/historical/lau-pa-sat/ (accessed 17 August 2023).

Yong, C. (2021) Humpty Dumpty eats kaya toast: Hawker culture in focus in three upcoming books. *The Straits Times.*

Yung, E.H., Chan, E.H. and Xu, Y. (2014) Community-initiated adaptive reuse of historic buildings and sustainable development in the inner city of Shanghai. *Journal of Urban Planning and Development* 140, 05014003.

7 Experiencing Cambodia's Living Arts Through Tourism

Bobbie Chew Bigby[1], Madura Thivanka Pathirana[2] and Sarin Chhuon[3]
[1]University of Waterloo, Canada; [2]Edith Cowan University, Sri Lanka Campus, School of Business and Law, Colombo, Sri Lanka; [3]Cambodia Living Arts, Phnom Penh, Cambodia

Summary
Experience Cambodian Living Arts (CLA) was a nightly program showcasing traditional Cambodian arts to tourists while providing employment opportunities to artists. Produced by the Cambodia-based non-profit organization CLA for nearly a decade, this program used tourism as a platform for sharing traditional arts while also earning revenue to support other arts initiatives and providing employment to several of its arts students.

This case study offers a glimpse into the arts-based programming of CLA, an organization based in Cambodia that has worked to revitalize traditional arts and inspire younger generations to engage with the arts over the past 25 years. 'Experience CLA' was one of the organization's programs that offered nightly performances of traditional dance, music and performing arts between 2011 and 2020. These performances stood at the heart of CLA's efforts in sharing traditional arts with international visitors through tourism, providing a platform for employment to young Cambodian artists, as well as helping the organization to earn revenue. However, due to the impact of COVID-19 on tourism as well as CLA adopting a new mission and vision that positioned it as more of an arts catalyst, rather than as an arts provider, the 'Experience CLA' program came to an end in 2020 as the organization began to shift its role in the Cambodian arts community.

The Value and Interest of the Case Study

The insights offered by this Cambodian case study help to paint a rich picture of the tourism offerings from an arts-based, non-profit organization operating in Southeast Asia. This case study is distinctive for its focus on the intersection of Cambodian arts, tourism, and the non-profit or non-governmental organization (NGO) sector. This study has potential relevance to other post-conflict nations that look to cultural and arts-based tourism for economic development and safeguarding traditional heritage. Additionally, this case study provides a unique example of a successful arts-based tourism model run by a local NGO and also one that comes to a natural conclusion based on both the organization's shifting roles and priorities, as well as the wider impacts of the COVID-19 pandemic upon tourism.

Background

Cambodian history

Cambodia is a Southeast Asian nation with an extremely rich cultural and artistic heritage. The country is recognized best for its temples of the Angkor Empire constructed between the 9th and 15th centuries and serving as a key draw card for millions of tourists annually. However, Cambodia is increasingly becoming known for its vibrant living heritage and art traditions, ranging from its traditional silk weaving and shadow puppets to classical dance and theater (Chheang, 2008; Tuchman-Rosta, 2014). This extensive cultural heritage and the art forms, both tangible and intangible, that continue into the present day are a product of the Khmer empire of Angkor that covered a large portion of mainland Southeast Asia from 802 AD to 1431 AD (Chandler, 2008). Beginning in 1863, Cambodia became a protectorate of France until it gained its full independence in 1953.

The period of the 1940s through to the 1960s in Cambodia saw intensive and exciting local investment in Khmer cultural expressions and activities across all domains of the arts and creative industries. Under the leadership of the Cambodian Royal Family, Queen Kossamak is credited with helping to rekindle and elevate the classical dance form into the esteemed artform it is today, while King Norodom Sihanouk served as a knowledgeable patron of the arts (Phim and Thompson, 1999; Heywood, 2008). Classical dance, also known as the Royal Ballet of Cambodia, drew on the proud legacy of the Angkor Empire, while also providing a strong sense of Khmer cultural and national identity for a newly independent country in the 1950s and beyond.

While Cambodia under the leadership of King Norodom Sihanouk aimed to maintain neutrality in the 1950s and 1960s with conflict growing and spreading throughout neighboring nations, by the 1970s the country was drawn deeply into the regional conflict. On April 17, 1975, the newly established Khmer Rouge regime assumed control of the capital, Phnom Penh, and forcibly ordered all Cambodians to move to the countryside and work growing rice. The Khmer Rouge sought to transform the Cambodian nation into a self-sufficient agrarian utopia and encouraged the abandonment of all traditional culture and modern education. The result of this genocidal period from 1975 to 1979 was the death of nearly 2 million Cambodian people, with an estimated 90% of artists—dancers in particular – not surviving this period (Heywood, 2008). The Khmer Rouge regime was brought to an end by Vietnamese occupation. However, civil war and deep internal conflicts continued in Cambodia up until the early 1990s.

The Paris Peace Accord of 1991, together with the Cambodian general elections overseen with the help of UN Peacekeeping Forces in 1993, helped to usher in a new era of post-war political and economic stability in Cambodia. From the late 1990s through the 2000s, Cambodia's economy grew by about 8%, relying most strongly on the textile and tourism sectors (Chheang, 2008). At present, the tourism, garment, and agricultural sectors remain the strongest economic pillars of the Cambodian economy (Winter, 2008).

Study methodology

The research informing this case study is centered upon the self-reflexive and autoethnographic conversations, recollections and participant observation of two of the study's co-authors, Sarin Chhuon and Bobbie Chew Bigby. For the duration of the Experience CLA Program, Sarin worked as the Program Manager of Experience CLA, assuming all responsibilities for producing the nightly cultural performances, as well as providing support for additional workshops and cultural activities that were offered to visitors by CLA. Bobbie was based in Cambodia from 2009 to 2011, first as a Fulbright Research Scholar hosted by CLA and later as an employee of CLA. Some of Bobbie's responsibilities included assisting with CLA arts festival production, as well as providing support to CLA's tourism and language education efforts. Apart from the primary method of autoethnography that is utilized in this case study (Coghlan, 2012), additional information is sourced from some of the promotional documents, CLA website pages and work notes that were used in the production of Experience CLA.

The Beginnings of CLA

CLA is a non-profit and non-governmental organization (NGO) based in Cambodia that was created in 1998 by Arn Chorn-Pond, a Cambodian survivor of the Khmer Rouge genocide (McCormick, 2012). As a young boy at the start of the Khmer Rouge period, Arn was separated from his family when he was conscripted as a child soldier. He was forced to play the flute for Khmer Rouge soldiers and generals – a task which he believes saved his life and forever cemented his love for music and the arts. After being adopted by an American family from a Thai refugee camp in 1980, Arn moved to the United States where he eventually began speaking to groups of people about his experiences during the Khmer Rouge period. In 1996, Arn returned to Cambodia to seek connection with family and members of the arts community. What he found were the few remaining artists who had survived the genocide and who were now living in poverty.

Beginning in 1998, Arn brought five inaugural students together to learn from master artists willing to teach, and CLA was born (CLA, n.d.-b). First focused on the preservation, transmission and teaching of Cambodia's rich, yet threatened performing arts legacy, the organization quickly grew in terms of the number of students served and provinces reached, impacting young arts learners in both urban and rural Cambodia. At this early stage, CLA's Master Performer Program worked with master arts instructors across seven provinces and urban areas. Some of the performing arts initially supported in the classes included folk and classical dance, Yike opera and chapei playing, a traditional Cambodian, long-necked stringed instrument. By 2012, the organization was offering 27 different classes across 12 different traditional art forms. These classes reached approximately 1000 students and were held across nine different provinces (CLA, n.d.-a).

As the organization has developed and grown, its work and mission have also evolved beyond providing art classes alone. CLA has also come to be known nationally and internationally for its art festivals and commissioned performances that it has produced, both within and outside of Cambodia, such as the 2013 Season of Cambodia festival that took place in New York City over a month. These Cambodian arts festivals, performances, workshops and other programs have long been at the heart of what CLA does not only as a means of employing master and young Cambodian artists alike but as a key conduit for reaching non-Cambodian audiences of tourists and visitors to share traditional Khmer arts.

At present, CLA is guided by the vision that 'the arts are at the heart of a vital society' and a mission, 'to be a catalyst in a vibrant arts sector, inspiring generations.' CLA has grown in its scope and reach for Cambodia's younger generations (CLA, n.d.-c). The organization has gradually transitioned away from serving as a direct arts instruction provider and producer. Instead, CLA now focuses on developing the capacity of Cambodians in the arts through offering scholarships, fellowships and other training; facilitating research, networking and international exchanges; as well as offering employment to students of the arts through regular festivals and performances.

CLA's Tourism Intersections Through 'Experience CLA'

While the activities of CLA in its early days were centered around arts training between masters and young students, over the years CLA's engagement with tourism, and specifically with performances and festivals, has grown and become a mainstay of the organization's programming. From the early 2010s up until the start of the COVID-19 pandemic in 2020, CLA produced 'Experience CLA,' a nightly performance of cultural dances and music at the National Museum (see Fig. 7.1) in Phnom Penh that was geared towards international tourists who paid admission (CLA, n.d.-d). In addition, 'Experience CLA' allowed visitors to take part in arts workshops, or arrange and book in advance demonstrations and smaller performances. These arts workshops and demonstrations could be held in Phnom Penh, Siem Reap, or other provinces and communities throughout Cambodia, often depending on where the art master and student community of a particular art form reside. For example, specific art forms such as Smot, or traditional Buddhist chants associated with funerals, were only located in Kampong Speu province, while support for specific Indigenous artforms were directed towards those communities in Rattanakiri and Mondulkiri provinces.

In the first year of Experience CLA, the show was held once a week in front of the

Fig. 7.1. The front of the National Museum in Phnom Penh. Courtesy of Jacky Desbiolles.

National Museum. Cambodia's National Museum was chosen as the venue for these performances for a number of reasons. For one, the National Museum comes under the authority of the Cambodian Ministry of Culture and Fine Arts, a National Ministry with which CLA has always had a consistent relationship and strong partnership. The idea of hosting performing arts shows was also attractive to the Museum as its leadership was keen to show that it could be a venue for living and performing cultural heritage, in addition to housing precious artifacts and sculptures. Finally, the National Museum's location was seen as a perfect hub for arts and culture given that it is close to Cambodia's other major artistic centers, including the Royal University of Fine Arts and numerous art galleries along Street 178, as well as being near other major tourism draws, including the Royal Palace and the Riverside Promenade.

After a more permanent stage was constructed at the Museum's side entrance, and due to the growing demand for performances, the show was put on three nights per week for another 2 years. Finally, with the demand continuing to grow, particularly during the high tourist season between October and March, Experience CLA was offered every night at the Museum up until the start of the COVID-19 pandemic. During these tourist high seasons, nightly audiences ranged from 70 to 100 people. Throughout the low season, from April to September, when Cambodia experienced its rainy season, the average attendance numbers were approximately 40–60 people per night.

These nightly performances generally involved between 30 and 50 artists and aimed to provide sustainable and fairly paid jobs for artists and technicians. The performing artists, including dancers, singers and musicians, were comprised in large part of arts students who had been trained through several of CLA's programs. After having completed the arts training offered by CLA, some of the CLA arts alumni who performed had also received training at the Royal University of Fine Arts in Phnom Penh.

At the beginning of Experience CLA, the organization worked primarily with the 'Children of Bassac' folk and classical dance class troupe led by Master artist Ieng Sithul. However, once the permanent outdoor stage was constructed and nightly performances commenced Experience CLA began to draw on talent from multiple other groups. These included a Yike opera troupe, a large shadow puppet troupe, a creative theater troupe, and other folk and classical dance groups.

While visitors to the program might have been entertained by slightly different artistic offerings depending on the night, Experience CLA shows almost always offered a mix of traditional pinpeat music, classical Cambodian dance and folk dances from Cambodia's different provinces and ethnic communities. The pinpeat orchestra and singers located at the side of the stage would provide the musical accompaniment and backdrop for the evening while helping to move the program from one dance to the next. Select pieces from the classical Cambodian dance repertoire, reflecting the highly refined tradition of the Khmer Royal Ballet, were a key program offering. Two pieces commonly performed were 'Robam Apsara,' or the 'Apsara Dance,' as seen in Fig. 7.2 and 'Robam Sovannmacha,' known commonly as the 'Dance of the Monkey and Golden Mermaid.' The Apsara Dance features several women who dance the role of the 'apsaras,' or female celestial beings, who play a central role in the creation story of the Khmer people and the identity of the Cambodian Nation. 'Robam Sovannmacha' is a story taken from the Khmer interpretation of the Hindu epic, The Ramayana, known in Khmer as 'Reamker.' The dance-drama piece tells the story of the monkey King, Hanuman, and his efforts to seduce the Golden Mermaid who is intent on destroying a vital stone bridge being constructed. After a flirtatious exchange, Hanuman is finally successful in his efforts to stop the Mermaid and her companions from undoing the bridge construction.

Interspersed throughout these pieces of classical dance were a number of lively, colorful and much faster-paced folk and traditional dances. These pieces were based on certain folk traditions of the rural areas in addition to specific ethnic or Indigenous communities living throughout Cambodia's various provinces. Some of the commonly performed dances included the Fishing Dance, the Buffalo Sacrificing Ceremony, the Dance of Crushing Dye, the Harvest Dance, the Cardamom Picking Dance (as seen in Fig. 7.3), and the Pestle Clapping Dance. Particular dances and ceremonies belong to specific Indigenous communities, with the Buffalo Sacrifice Ceremony an activity of the Phnong Indigenous people of eastern Cambodia and the Harvest Dance belonging to the Suoy people of Kampong Speu

Fig. 7.2. Dance students of the 'Children of Bassac' troupe performing Robam Apsara. Courtesy of Jacky Desbiolles.

province. On the other hand, dances such as the Fishing Dance or Cardamom Picking were representative of rural and folk activities common to many Cambodian communities, or those villages that are located near resources such as waterways or mountains where cardamom grows. Unlike the classical dance pieces that required dancers to be adorned in costumes of elaborate silk, jewelry, and gold crowns, the folk and traditional dancers would have diverse, colorful costumes that often included specific hats, scarves or props that matched the theme of the dance.

The revenue generated from the shows and workshops that comprised the Experience CLA program helped to provide a steady income for students but also served as an important revenue source for CLA to help fund its other programs that were not necessarily revenue-earning. From 2014 to 2020, the revenue generated by Experience CLA contributed approximately one-third of the overall CLA organizational budget, which stood at approximately US$2 million in early 2020. This arts tourism-based income was a very helpful source of support for an organization whose budget draws heavily on grants and donations. Experience CLA contributed between US$50,000 and US$150,000 to other CLA programs annually, while in some years revenue was reinvested back to Experience CLA in order to make program improvements such as adding seats or constructing a roof for rainy season performances.

CLA as an Arts Catalyst and COVID-19 Impacts

As the spread of COVID-19 began to intensify in the early part of 2020, few people initially realized the impacts of the disease, which would soon grow to be a pandemic, not only on a program like Experience CLA, but on tourism

Fig. 7.3. 'Children of Bassac' troupe performing Robam Beh Krawine, or the Cardamom Picking Dance. Courtesy of Jacky Desbiolles.

throughout Cambodia and globally. However, the months of February and March 2020 would be the final window of opportunity for the Experience CLA performances to be staged. Fig. 7.4 shows the final billboard for Experience CLA outside of the National Museum before the show ended. By the end of March 2020, tourism flows into Cambodia were halted and further performances of Experience CLA were cancelled.

Throughout the most severe period of the global COVID-19 pandemic between 2020 and 2021, performing artists across Cambodia, like so many artists and other professions throughout the country, suffered from the lack of tourism and other employment options (Prasad, n.d). Cambodia was certainly not unique in this regard. Globally, arts-based tourism, like most other forms of tourism, along with arts and cultural-based economies more generally, felt massive negative impacts and revenue losses. Several of the research studies of the impact of the COVID-19 pandemic on arts tourism and the creative sectors show not only that impacts were severe, but that recovery was also slower in these sectors (Flew and Kirkwood, 2021; Cohen, 2022). As COVID-19 continued to rage on globally, Cambodia faced the lack of consistent tourism streams to provide audiences for Experience CLA. Additionally, with the rapidly evolving local government regulations on public gatherings and crowd sizes to prevent the spread of COVID-19, Experience CLA had no pathway to continue during this pandemic time.

Running parallel to the massive transitions imposed upon CLA as an organization by the pandemic and the COVID-19-driven impacts on cultural economies and tourism, in this period, CLA began to reconsider and reevaluate its role, vision and mission within the arts ecosystem of Cambodia. The organization's growth over the past decades had resulted in larger numbers of students and communities being reached. Yet for many of the arts students who had been trained in these

Fig. 7.4. The billboard for Experience CLA outside of the National Museum, taken in February 2020 at the start of the COVID-19 pandemic. Courtesy of Bobbie Chew Bigby.

traditions, further opportunities and funding for professional growth in the broader environment had remained elusive and out of reach for some. Given this context of wanting to be able to support emerging artists through opportunities outside of arts training itself – including through grant-making and arts administration professional development – CLA began the process of revising its vision and mission statements. In this latest transition, CLA has been developing and evolving in its role as an art 'catalyst' in Cambodian society by providing the support, capacity building and opportunities for artists that might not be readily found elsewhere. With this new outlook, Experience CLA was recognized as a program that no longer fit directly with the host of current initiatives driven by this 'catalyst' mission. As a consequence, Experience CLA was discontinued as a result of both the COVID-19 pandemic and a change in the organization's priorities. Yet this change did not diminish the important role and legacy of Experience CLA that lasted nearly a decade.

Conclusion

Experience CLA is undeniably a strong example of a local arts-based tourism program that flourished over multiple years, provided employment to performing artists and revenue to an organization and was able to share classical and traditional Cambodian dances with countless audiences and tourists. Yet through the program's story, it is clear that the success of the program is closely tied to both the wider context of the Cambodian arts environment and the organization's own priorities, role and mission within Cambodian society. As the organization's own vision and role began to adapt and change, first in line with the broader society and secondly, in response to the unprecedented impacts of COVID-19, CLA similarly transitioned and adapted. Thus, in spite of the fact that Experience CLA is no longer a program offered to inspire, employ and entertain through arts-based tourism, the meaningful roles it played for artists, tourists and community members over many years can point to success in its impacts.

References

CLA (n.d.-a) About. Available at: https://www.cambodianlivingarts.org/en/about/ (accessed 11 August 2023).

CLA (n.d.-b) Arn Chorn-Pond. Available at: https://www.cambodianlivingarts.org/en/arn-chorn-pond/ (accessed 11 August 2023).

CLA (n.d.-c) Cambodian Living Arts. Available at: https://www.cambodianlivingarts.org/en/ (accessed 11 August 2023).

CLA (n.d.-d) Experience CLA. Available at: https://www.cambodianlivingarts.org/en/uncategorized/experience-cla/ (accessed 11 August 2023).

Chandler, D. (2008) *A History of Cambodia. (Fourth Edition)*. Westview Press, Boulder.

Chheang, V. (2008) The political economy of tourism in Cambodia. *Asia Pacific Journal of Tourism Research* 13(3), 281–297. DOI: 10.1080/10941660802280414.

Coghlan, A. (2012) An autoethnographic account of a cycling charity challenge event: Exploring manifest and latent aspects of the experience. *Journal of Sport & Tourism* 17(2), 105–124. DOI: 10.1080/14775085.2012.729899.

Cohen, R. (2022) COVID-19's pandemic's impact on the arts: Research update. COVID-19's pandemic's impact on the arts: Research update May 12, 2022. Americans for the Arts.

Flew, T. and Kirkwood, K. (2021) The impact of COVID-19 on cultural tourism: Art, culture and communication in four regional sites of Queensland, Australia. *Media International Australia Incorporating Culture & Policy* 178(1), 16–20. DOI: 10.1177/1329878X20952529.

Heywood, D. (2008) *Cambodian Dance: Celebration of the Gods*. River Books, Bangkok.

McCormick, P. (2012) *Never Fall Down: A Novel, Based on the True Story of Arn Chorn-Pond*. Balzer + Bray, New York.

Phim, T.S. and Thompson, A. (1999) *Dance in Cambodia*. Oxford University Press, Oxford, UK.

Prasad, A. (n.d) The COVID-19 impact on cambodian art & culture. *The Borgen Project* Available at. Available at: https://borgenproject.org/cambodian-art/ (accessed 11 August 2023).

Tuchman-Rosta, C. (2014) From ritual form to tourist attraction: Negotiating the transformation of classical Cambodian dance in a changing world. *Asian Theatre Journal* 31(2), 524–544. DOI: 10.1353/atj.2014.0033.

Winter, T. (2008) Post-conflict heritage and tourism in Cambodia: The burden of Angkor. *International Journal of Heritage Studies* 14(6), 524–539. DOI: 10.1080/13527250802503274.

8 Aikido Tourism Development in Tanabe, Wakayama, Japan

Eiji Ito
School of Health and Sport Sciences, Chukyo University, Toyota, Aichi, Japan

Summary
This case study illustrates the successful promotion and development of budo (martial arts) tourism by the Japan Sports Agency. It focuses primarily on aikido tourism and provides insights into strategies and initiatives that can be implemented to leverage the region's unique cultural and sport-related heritage for tourism purposes. Furthermore, it highlights the importance of authenticity in sport tourism experiences, emphasizing Tanabe's role as the birthplace of aikido's founder and the involvement of the sensei (instructor), and its possession of a rich historical background and competent guides/translators. These aspects ensure authentic and meaningful encounters with tourists. Additionally, this case study emphasizes the significance of industry-government-academia collaboration in enhancing sustainable tourism development. Collaboration between the Japan Sports Agency, local tourism agencies, aikido practitioners, and the community demonstrates the value of leveraging resources and expertise to create meaningful tourism experiences.

The Value and Interest of the Case Study

This case study provides valuable insights into the planning, development, and promotion of budo tourism by illustrating effective strategies for harnessing a region's cultural heritage and engaging with domestic and international tourists. It serves as a practical example for destinations and organizations seeking to create authentic and sustainable sports tourism offerings.

Background

The Japan Sports Agency, a governmental organization under the Ministry of Education, Culture, Sports, Science, and Technology in Japan, has been actively promoting sports tourism since its inception in 2015. TheJapan Sports Agency (2019) announced the *New Strategy for Expanding Sports Tourism Demand* in 2018, focusing primarily on promoting two types of sport tourism: budo (martial arts) tourism and outdoor sport tourism. This strategy has leveraged popular tourism attractions in Japan. Budo tourism has a significant advantage as a highly recognizable and respected sports tourism brand, being the birthplace of budo disciplines such as judo, karate, and aikido. According to a Japan Sports Agency's survey Japan Sports Agency (2018), inbound tourists from seven countries/regions (America, Australia, China, Hong Kong, South Korea, Taiwan, and Thailand) expressed a strong interest in watching traditional

Japanese sports. Budo emerged as the top choice, with 31.9% of respondents showing a keen interest. Sumo, a traditional Japanese wrestling sport, followed closely with 28.5%. These two groups garnered more interest than baseball (26.8%), snow sports (24.0%), and football (23.8%).

The Japan Sports Agency initiated the *Sport Tourism Attraction Development Project* in 2021 to promote sports tourism in Japan, with the primary objective of developing attractive sport tourism packages for both domestic and inbound tourists by utilizing local attractions (Japan Sports Agency, 2023a). Four of the seven projects that commenced in 2022 focused on developing budo tourism (Japan Sports Agency, 2023b). One project focused on the development of aikido tourism in Tanabe, Wakayama. Aikido is popular in over 140 countries and originated in Tanabe, the birthplace of its founder, Morihei Ueshiba (Aikikai Foundation, n.d.). This region has been witness to some of the aikido tourism phenomena, including visits to Ueshiba's cemetery (heritage sport tourism) and the 2008 International Aikido Congress (event sport tourism), which attracted both domestic and international aikido practitioners to Tanabe. However, these phenomena appear to occur naturally rather than strategically, potentially endangering the sustainability of sport tourism development (Hinch and Ito, 2018). This indicates the importance of diligent planning (Higham and Hinch, 2018). To address this issue, a tourism agency called Tabikyo Japan Co., Ltd initiated a project concerning aikido tourism development in collaboration with the Japan Sports Agency and local stakeholders, such as the Tanabe Aikido Dojo and the Tanabe Tourism Association (Japan Sports Agency, 2023b). They subsequently developed and commercialized aikido tourism itineraries (Kumano Travel, n.d.-a, Kumano Travel, n.d.-b).

This case study underscores not only the significance of industry-government-academia collaborations in the sport tourism sector for sustainable development but also explores the challenges associated with the commercialization (or commodification) process and its impact on authenticity (Higham and Hinch, 2018). It also aims to strike a balance between commercializing aikido and preserving its essential and authentic aspects.

The author served as a committee member for the Japan Sports Agency's *Sport Tourism Attraction Development Project* in 2022. In his role as an academic adviser for aikido tourism development, he utilized his expertise in sport tourism and 6 years of experience living in Wakayama while working for the Faculty of Tourism at Wakayama University. The information for this case study was primarily gathered through direct observations of project activities and informal discussions with stakeholders.

Authenticity

Tourists who actively engage as spectators or participants in a sport consider their experiences authentic, 'regardless of how others may assess the situation' (Higham and Hinch, 2018, p. 73). Unlike other tourist attractions, sports uniquely emphasize engagement (Higham and Hinch, 2018), with this activity-related authenticity referred to as existential authenticity (Wang, 1999). Existential authenticity refers to the potential existential state of being activated through tourist activities (Wang, 1999), which is relevant to tourism's experiential focus (Higham and Hinch, 2018).

Tanabe is one of the most crucial elements contributing to authenticity in aikido tourism development. Tanabe is home to the founder's cemetery, representing a tangible, immovable, sport-related heritage (Ramshaw and Gammon, 2005) that cannot be replicated elsewhere. This unique setting holds a deep meaning for aikido practitioners and attracts enthusiasts from across the globe. By leveraging this valuable asset, Tanabe built a new budo arena in 2020, thereby providing opportunities for both locals and tourists to participate in aikido. Furthermore, the arena proudly displays a statue of Morihei Ueshiba at its entrance (Fig. 8.1), and the Morihei Ueshiba Memorial Museum is located within the facility (Fig. 8.2). Although these elements are tangible, movable sport-related heritage (Ramshaw and Gammon, 2005), the statue and museum in Tanabe strengthen the potent authentic component linked to the founder's birthplace, ensuring authentic experiences for aikido tourists.

Fig. 8.1. Statue of Morihei Ueshiba at the entrance of Tanabe Budo Arena (courtesy of the author).

Another essential element of authenticity is the sensei (instructor), who facilitates the aikido experience. Junichi Gomita, the 5th Dan practitioner, is significant in this regard. Born into an aikido family, his parents were practitioners and he commenced aikido training during early childhood. His father, Seiji Gomita, is a direct student of Morihei Ueshiba and currently serves as the head of the Aikido Tanabe Dojo. Junichi Gomita, head of the Tanabe Aikido Dojo and deputy head of the Morihei Ueshiba Memorial Museum, has a rich historical background

Fig. 8.2. Morihei Ueshiba Memorial Museum with local tour guides (courtesy of Tabikyo Japan Co., Ltd).

(Kumano Travel, n.d.-a) that contributes to the provision of authentic experiences for beginners and experienced aikido practitioners.

Finally, guides/translators fluent in both Japanese and English and in possession of extensive knowledge of aikido play a crucial role in delivering authentic experiences to inbound tourists. Finding suitable guides/translators for the development of budo tourism remains challenging. More specifically, it is difficult to find individuals capable of accurately translating physical movements and skills as well as the essence and philosophy of budo. Without such competent guides/translators, inbound tourists may not fully appreciate the aikido experience, despite the authenticity of the place and the sensei. Another noteworthy aspect of the development of aikido tourism is the initiative to train future aikido guides/translators. This aligns with Higham and Hinch's (2018) perspective, emphasizing the significance of a process-oriented approach over a goal-oriented one in achieving sustainable development. This endeavour aims to foster the long-term sustainability of aikido tourism by providing training sessions for local tour guides (Fig. 8.3).

Commercialization

Authenticity is intricately linked to commercialization, which introduces economic relations into an area where it did not exist earlier (Higham and Hinch, 2018). Pricing is a challenge in the commercialization of budo tourism. Most budo practitioners engage in budo for reasons other than financial gain, which can make them uncomfortable when their budo activities are priced. Thus, effective communication between tourism agencies and budo practitioners is crucial in addressing this challenge. For this aikido tourism project, Tabikyo Japan Co., Ltd – a tourism agency – worked closely with aikido stakeholders (e.g. Aikido Tanabe Dojo, Aikikai Foundation), local

Fig. 8.3. Training session for local tour guides (courtesy of Tabikyo Japan Co., Ltd).

stakeholders (e.g. tour guides, local companies/communities), Tanabe City (e.g. Tourism Division, Kumano Tourism Bureau), and the Japan Sports Agency. Mr Haneda, a tourism agent at Tabikyo Japan Co., Ltd, played a vital role in the development of aikido tourism by associating with stakeholders. As an outsider to the destination, he stated that pricing budo tourism is feasible without feeling restricted by complex ties (personal communication, 17 November 2022). Based on his extensive experience as an outdoor tour guide – both domestically and internationally – he effectively communicated with stakeholders while highlighting the importance of commercializing aikido in Tanabe from the perspective of tourism agents. Exceptional collaboration among these stakeholders is exemplified by a local company president donating a set of aikido-gi (training uniforms) to the aikido tourism project.

Consequently, two aikido tourism packages were developed through a series of activities, including a 3-day and 2-night monitor tour, a monitoring session of aikido experiences, and stakeholder meetings. The first package is an all-inclusive 3-day and 2-night *Aikido, Kumano Kodo Experience* (Kumano Travel, n.d.-a). As the name suggests, this tourism plan capitalizes on the popularity of Kumano Kodo among inbound tourists and its compatibility with aikido in terms of Japanese culture. Kumano Kodo is a network of ancient pilgrimage routes located on the Kii Peninsula of Japan (including Tanabe City) which is recognized as a UNESCO World Heritage Site and has significant cultural and spiritual value in Japanese history (UNESCO, n.d.). During this trip, tourists have the opportunity to engage in aikido experiences, embark on walks along Kumano Kodo, indulge in local cuisine, participate in mikan orange picking, relax in onsens (hot springs), and visit the Kumano Hongu Taisha Grand Shrine with a knowledgeable tour guide. Kumano Travel, n.d.-a offers this tourism package at a price of JPY348,000 (approximately US$2,400) for individuals and JPY398,000 (approximately US$2,700) for two guests.

Another offering is a private 3-hour aikido session referred to by the name '*Aikido Experience Basic Course*' (Kumano Travel, n.d.-b). The minimum number of participants is one, with a cost of JPY24,400 (approximately US$165) per adult, whereas the maximum number of participants is five, with a cost of JPY9,700 (approximately US$65) per adult. Although this course serves as a basic introduction to aikido, requests from more experienced visitors are acceptable. By offering a standalone aikido session, tourists who already have their own plans can participate in the aikido experience in Tanabe. Market diversification plays a crucial role in product/market development (Higham and Hinch, 2018). This idea originated from committee members (e.g. professors, tourism experts) involved in the *Sport Tourism Attraction Development Project* led by the Japan Sports Agency. This illustrates the significance of industry-government-academia collaboration in tourism development.

Promotional materials were enhanced as part of this project. A promotional video entitled '*Step into the land of aikido*' (Tabikyo Japan Co., Ltd, 2023) was uploaded on YouTube to showcase aikido tourism using English subtitles (Fig. 8.4). Considering the limited availability of online aikido videos specifically tailored for inbound tourists compared to judo and karate, this video holds significant value in promoting not only the two aikido tourism packages but also aikido in Tanabe as a whole. This significance is underscored by data from the aforementioned Japan Sports Agency survey Japan Sports Agency (2018), which revealed that approximately 20% of inbound tourists rely on YouTube and other video-sharing platforms for sport tourism information in Japan. The breakdown by country/region is as follows: America (21.9%), Australia (15.6%), China (22.0%), Hong Kong (26.5%), South Korea (20.0%), Taiwan (28.4%), and Thailand (36.6%). Brochures detailing aikido tourism in Tanabe were also developed in English and French.

Although commercialization may jeopardise authenticity (Higham and Hinch, 2018), the development of aikido tourism in Tanabe has preserved the authentic aspects of the process. As the relationship between sport tourism development and existential authenticity has been highlighted by Higham and Hinch (2018), the case of aikido tourism development aligns with four sub-categories of existential authenticity (Wang, 1999). First, the core product of tourism packages is aikido engagement, which represents intra-personal authenticity through bodily feelings. Second, aikido practitioners

Fig. 8.4. Screenshot of the promotional video (courtesy of Tabikyo Japan Co., Ltd).

can reaffirm their self-identity through aikido engagement in the founder's birthplace, representing the second sub-category, intra-personal authenticity, as manifested in self-identity. Third, inter-personal authenticity, which enhances family ties, is emphasized by targeting inbound family tourists. Finally, engaging with vibrant communities through aikido activities in Tanabe demonstrates inter-personal authenticity because several local stakeholders and people are involved in tourism development. This case serves as a practical example for destinations and organizations seeking to create authentic and sustainable sports tourism offerings while preserving their authentic aspects.

Conclusions

The Japan Sports Agency has been promoting sport tourism extensively, with a primary focus on budo tourism. Budo tourism has several advantages, such as possessing strong tourism attractions and benefitting from the authenticity of the birthplace of martial arts disciplines such as judo, karate, and aikido. *Sport Tourism Attraction Development Project* conducted by the Japan Sports Agency has contributed to the development of aikido tourism in Tanabe, highlighting the importance of meticulous planning and collaboration among stakeholders. The significance of Tanabe as the birthplace of the founder of aikido and the collaborations with sensei, in possessing a rich historical background and competent guides/translators all enhanced the authenticity of this project. The challenges of commercializing aikido tourism, including its pricing, have been addressed through effective communication and collaboration among various stakeholders. The project has resulted in the creation of two aikido tourism packages, one of which – the all-inclusive *Aikido Kumano Kodo Experience Plan* – capitalizes on the popularity of the historic Kumano Kodo pilgrimage routes. The promotion of aikido tourism through videos with English subtitles and brochures in multiple languages has expanded its reach.

One of the key learnings for the organizations involved is the potency of collaboration. Aikido practitioners may not fully grasp the significance of aikido as a tourism attraction because it is a natural and integral part of their everyday lives. Therefore, gaining an outsider perspective is essential for the development of tourism. However, it is challenging for outsiders to understand the sentiments and situations of aikido practitioners and locals. Identifying individuals from outside a community who are passionate about local development through tourism can be challenging. Fortunately, in the case of aikido tourism development, a fortuitous encounter occurred between the locals and Mr Haneda from the tourism industry, leading to a mutual understanding. Engaging with the government and academic sectors is another critical success factor that benefits this project. Individuals from these sectors – including the author – actively participated in monitoring sessions on aikido experiences and stakeholder meetings. They provided constructive feedback from an outsider's perspective by leveraging their expertise. This project exemplifies the significance of industry-government-academia cooperation in achieving sustainable tourism development in the sport tourism sector.

With that said, aikido tourism development in Tanabe is not a concluded chapter, but rather the beginning of an exciting journey. Although two aikido tourism packages have been developed, it is imperative to promote and sell them effectively. The creation of additional aikido tourism packages is essential for long-term sustainability. In 2024, another aikido tourism development project, led by Tabikyo Japan Co., Ltd, has been accepted by the Japan Tourism Agency's *Restarting Tourism Project*. This endeavour will continue to enhance and refine aikido tourism development in Tanabe. The Japanese saying 'senri no michi mo ippo kara' ('A journey of a thousand miles begins with a single step') aptly captures the essence of aikido tourism development as well.

References

Aikikai Foundation (n.d) What is Aikido? Available at: http://www.aikikai.or.jp/eng/aikido/index.html (accessed 11 August 2023).

Higham, J. and Hinch, T. (2018) *Sport Tourism Development*, 3rd edn. Channel View Publications, Bristol, UK.

Hinch, T. and Ito, E. (2018) Sustainable sport tourism in Japan. *Tourism Planning & Development* 18(1), 96–101. DOI: 10.1080/21568316.2017.1313773.

Japan Sports Agency (2018) Supotsu tsurizumu ni kansuru kaigai maketeingu chousa houkokusho. [Inbound Marketing Survey Report about Sport Tourism]. Available at: https://www.mext.go.jp/sports/content/20200330-spt_stiiki-000005408-12.pdf (accessed 15 September 2023).

Japan Sports Agency (2019) Regional Revitalization Through Sports Tourism. Available at: https://www.mext.go.jp/sports/en/b_menu/policy/economy/rrstourism.htm (accessed 11 August 2023).

Japan Sports Agency (2023a) Reiwa 5 nendo 'supotsu tsurizumu kontentsu soushutsu jigyou' no koubo ni tsuite. [2023 Sport Tourism Attraction Development Project]. Available at: https://www.mext.go.jp/sports/b_menu/boshu/detail/jsa_00207.html (accessed 11 August 2023).

Japan Sports Agency (2023b) Supotsu tsurizumu kontentsu soushutsu jigyou seika happyoukai no kaisai ni tsuite. [Workshop of the Sport Tourism Attraction Development Project]. Available at: https://www.mext.go.jp/sports/b_menu/houdou/jsa_00130.html (accessed 11 August 2023).

Kumano Travel (n.d.-a) Aikido, Kumano Kodo Experience. Available at: https://www.kumano-travel.com/en/tours-activities/local-package-tours/3565 (accessed 11 August 2023).

Kumano Travel (n.d.-b) Aikido Experience Basic Course. Available at: https://www.kumano-travel.com/en/tours-activities/aikido-experience/3567 (accessed 11 August 2023).

Ramshaw, G. and Gammon, S. (2005) More than just nostalgia? Exploring the heritage/sport tourism nexus. *Journal of Sport & Tourism* 10(4), 229–241. DOI: 10.1080/14775080600805416.

Tabikyo Japan Co., Ltd (2023) *Step into the Land of Aikido* [Video]. YouTube. Available at: https://www.youtube.com/watch?v=Ke1QMmDzqGk

UNESCO (n.d.) Sacred Sites and Pilgrimage Routes in the Kii Mountain Range. Available at: https://whc.unesco.org/en/list/1142/ (accessed 11 August 2023).

Wang, N. (1999) Rethinking authenticity in tourism experience. *Annals of Tourism Research* 26(2), 349–370. DOI: 10.1016/S0160-7383(98)00103-0.

9 Forging Social Identity: The Making of a Destination – A Case of the Bullet Baba Shrine in Marwar, Rajasthan, India

Abhisek Porya[1], Sweety Mishra[1], Sunayana[2] and Nimit Chowdhary[3]

[1]Department of Tourism and Hospitality Management, Jamia Millia Islamia (A Central University), New Delhi, India; [2]Department of Management Studies, FMS, Jamia Millia Islamia (A Central University), New Delhi, India; [3]Jamia Millia Islamia (A Central University), New Delhi, India

Summary

The Om Banna Temple (also called the Bullet Baba Temple) is a remarkable site in Rajasthan's arid Pali district. It defies traditional spiritual norms, merging supernatural tales with the world of motorcycle enthusiasts and explorers. The temple honours Om Banna, a local deity and guardian for bikers. Om's story began with a tragic motorcycle accident in 1988, after which his motorcycle mysteriously reappeared at the accident site, defying explanation. This phenomenon led to the temple's creation, representing a belief in a protective force residing in Om's bike. This temple's popularity has grown, becoming an iconic and unconventional tourist attraction. Travellers from various backgrounds are drawn to its unique blend of spirituality and reverence for motorcycles. The temple signifies faith, protection, and an extraordinary connection to humanity. The case study explores cultural exchange in tourism, where visitors engage with the temple's rituals and stories, leading to mutual learning between tourists and locals. Social Identity Theory is applied to show how both groups develop distinct identities shaped by factors like nationality and interests and how these identities influence their interactions. The temple's identity is built on uniqueness, storytelling, emotional connection, word-of-mouth marketing, niche appeal, protective symbolism, and cultural interaction. This captivating identity attracts diverse visitors, from motorcycle enthusiasts to those seeking cultural immersion, turning the temple into an unforgettable and unconventional destination that bridges spirituality, identity, and shared experiences.

The Significance and Appeal of the Case Study

The Om Banna Temple case study holds significant value and interest due to its unique blend of spirituality and culture. This is not just another story but a journey that uncovers the essence of cultural exchange. The Om Banna Temple case study holds a treasure trove of insights, showcasing the dynamic power of shared experiences in tourism. It is a beacon for those seeking to bridge cultural gaps, revealing the transformative potential of engaging with local practices and beliefs. It highlights the power of cultural exchange in tourism, where visitors engage with local practices, fostering mutual understanding. The application of Social Identity Theory sheds light on how tourists and locals develop distinct identities that influence their interactions.

Two authors visited the temple (10–13 June 2023) and encountered travellers, unravelling their social identity dynamics. Collaborative discussions about the temple's identity, the congregation in the temple for prayer, and the gradual development of a tourist destination within the research team were followed by analysis of shared beliefs. This was further deepened by triangulating the understanding with the material available in local, national, and international magazines or news. The combination of firsthand accounts and archival data offers a nuanced perspective, allowing us to investigate the intersection of Social Identity Theory and the unique cultural phenomenon of the Om Banna temple in Rajasthan.

The temple's identity, rooted in a captivating story and protective symbolism, showcases the appeal of authenticity, emotional connection, and niche attraction in tourism. This case study offers insights into creating memorable and meaningful destinations that resonate with diverse audiences while bridging cultural gaps. It exemplifies how an extraordinary narrative can transform an ordinary site into an iconic and offbeat tourist attraction.

The Mystical Realm of the *Om Banna* Temple

Amidst the arid expanse of Rajasthan's Pali district, where golden sand dunes stretch as far as the eye can see, lies a temple that defies the conventional norms of spirituality. Welcome to the captivating realm of the Om Banna Temple, fondly known as the Bullet Baba Temple (see Fig. 9.1), where a tale of the supernatural intertwines with the rumble of motorcycle engines and the hopes of weary travellers (see Fig. 9.2).

Nestled within this desert landscape is the shrine dedicated to Om Banna, a local deity who takes on the role of a guardian for bikers and explorers. However, there is more to his name than meets the eye – 'Om' pays homage to the person he once was, while 'Banna' conjures the image of a spirited, unmarried youth, echoing through the labyrinthine alleys of Rajasthan's dialect.

Let us rewind to a fateful day – 23 December 1988. Om Singh Rathore, a mere 23 years old, met his tragic end in a motorcycle accident. Little did anyone know that his story was far from over. In an eerie twist, the police took custody of his motorcycle. This prized possession had carried him into the pages of history. However, as the dawn broke, the bike vanished

Fig. 9.1. The Image of OM Banna or Bullet Baba in the temple.

Fig. 9.2. The image of Bullet Baba's motorcycle (Royal Enfield Bullet 350) placed inside the temple.

from the station, reappearing at the very spot of the accident. Chains, locks, and reason could not hold back this mysterious force.

The bewildering spectacle of the motorcycle's nightly escapades from the police station for two consecutive nights intrigued the villagers, fuelling whispers of divine intervention. Thus, the Om Banna Temple emerged (see Fig. 9.1), embodying their belief in a higher power, a protector residing in the metal and gears of Om Singh Rathore's beloved bike (see Fig. 9.2). The tale unfolded like a mesmerizing melody, drifting beyond the village's borders, drawing travellers into its enigmatic web.

However, this was no ordinary road, the NH-62, Jodhpur-Pali expressway. It had witnessed its fair share of tragedies before Om Banna's encounter with destiny. Locals whispered that the spirits of the land were restless, and it took the spirit of a young boy on a motorcycle to bring them solace. Moreover, it did bring solace as accidents dwindled, and tales of miraculous escapes spread like wildfire.

Venture into the temple, and you will find it, the motorcycle that defies reason and blurs the line between the tangible and the ethereal. It is said that Om Banna's spirit finds its eternal home here, forever cradled in the curves of the bike's frame. Legends whisper that he still roams these roads, occasionally stepping out of the realm of the unseen to lend a hand to those in need.

As the sun dips below the horizon, travellers pause in reverence. A pilgrimage to this temple is more than just a journey, it is a ritual of safety, a pact with the divine. As the villagers say, passing by without paying respects is an invitation to fate's capriciousness. With the temple guarding the path (NH-62, Jodhpur-Pali expressway), who would dare to take fate for granted?

However, it is not just prayers that grace the temple's hallowed halls. It is an assembly of gratitude, a sea of motorcycles that stand as testaments to lives saved and stories shared. Each two-wheeled offering is a heartfelt ode to the interplay between the human spirit and the supernatural. Thus, against Rajasthan's sun-kissed sands, the Om Banna Temple is a testament to the extraordinary, a magnet for the curious and the intrepid. The Bullet Baba phenomenon showcases India's rich tapestry of beliefs and traditions, where religious and cultural practices can take on diverse and unexpected forms.

The story of Bullet Baba has gained popularity over the years. The temple has become an iconic and offbeat tourist attraction in Rajasthan. Visitors come far and wide to witness this unusual blend of spirituality and reverence for a motorcycle. The temple symbolizes faith, protection, and a unique connection between humans and machines. Here, the ordinary is transformed into the extraordinary, where a motorcycle is not just a machine. It is a conduit between realms, and a journey becomes more than a destination – it is an encounter with the divine.

Transformative Experiences

The case of Om Banna Temple delves into the concept of cultural exchange within tourism, where visitors from diverse backgrounds engage with the shrine's rituals, stories, and practices, gaining insights into the local culture and beliefs.

The narrative of the tale woven around the Om Banna temple highlights how tourists, arriving with their cultural perspectives, interact with the temple as a platform for exchanging ideas, stories, and experiences. This interaction leads to a mutual learning process where visitors and locals share and understand different ways of life. The temple serves as a bridge between cultures, fostering a deeper appreciation for diversity.

Intertwined with this cultural exchange is the concept of Social Identity Theory, which psychologist Henri Tajfel developed in the 1970s (Tajfel and Turner, 1979). It provides insights into how individuals categorize themselves and others into groups. These group memberships have an impact on our behaviour and attitudes. It is natural for people to classify themselves and others based on shared characteristics such as ethnicity, nationality, or hobbies, creating a sense of identity associated with these groups. This results in the formation of 'in-groups' (groups to which individuals belong) and 'out-groups' (groups to which they do not belong), often leading to a preference for the in-group. Individuals tend to compare their group with others in search of distinctiveness that boosts their self-esteem by perceiving their group as superior. Unfortunately, this favouritism can contribute to intergroup bias conflicts between groups and even instances of discrimination. Social Identity Theory offers a framework for comprehending social dynamics like prejudice, stereotypes, and the influence of group membership on our perceptions and behaviour in various social contexts. For the local community, the temple becomes a stronghold of their collective identity and cultural heritage. Meanwhile, their motivations shape tourists' identities – motorcycle enthusiasts or curious explorers.

Cultural exchange in the context of tourism refers to the interaction and sharing of cultural elements between tourists and the host community. In the case of the Om Banna Temple, tourists from various backgrounds come to the shrine to witness and participate in a unique local tradition. They engage with the rituals, stories, and practices associated with the temple, which provides them with insights into the local culture and beliefs. Tourists bring their cultural perspectives, beliefs, and practices, and their interaction with the locals can lead to exchanging ideas, stories, and experiences. The temple becomes a platform for cultural exchange as visitors learn about and appreciate a different way of life. At the same time, locals also get exposure to various cultures through interactions with a diverse range of tourists.

Social Identity Theory, conversely, suggests that people categorize themselves and others into groups based on various social factors. Social identity becomes essential to how individuals define themselves and relate to others. In the context of the Om Banna Temple, both tourists and locals have social identities that shape their interactions and experiences.

The Bullet Baba Temple holds deep cultural and spiritual significance for the local community. It reinforces their sense of identity as the guardians of this unique tradition. The temple's practices and beliefs reinforce their social cohesion and collective identity. Whereas tourists who visit the temple come from different social and cultural backgrounds. Some may be motorcycle enthusiasts seeking to connect with a shared passion. Others might be curious travellers interested in experiencing the local culture and rituals. Visiting the temple becomes a part of their travel identity, shaped by their motivations and experiences.

Further, the interaction between tourists and locals at the temple brings these social identities into contact. Tourists learn about and engage with the local culture, potentially challenging or enriching their social identity perceptions. Locals, in turn, witness the interest and respect tourists show towards their traditions, possibly reinforcing their identity as the preservers of this cultural practice.

Travelling to Explore Culture and Identity

Travel has an extraordinary power to bridge cultural gaps, and kindle shared connections through experiences and beliefs. This magic lies at the heart of cultural exchange, a vital force within travel and tourism. Imagine people from different walks of life converging, each carrying their unique stories, and in their interactions, ideas, customs, and perspectives are woven into a rich tapestry of shared understanding. McIntosh *et al.* (2002) conducted a study that underscored how travel can become a conduit for cultural knowledge, leading to a profound appreciation of diverse cultures. This journey often redefines personal beliefs as travellers encounter viewpoints that challenge their preconceived notions.

Cultural exchange, a cornerstone of travel, finds a kindred spirit in the 'social identity theory' proposed by Tajfel and Turner (1979). This theory unveils the innate human tendency to classify ourselves into social groups, an instinct driven by the desire for self-esteem and a sense of belonging. This inclination springs to life as travellers embark on adventures, affiliating with tour groups to forge a social identity beyond mere companionship. These identities extend into roles that define their journey, cultivating self-worth and a shared sense of purpose among fellow travellers (Twigger-Ross and Uzzell, 1996).

The echoes of social identity ripple across the tourism landscape, giving birth to tailored experiences that cater to different tour groups' distinct needs and preferences. Social identity is at the heart of this, reflecting how individuals see themselves by comparing attributes with those who share similarities (Tajfel, 1972; Wang *et al.*, 2014). A fascinating twist emerges as travellers immerse themselves in new cultures, finding comfort in relating to local traditions through participation in cultural events. Here, new social identities evolve, shaped by shared ideals and experiences. This intricate interplay between individual identity and group affiliation rests on the bedrock that who we are is intertwined with the groups we align ourselves with. Social Identity Theory posits that humans yearn to classify – seeking common threads like race, religion, and nationality to create order and foster connections.

In the realm of social psychology, Social Identity Theory stands as a guiding light (Hornsey, 2008). Scholars like Liu and László (2007) venture into its cognitive, interactive, and social dimensions. Hornsey (2008) emphasizes that dissecting individual actions or interactions cannot fully grasp the collective fabric. Social Identity Theory delves into the intricate dance between self-perception and group dynamics, embracing group engagement, intergroup bonds, and social affiliations (Yuki, 2003). A structured three-step process guides group behaviour: social categorization, social identification, and social comparison. It begins with categorization, where norms shape how we see others. These norms can be as varied as the spectrum of humanity.

This case study bridges social categorization with tourism. It dives into functional and hedonic attributes, ultimately sorting individuals based on identity and paving the way for social comparison. As individuals compare favourably within their group, it bolsters self-esteem, creating a powerful affirmation cycle. Social Identity Theory's versatility transcends fields, from politics (Calhoun, 1994) to human resources (Hogg, 2001) and even education (Gurin *et al.*, 2002). In each domain, it unveils the hidden mechanisms of group dynamics. With this lens, we journey into the Om Banna Temple, a tourist haven, peering through the prism of Social Identity Theory to uncover the secrets of its development.

Understanding this Temple Dynamic through Social Identity Theory

Our sense of self is a mosaic forged by memories, emotions, and interpretations woven through our journeys across diverse landscapes. Our identity transforms as we enter novel experiences, shaping our perception of who we are. The destinations we explore become brushes that paint our identity, a canvas etched with stories of our travels. Each place we visit leaves an indelible mark, crafting narratives that weave the tapestry of our identity. In tourism, this process is a dance with Social Identity Theory. This concept whispers how we categorize ourselves and others, nestling ourselves within the embrace of different groups. Let us unveil this theory through the captivating lens of the Om Banna Temple:

As pilgrims walk the hallowed ground of the Om Banna Temple, they undergo a subtle transformation – an act of *social categorization*. Within these sacred walls, they bind themselves to the spirit of the Royal Enfield Bullet motorcycle, drawing a shared identity. Devotees, united by their belief in the motorcycle's divine guardianship, embrace a standard narrative that orbits around faith, reverence, and the supernatural.

The temple's aura resonates through *social identification*, forging links between souls. Visitors, who once were travellers passing through, now find themselves part of a tribe bonded by devotion. They are not just individuals, they are torchbearers of a shared belief. By bowing before the Bullet Baba and immersing themselves in ritual, they are painting strokes of identity that herald their place in this vibrant community.

Within the temple's embrace blooms *social comparison*, a dance of shared experiences. Devotees gather, weaving tales of their encounters with Om Banna's spirit. These shared narratives breathe life into faith, reinforcing the connection between individual experiences and collective belief as they recount stories of reduced accidents and bolstered safety, a symphony of shared belonging crescendos.

In the grand narrative, the Social Identity Theory unveils itself in this unique context:

- *Social categorization*: Devotees don the mantle of believers, entwined with the temple's aura and the motorcycle's mystique, seeking safety's embrace.
- *Social identification*: A community is born, shaped by shared worship, a collective heartbeat that thrums with faith and reverence.
- *Social comparison*: Conversations become a conduit for shared journeys, fortifying faith and nurturing a sense of unity within the temple's realm.

Through the prism of Social Identity Theory, the Om Banna Temple emerges as a living testament. It showcases the artistry of human identity, woven by affiliation with a temple that holds not just a motorcycle but the embodiment of belief – a belief that bridges gaps, stitches hearts, and shapes an identity forged in the fires of devotion.

Conclusion

The case of the Bullet Baba Temple exemplifies how forging a compelling identity is a cornerstone in developing an attractive destination for travellers, achieved through various pivotal mechanisms. In a world of commercialized tourist attractions, the Bullet Baba Temple's story and identity stand out as unique and authentic, catering to the growing demand for offbeat and genuine experiences. It embodies a distinct local history and cultural significance, offering a refreshing alternative to mainstream destinations.

The temple's narrative, entrenched in themes of tragedy, mysticism, and protection, serves as a powerful emotional catalyst for visitors. People are inherently drawn to stories that possess a human touch and foster a deeper connection. The temple's identity, intricately linked with the real-life story of Om Singh Rathore, weaves an emotional bond between visitors and the place itself, elevating the overall visitor experience and etching the destination into their memories. Word-of-mouth recommendations, often initiated by visitors emotionally engaged and captivated by the temple's identity, play a pivotal role in promoting the place. Personal stories and heartfelt endorsements contribute significantly to the temple's reputation, driving more people to yearn to experience it.

Further enhancing the temple's appeal is its ability to cater to a specific niche audience, particularly motorcycle enthusiasts and those with a penchant for unique cultural encounters. This tailored approach attracts a dedicated and passionate group of visitors who might never have considered the area otherwise, thereby establishing a distinct market for the destination.

The temple's identity as a protector of travellers and motorcyclists adds a layer of allure.

Many visitors are naturally drawn to places that offer them a sense of safety and blessings. The temple's reputation for safeguarding people on their journeys aligns perfectly with the desires of travellers who seek protection while on the road.

For those travellers who harbour a keen interest in cultural immersion and learning, the Bullet Baba Temple extends an invitation to engage with local beliefs and practices. Visiting the temple, delving into its history, and participating in its rituals becomes a rich cultural learning experience, imbuing the visit with depth and meaning for those who seek to connect with the local culture on a profound level.

The Bullet Baba Temple has skilfully metamorphosed into an alluring destination by crafting a robust and authentic identity interwoven with captivating stories and cultural traditions. This identity resonates profoundly, personally, and emotionally with travellers, catering to their interests and desires. It transcends conventional tourism, transforming the temple into a place of profound and meaningful experiences.

References

Calhoun, C. (1994) *Social Theory and the Politics of Identity*. Wiley-Blackwell Malden, MA.

Gurin, P., Dey, E., Hurtado, S. and Gurin, G. (2002) Diversity and higher education: Theory and impact on educational outcomes. *Harvard Educational Review* 72(3), 330–367. DOI: 10.17763/haer.72.3.01151786u134n051.

Hogg, M.A. (2001) A social identity theory of leadership. *Personality and Social Psychology Review* 5(3), 184–200. DOI: 10.1207/S15327957PSPR0503_1.

Hornsey, M.J. (2008) Social identity theory and self-categorization theory: A historical review. *Social and Personality Psychology Compass* 2(1), 204–222. DOI: 10.1111/j.1751-9004.2007.00066.x.

Liu, J.H. and László, J. (2007) A narrative theory of history and identity. In: Moloney, G. and Walker, I. (eds) *Social Representations and Identity*. Palgrave Macmillan, New York, pp. 85–107.

McIntosh, A.J., Hinch, T. and Ingram, T. (2002) Cultural identity and tourism. *International Journal of Arts Management* 4(2), 39–49.

Tajfel, H. (1972) Social categorization, english manuscript of 'La catégorisation sociale'. In: Moscovici, S. (ed.) *Introduction a La Psychologie Sociale*, Vol. 1. Larousse, Paris, pp. 272–302.

Tajfel, H. and Turner, J.C. (1979) An integrative theory of intergroup conflict. In: Austin, W.G. and Worchel, S. (eds) *The Social Psychology of Intergroup Relations*. Brooks/Col, Monterey, CA, pp. 33–48.

Twigger-Ross, C.L. and Uzzell, D.L. (1996) Place and identity process. *Journal of Environmental Psychology* 16(3), 205–220. DOI: 10.1006/jevp.1996.0017.

Wang, S., Zhou, L., Lee, S. and King, C. (2014) Analysis of residents' social identity, tourism engagement, and propensity for tourism advocacy. *Advances in Hospitality and Leisure* 10, 109–129. DOI: 10.1108/S1745-354220140000010006.

Yuki, M. (2003) Intergroup comparison versus intragroup relationships: A cross-cultural examination of social identity theory in North American and east asian cultural contexts. *Social Psychology Quarterly* 66(2), 166–183. DOI: 10.2307/1519846.

10 Lived Religious Practice and Tourism Perceptions of Southern Buddhist Monks in an Ethnic Tourism Community in Xishuangbanna, China

Yu Li
College of Ethnology, Guangdong Polytechnic Normal University, Guangzhou, Guangdong, China

Summary
The Dai Garden, situated in Menghan Town of Xishuangbanna, Yunnan Province, China, is an ethnic tourist destination operated by the local company of Jinghong city investment and development. This case study was created to illustrate the complexity of Buddhist tourism. It outlines how the tour guides utilize cultural differences to stigmatize Southern Buddhist monks and addresses the negative tourism perceptions of Southern Buddhist monks in tourism in China, which leads to social and cultural unsustainability in religious tourism. It also reveals the challenges of business, primarily in marketing and management. This study should serve as a reference for other religious tourism businesses seeking to attain a higher level of social/cultural sustainability.

The Value and Interest of the Case Study

The narrative discourse of tour guides often influences the perception of Southern Buddhist monks towards tourism. When there is a potential conflict between tourists and Southern Buddhist monks in the sharing of sacred and ritual space, the monks often strive to maintain a balance through temporal and spatial separation. In contrast, the tour guides, as intermediaries in tourism discourse and culture, are constantly exacerbating conflicts. This study enhances our understanding of the complexity of the Buddhist tourism situation by identifying the key factors that influence the tourism perceptions of southern Buddhist monks. It also provides valuable insights for the management of religious tourism.

Background

Southern Buddhism and the Southern Buddhism monks

'Southern Buddhism'[1], translated into Chinese as 'xiaochengfojiao' (小乘佛教), first spread from India to Sri Lanka, Myanmar, Thailand, Cambodia, Laos and Dai-inhabited areas of Yunnan Province, China. Since it spread southward from India, it is referred to as Southern Buddhism (Liang, 2019). In terms of doctrine, Southern Buddhism places great importance on the recognition of suffering and personal liberation, in direct contrast to Han Buddhism and Tibetan Buddhism, which seek to alleviate human suffering. Specifically, the theoretical basis of Southern Buddhism follows the basic doctrines of the 'Triple Gem' (the Buddha, Dharma and Sangha), 'the four Noble Truths' and 'the Eightfold Path doctrines'. According to Southern Buddhism, the cause of suffering is desire, and only by renouncing desire can we be free from pain (Song, 2016). Other important concepts include karma and compassion. So-called 'karma' literally means 'action', and it refers to the seed planted by a person's behaviour, especially their intention, that influences their future outcomes. Empathy and compassion are manifested through actions that prioritize helping others (Abe, 1993; Wisner, 2011).

Currently in China, the local government of Xishuangbanna regards Southern Buddhism as a potential resource for development, especially in terms of local economic growth and diplomatic relations (Li, 2021).

The religious practices of Southern Buddhist monks in Xishuangbanna are pluralistic, encompassing complex and diverse discourses, memories and histories from various historical periods. It mainly consists of two parts: First, the folk beliefs that the Dai people have retained for a long time are based on the concept of 'soul', and it is believed that all things possess animism. Second, Southern Buddhism was introduced to the region from Thailand. It experienced a process of moving from conflict to reconciliation with local folk beliefs after it was introduced into the Dai ethnic region.

In the Dai Garden, there is frequent interaction between Southern Buddhist monks and tourists. Previous studies on the interaction between Buddhist monks and tourists have shown that Buddhism and tourism are harmonious (Shackley, 2001; Wong *et al.*, 2013). The degree of faith plays an important role in the way Buddhist monks treat tourists (Wong *et al.*, 2013). However, this case study demonstrates that in the unique context of ethnic tourism in China, the way Buddhist monks treat tourists is not related to the degree of belief, but rather to the discourse of tourism, that is the narrative discourse of tour guides. According to social exchange theory, this behaviour is influenced not only by the act of exchange itself but also by the perceived fairness of the exchange (Kreiner *et al.*, 2015). Specifically, Southern Buddhist monks and tourists have expectations and participate in relationships based on exchange and reciprocity. Tourists seek spiritual guidance and cultural experiences while also elevating the social status and reputation of Southern Buddhist monks. However, the intervention of the tour guide's discourse disrupts the exchange relationship, shifting the focus to the fairness of Buddhism and the identity of Southern Buddhist monks.

The Dai garden in Xishuangbanna

Dai Garden is situated in Menghan Town (formerly known as Mengtani and also referred to as 'Ganlanba') and is 28 km from Jinghong, the capital city of Xishuangbanna in Yunnan Province, China (see Fig. 10.1). According to local legend, the Dai people used 'fahan' (rolled white cloth) to spread Buddhism all the way to Manchunman Village in Dai Garden when Sakyamuni went to Menghan to preach. Sakyamuni was deeply moved and promptly named the place Menghan.

The Dai Garden is comprised of five ancient villages of the Dai nationality – Manjiang, Manchunman, Manzha, Mantin and Manga. The Dai Garden displays the history, culture, religion, art, architecture, clothing, diet, production and living customs of the Dai people in Xishuangbanna and integrates nature and culture (Sun, 2009). In July 2020, there were 357 households with a total of 1764 people in the five villages. There are four Buddhist temples in the Dai Garden: Manchunman, Manting, Manzha and Manjiang. The monks come from the villages of the Dai Garden and surrounding areas. They are organized into distinct ranks based on their levels and positions within the Buddhist monastic system. There are six 'Thera', one 'Phra' and ten 'Samanera' monks[2]. While the monks in the temples do derive part of their income from tourists, their main reliance is on the residents of their own villages.

Methods

The information for this case study was gathered through a series of non-participatory observations and in-person interviews with eight Southern Buddhist monks, ten villagers and six tourists of the Dai Graden in Menghan Town, Xishuangbanna, Yunnan Province, China. Research on the lived religious practice

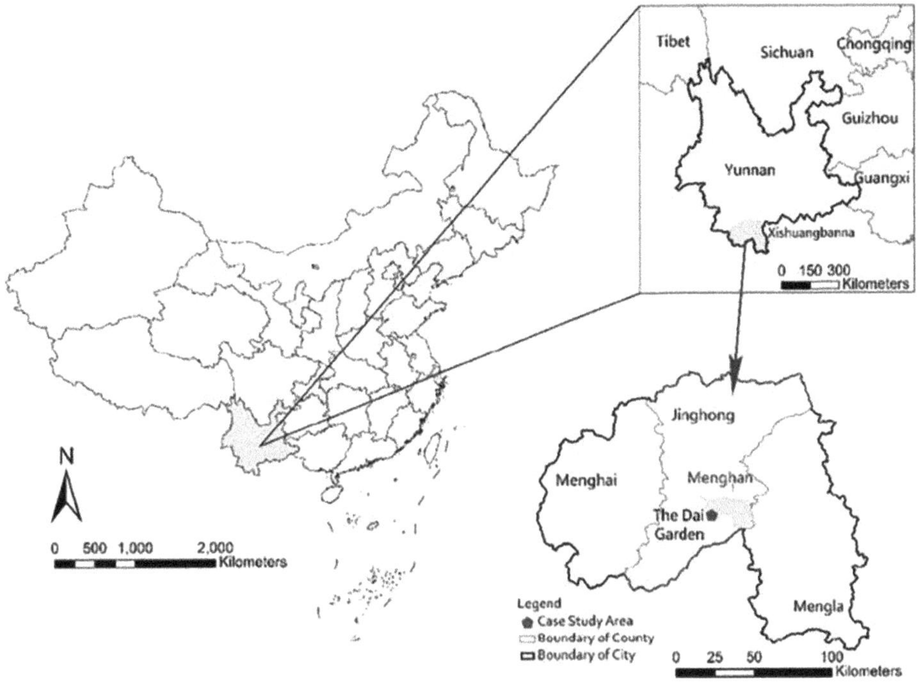

Fig. 10.1. Location of the Dai Garden in Menghan Town in Xishuangbanna, Yunnan, China.

and tourism perceptions of Southern Buddhist monks was also conducted on-site by the author over a 3-month period in 2020.

Lived Religious Practice of Southern Buddhist Monks

In the Dai Garden, tourists can be classified into two types based on their purpose of visit. First, there are those who are interested in Dai culture and landscapes. This type of tourist rarely participates in the worship activities in Southern Buddhist temples. They may have different beliefs, such as Christianity and Catholicism, or some tourists believe that the beliefs of ethnic minorities differ from those of Han people, so they cannot worship casually. The second type of tourist integrates meditation and leisure tourism to attain a state of tranquillity and alleviate stress. They are often interested in the culture and landscape of the Dai Garden and are highly enthusiastic about participating in pilgrimage and religious activities. This aligns closely with the concept of 'pilgrimage-tourist' (Eade and Sallnow, 1991). As Turner and Turner (1978, p. 20) proposed, 'a tourist is half a pilgrim, if a pilgrim is half a tourist'. The lived religious practices of Southern Buddhist monks are closely connected to pilgrimage-tourists, encompassing both shallow material communication and deep spiritual interchange.

Shallow material communication

The interaction between pilgrimage-tourists and Southern Buddhist monks based on religious material often takes the form of a nonverbal gaze, which is influenced by the atmosphere, religious material and embodied practice. Monks engage in religious bodily practices, such as chanting scriptures, performing dripping water rituals, and taking part in ceremonial practices. Furthermore, the existence of religious landscapes such as Buddhist temples, pagodas, and statues, captures the interest of tourists

Fig. 10.2. Buddhist temples, pagodas and statues, capture the interest of tourists.

and enhances their experiential travels (see Fig. 10.2).

In addition, the material communication also reflects the interaction between Southern Buddhist monks and pilgrimage-tourists in the religious economy. As Morgan (2010) noted, the structure of religious practice is remarkably economical, with objects and images serving as symbols in the circulation of the economy. The Southern Buddhist monks conducted negotiations through the practice of giving or donating, begging or praying, and exchanging. During the process, each negotiation has different expectations regarding exchange and favours. For example, Southern Buddhist monks who manage temples sell incense to pilgrimage-tourists according to their needs at the Manchunman Buddhist temple in the Dai Garden. Some pilgrimage-tourists purchase incense and proceed to make wishes and kowtow in front of a Buddha statue, while others donate money and then write their basic information and wishes on a donation pad to wish for good health and prosperity for their families. Based on my investigation, the 'Tying the thread'[3] practice of the Dai people has been endowed with more extensive functions and meanings, such as 'blessing you with good health and good luck' (interviewee-monk). Pilgrimage-tourists often ask Southern Buddhist monks to help tie the thread, which makes their religious practice more valid.

Deep spiritual interchange

The interaction between Southern Buddhist monks and pilgrimage-tourists involves cultural exchange and meditation. In the Dai Garden, pilgrimage-tourists often visit the temples for meditation and spiritual practice. During the meditation period at Manting temple, the religious practice of Southern Buddhist monks mainly includes chanting sutras, teaching believers basic etiquette and Buddhist knowledge

Fig. 10.3. The local Dai knelt outside the Buddhist temple to scatter flowers as a warm welcome to Master Haitao.

from the Southern Buddhist sutra, and guiding believers in meditation. In addition, pilgrim-tourists from all over the country come to engage in Buddhist cultural exchange. For example, during the investigation, on 13 July 2014, the author happened to encounter Master Haitao and his disciples at the Manchunman Buddhist temple in the Dai Garden for a cultural exchange that included a ceremony to offer Buddhist robes. The purpose of the trip of the pilgrim-tourists was to integrate meditation with leisure travel to promote relaxation and stress relief. For the worshippers of the Dai people residing in the Dai Garden, participating in such Buddhist events holds great significance, akin to accumulating merits and virtues. The local Dai knelt outside the Buddhist temple to scatter flowers as a warm welcome to Master Haitao and brought fruits from their homes to offer to the visiting pilgrims (see Fig. 10.3).

During the ceremony, DBHT, the host of the Manchunman Buddhist temple, first provided a general introduction to Southern Buddhist beliefs in Xishuangbanna, the history of the temple, and the basic doctrines of Southern Buddhism. Second, Master Haitao and DBHT, the Buddha of the Manchunman Buddhist temple, presided over the ritual of offering Buddhist robes, which involves teaching the five precepts, offering Buddhist robes (see Fig. 10.4), and the reciprocation of blessings through the ritual of dripping water. The pilgrim-tourists and Dai worshippers learned about the culture of Southern Buddhism through the verbal and embodied ritual practices of Master Haitao and DBHT, the Buddha of the Manchunman temple. A nun from Taiwan with Master Haitao noted, 'We met here, and our spiritual journey has helped us expand our knowledge and promote the Dharma'.

Tourism Perceptions of Southern Buddhist Monks

The Southern Buddhist monks believe that the arrival of tourists increases income for the

Fig. 10.4. Master Haitao and DBHT, the Buddha of the Manchunman Buddhist temple, presided over the ritual of offering Buddhist robes.

temples and the community. However, they also hold more negative perceptions of the influences of tourism brought about by tour guides. This issue mainly manifested in two aspects: Misunderstanding and stigmatization from Han tour guides, as well as mixed reviews from local tour guides, leading to detrimental consequences for the tourism industry.

Misunderstanding and stigmatization from the Han tour guides

In mass tourism, Han tour guides play a prominent role in facilitating host-guest interactions, leading to a communication gap between Southern Buddhist monks and tourists. Some Han tour guides often use the terms 'xiaochengfojiao' (小乘佛教) and 'dachengfojiao' (大乘佛教) to differentiate the belief range between Han Buddhism and Southern Buddhism. They describe Southern Buddhist monks as 'Flower monks' (interviewee-villager), such as Lu Zhishen, a historical figure in China who did not obey the Buddhist rules and had a bad reputation. For example, in China, Han Buddhist monks usually eat vegetarian food instead of meat. In contrast, Southern Buddhist monks do not have strict dietary restrictions regarding meat consumption, and those who return to secular life are allowed to marry. The essence of stigma is to produce biased cognition and enlarge the differences between groups. The guides exploit the inherent differences in Buddhist precepts to arouse tourists' interest, thus deliberately belittling and stigmatizing Southern Buddhist monks.

The irresponsible behaviour of tour guides often has negative consequences:

1. Damaging the image of the tourism destination: As cultural intermediaries of the tourism destination, tour guides' discourse directly affects tourists' impression of the destination. The stigmatization of Southern Buddhist monks by tour guides creates an impression of unprofessionalism and prejudice, which negatively impacts tourists'

perception and their objective evaluation of the tourism destination.

2. Hindering the development of tourism: The stigmatization of monks by tour guides often sparks controversy and diminishes tourists' interest in the destination. This has a negative impact on tourism development, resulting in decreased revenues and employment opportunities for the local population.
3. Impairing the tourism experience: One of the main responsibilities of tour guides is to offer accurate and comprehensive explanations to tourists, helping them to gain a better understanding of Dai culture and customs. When Han tour guides engage in stigmatizing behaviour, tourists may not receive genuine and objective information, thus diminishing the cultural value of their tourism experience.

Therefore, the tourism company of the guides should collaborate closely with local religious institutions and communities to establish appropriate standards and guidelines for tour guides. Additionally, the company should organize regular training sessions on Southern Buddhist knowledge for newly recruited tour guides.

Mixed reviews from local tour guides

The organization of tour routes and the involvement of tour guides has led to cultural segregation and social distancing between the tourists and the Southern Buddhists. However, despite differences in language, customs, religious beliefs and values, there are no conflicts between Southern Buddhist monks and tourists. The local Dai tour guide's explanations are very perfunctory due to a lack of religious knowledge. This has led to dissatisfaction among the monks, prompting them to resist and refuse to receive tourists. As noted by a Buddhist monk at the ManJiang temple, 'their guide said that all Buddhist temples are the same, with only one Buddha statue. The Buddha statues here are all of Sakyamuni, and what they said is generally nonsense'.

In this regard, Buddhist monks usually attribute the negative impact of tourism to the negligence of the company management of Dai Garden. A monk at Manting temple stated, 'The tour guides inform tourists that one temple is worth visiting while another temple is not in order to steer tourists towards the designated temples within the travel itinerary. We believe that any matters related to Buddhist temples should be discussed and negotiated with us. If the company provides professional training for its employees, they should consider creating a guidebook for the new tour guides to outline what they should and should not say'.

As a result, tourists and Buddhists hesitate to initiate communication due to the cognitive differences and obstacles caused by tour guides' lack of local knowledge. In particular, the lack of professional knowledge and irresponsible conduct by tour guides hinders communication and interaction between tourists and Buddhists, making it difficult to cultivate mutual understanding and respect. This may lead to cultural conflicts and social segregation between tourists and Southern Buddhist monks. Tour guides should show respect and deepen their understanding of the Southern Buddhist culture in the Dai religion in order to provide professional services and create an atmosphere of respect and inclusivity within the tourism industry. Additionally, local governments should prioritize this issue by systematically training local tourism service personnel in Buddhist knowledge. They should actively guide the positive development of the tourism market and provide enhanced services to eliminate cultural conflicts and social segregation. The tourism company in the Dai Garden should improve the tourist experience by offering products such as educational Buddhism tours and Zen meditation. This will help foster interaction and exchange opportunities between monks and tourists, promoting a relationship of equal dialogue and mutual understanding. The ultimate objective is to guarantee the sustainable development of culture and tourism.

Conclusions

In summary, during the development of tourism, the daily religious practices of Southern Buddhist monks have extended from serving the community to engaging with pilgrimage tourists. This reflects the evolution of interactive relationships among religious material, the religious economy, and religious culture, transitioning from shallow to deep, and becoming multidimensional and multilevel.

During the interaction between Southern Buddhist monks and pilgrimage-tourists, it is evident that Buddhist monks strive to maintain balance based on economic considerations, while tour guides, as promoters of tourism discourses and cultural introducers, are constantly intensifying conflicts. The tour guides introduce tourists to the denigration, deliberate disparagement, and peculiar aspects of Southern Buddhism, the negative evaluation of Buddhist temples, and the disrespect for monks, known as 'flower monks'. This has led to a negative perception among Southern Buddhist monks in the context of tourism in Southern Buddhism.

In the long term, religious interactions between Southern Buddhist monks and tourists are not static but changeable and overlapping. In other words, the negative perception and attitude of monks towards tourism discourse is not static and may change dynamically in the future through tour guide management and standardized Southern Buddhist knowledge education of Dai Garden managers in Southern Buddhist knowledge. However, it is important to further refine the understanding of the complex dynamics and power relations between Southern Buddhist monks and tour guides during these interactions. By delving deeper into the underlying reasons for the negative portrayal of Southern Buddhism and addressing the issue of fairness in tourism practices, a more comprehensive analysis can be achieved.

Finally, this study contributes to the understanding of the complexity of Buddhism in tourism. The study of the relationship between tourism and religion encompasses various contexts, including Europe, America and Asia. It focuses on religious and cultural conflicts in tourism, particularly involving Islam, Christianity, Judaism and Hinduism. While a few studies have focused on the interactions between Buddhist monks and tourists and their perceptions (Timothy and Olsen, 2006), there has been a lack of attention to the nuances of the Buddhist system. This study highlights the subtle differences in doctrines, precepts, and appellations between Southern Buddhism and Han Buddhism, particularly in the context of ethnic tourism. Tour guides' discourse gives rise to certain contradictions and conflicts that indicate the necessity of re-evaluating the role of travel guides in the sustainable development of religious tourism and broader global religious equality.

Funding

National Science Fund for Young Scholars in China, (Grant / Award Number: '41801133')

Notes

1 From the perspective of the Dai monks, the term '小乘佛教' (Xiǎo chèng Fó Jiào), used to translate 'Southern Buddhism' into Chinese, is seen as carrying a negative connotation. Actually, in the academic discourse system of Chinese, the terms 'large' and 'small' are commonly used to distinguish the belief range between Han Buddhism and Southern Buddhism. However, the Dai monks believe that all beings are equal and have no distinction between 'large' and 'small'. Therefore, in this study, the author adopts the local view, using the term 'Southern Buddhism' (Nán chuán fó jiào '南传佛教').

2 The terms 'Thera', 'Phra', and 'Samanera' in Southern Buddhism are appellations that categorize monks into different levels and positions within the Buddhist monastic system. For example, 'Thera refers to senior monks in Southern Buddhism who typically possess extensive experience in practice, profound knowledge of Buddhist teachings, and abundant practical experience. Phra denotes intermediate-level monks in Southern Buddhism who have received full ordination (novice ordination and higher ordination) and have begun participating in the formal monastic community. Samanera refers to novice monks who have received novice ordination (preliminary ordination), usually at a younger age' (interviewee-monk).

[3] 'Tying the thread' (拴线), which is called 'shuhuan' (树欢) in the Dai language, signifies 'tying the soul'. It is a traditional blessing ceremony of the Dai people in Xishuangbanna. The ceremony originated from the belief in 'soul immortalization' in the Dai native religion. 'The 'tying the thread' ritual is typically performed on four different occasions in the lives of the Dai people: during the first month after birth, on one's wedding day, during evocation, and on one's funeral day' (interviewee-villager).

References

Abe, M. (1993) Buddhism. In: Sharma, A. (ed.) *Our Religions*. Harper Collins, New York.

Eade, J. and Sallnow, M. (eds) (1991) *Contesting the Sacred: The Anthropology of Christian Pilgrimage*. Routledge, London and New York.

Kreiner, N.C., Shmueli, D.F. and Gal, M.B. (2015) Understanding conflicts at religious-tourism sites: The Baha'i World Center, Israel. *Tourism Management Perspectives* 16, 228–236.

Li, Y. (2021) *Folk Tourism and Reconstruction of Local Culture*. Social Science Press, Beijing, China.

Liang, X.F. (2019) Textual research on the early spread of Theravada Buddhism in Yunnan. *Religious Studies* 4, 150–158.

Morgan, D. (2010) The material culture of lived religion: Visuality and embodiment. In: Vakkari, J. (ed.) *Mind and Matter: Selected Papers of Nordic Conference*. Society of Art History, Helsinki, pp. 14–31.

Shackley, M. (2001) *Managing Sacred Sites: Service Provision and Visitor Experience*. Continuum, London and New York.

Song, L.D. (2016) The righteousness and characteristics of Southern Buddhism. *Chinese Religion* 4, 64–66.

Sun, J.X. (2009) *Community Tourism and Community Participation in Tourism Anthropology*. Commercial Press, Beijing, China.

Timothy, D.J. and Olsen, D.H. (2006) *Tourism Religion and Spiritual Journeys*. Routledge, London.

Turner, V. and Turner, E. (1978) *Image of Pilgrimage in Christian Culture*. Columbia University Press, New York.

Wisner, B.L. (2011) Exploring the lived religion of Buddhists: Integrating concepts from social work and religious studies. *Journal of Religion and Spirituality in Social Work: Social Thought* 30(4), 385–404.

Wong, C.U.I., McIntosh, A. and Ryan, C. (2013) Buddhism and tourism perceptions of the monastic community at Pu-Tuo-Shan, China. *Annals of Tourism Research* 40, 213–234.

Section 3: Empowerment Through Tourism

In this section the discourse shifts to community-based tourism initiatives and their role in empowering local communities, especially women, while promoting cultural and environmental preservation. Cases from the Philippines, Maldives, India and Thailand showcase how tourism can be a tool for economic empowerment and social change.

11 Community-Based Ecotourism and Conservation in Tboli, Philippines

Ramon Benedicto A. Alampay
Asian Institute of Tourism, University of the Philippines Diliman, Quezon City, Philippines

Summary
This case study narrates the story of Lake Holon and the choices made by the Municipality of Tboli (and the indigenous people whose name it bears) in opening their sacred lake to tourism. Until 2013, tourism to the lake had been limited to occasional expeditions and outreach activities by local mountaineering groups. The municipal government chose to adopt a community-based tourism strategy as a means of empowering the resident Tboli – especially those in the villages leading to, and around, Lake Holon. From 2014 to 2018, the municipality worked with five rural villages to develop ecotourism attraction sites, while also preparing the residents to operate and manage the attractions themselves.

This narration of the Tboli experience comes primarily from direct interviews and published quotes from two persons directly involved in establishing the tourism program: Rodel Hilado, the municipal tourism officer of Tboli since 2013, and Nelly Nita Dillera, regional director for the Department of Tourism (DOT) from 2013 to 2019. Both have graciously granted permission for their names and statements to be cited in this case study. Additional details on the program and the subsequent outcomes come from local government documents and other secondary sources.

The Value and Interest of the Case Study

Readers of this case study will appreciate the fine line that Tboli – the municipality and the people – have chosen to walk, using community-based ecotourism as a strategy for inclusive and sustainable development. The case presents the choices made by the Tboli in balancing sometimes competing interests of local economic empowerment, environmental sustainability, and cultural rejuvenation. The narrative also provides insight into how Tboli combined its natural comparative advantages with its cultural heritage to establish a unique tourism identity.

Background

In July 2013, Mayor Dibu S. Tuan of the municipality of Tboli in the Province of South Cotabato, Philippines received a visit from Nelly Nita Dillera, regional director for the Department of Tourism (DOT) for Region 12. She told Mayor Tuan, who had recently been elected to his first term of office, that she was pleasantly surprised to see how the town's urban center had been rapidly urbanizing with the discovery of gold deposits in the area, and the entry of multinational companies engaged in planting and processing bananas, pineapple, palm oil, and other crops.

The DOT director wondered how tourism would fit into the town's current direction. Dillera recalled Mayor Tuan's response to be something like:

> Your observation is correct. We have many plantations. In fact, many of the top managers and employers come from outside Tboli. If we keep to the current path, the Tboli will remain laborers for the rest of their lives. They will always be working for outsiders. I have thought of tourism [as the strategy], but please help us with how to do it.

Choosing a Different Tourism Path for Tboli

Mayor Tuan saw tourism as a way to empower the Tboli people to become entrepreneurs and managers, not only of their livelihoods but, more importantly, of their heritage. The Tboli, or T'boli, are one of the 18 ethnolinguistic groups (collectively called *lumad*) in the Philippines' southernmost island of Mindanao, who have held on to their traditional beliefs, practices, and traditions (Bengwayan-Anongos, 2022). The name is derived from '*tao belil*,' meaning 'man living in the hills' in the indigenous language.

Though there are no official figures for the numbers of indigenous people in the Philippines (Perez-Brito, 2021), the 2015 Census showed that almost two-thirds of the town's 91,000 residents spoke Tboli as their mother tongue, with another 8% speaking Blaan, another indigenous language. Notwithstanding the rapid growth of the town, 85% of Tboli's population continued to live in rural barangays[1] outside the urban center (Municipal Government of Tboli, 2021) (Fig. 11.1).

The mayor and other Tboli stakeholders were wary of the kind of tourism development they had seen in the neighboring town of Lake Sebu. Like the municipality of Tboli, Lake Sebu was also home to many indigenous people of Tboli heritage, including master weavers and artisans who made *T'nalak* cloth, beadwork, brassware, and other indigenous crafts. Lake Sebu gained more visitors when the Provincial

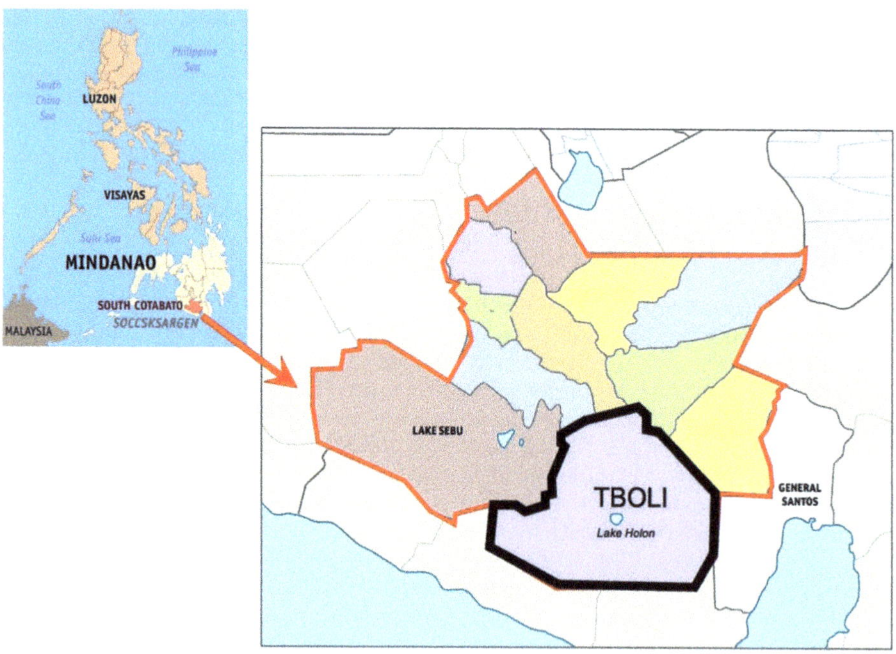

Fig. 11.1. Location map of Tboli and the Province of South Cotabato in the Philippines.

Government of South Cotabato developed the Seven Falls Park in Lake Sebu, whose feature attraction was a more than 1 km long zipline that soared 60m above the waterfalls. The combination of Instagram-able lake settings and Tboli culture, plus the promise of fun and adventure at the Seven Falls zipline soon made Lake Sebu one of the top destinations in the province.

However, the rapid growth of Lake Sebu had come with a price. Uncontrolled development of commercial *tilapia* fish pens resulted in cyclical bouts of fishkill every few years because of fish pen overfeeding and pollution. With Lake Sebu's increasing popularity, investors from outside the town had also begun to put up hotels and resorts alongside the fish pens. For Tboli stakeholders, such as Tboli's Municipal Tourism Officer Rodel Hilado, this was not the kind of tourism development they aspired to.

> Lake Sebu is more on (commercial) investments.... As you observe there are no big restaurants or establishments near the [T'boli tourism] sites. We are more into community immersion. You will experience not just the sites but also the culture and traditions of the Tbolis (Alfasain, 2016).

Developing the Tboli Tourism Product

For Tboli, Lake Holon was the most obvious building block for tourism. Lake Holon is a 304-ha lake, located some 1756m above sea level that was created from the eruption of Mt. Melibingoy (formerly Mt. Parker) on January 4, 1641. With a positive reputation stemming from being declared 'the cleanest inland body of water' by the national government in 2003 and 2004, it was also culturally important to the Tboli who believed it to be the dwelling place of the deity *Tud Bulol*. According to tradition, the Lake was protected by 15 guardians who could be seen as peaks of Mt. Melebingoy surrounding it (Fig. 11.2).

Fig. 11.2. Lake Holon in the mouth of Mt. Melebingoy's caldera (Photo courtesy of Louie Pacardo).

Lake Holon was largely undeveloped for tourism, especially in comparison to Lake Sebu. The only visitor activities consisted of sporadic treks or outreach activities to Tboli rural communities that were organized by the local government or by local mountaineering groups.

In June 2014, Mayor Tuan announced that Lake Holon would be closed to visitors for 9 months (Philippine News Agency, 2014). During this time, they would work to rehabilitate and prepare the lakeside location that had become the *de facto* campsite for Lake Holon. The time would also be used to construct visitor centers, viewing decks, and way stations of light materials along the trails that visitors would pass, and to train communities around Lake Holon to be hosts and guides.

On March 10, 2015, the local government declared Lake Holon to be 'The Crown Jewel of the South' and reopened it to visitors. To mark the occasion, the Tboli Tourism Office organized the 'Great Lake Holon Trek.' Limited to a maximum of 150 participants, the overnight camp would become the first 'Pilgrimage to Lake Holon' that is now a regular event in the annual Seslong Festival. It also established the Holon trekking experience that is in place now.

The trek to Lake Holon starts at either of the two alternative trails leading to the campsite at Lake Holon, both about an hour's ride from the center of town. Most visitors take the more challenging, but more scenic *Kule* Hunter's Trail to the lake (Fig. 11.3). They then go back using the easier *Salacafe* or Monkey Trail after spending at least one-night camping on the lakeside (Fig. 11.4). Alternatively, visitors can also spend the night in the houses of Tboli families around the Sitio Kule visitor center.

Following the reopening of Lake Holon, the Tboli gradually introduced new visitor attractions and homestays in other areas of the town both to diversify its tourism product portfolio and to spread the benefits of tourism to other peripheral barangays of the municipality.

Fig. 11.3. Hikers taking the Kule Hunter's Trail on Mt. Melebingoy, going to Lake Holon (Photo courtesy of Louie Pacardo).

Fig. 11.4. Designated campsite on the banks of Lake Holon (Photo courtesy of Louie Pacardo).

The *El Kini* Hot Springs in Barangay Lamhaku opened in June 2015. The *Bakngeb* River Cave – offering canyoneering, diving, and river rafting, as well as volunteer bamboo and tree planting activities – opened in January 2016. *Hidak* Falls (which refers to only the first of a cluster of three different waterfalls) in the village of Kematu, quickly followed in March 2016 (Fig. 11.5).

The last major piece to the Tboli's initial push into tourism came in February 2017 with the opening of the *Lemsnolon Cultural Village*, the only site, so far, which is not nature-based. The main feature of the cultural village is a traditional Tboli long house or *Guno Bong* which can accommodate up to 100 visitors (Miranda, 2023). The *Guno Bong* also offers sleeping spaces where guests can stay and immerse in Tboli traditions, arts, music, and history. It is also intended to be a space for cultural masters to pass on the traditions of livelihood, art, and performance to younger generations of Tboli.

Implementing Community-Based Sustainable Tourism in Tboli

Rodel Hilado, Municipal Tourism Officer of Tboli, explained that Lake Holon and the other attractions were planned to be community-based tourism enterprises from the outset. By his estimate, the local government invested at least 4 years from the start of Mayor Tuan's first term in 2013 to prepare the communities to become tourism hosts.

> We have included the community from the beginning. We have observed that projects in other parts of the Philippines failed because the community was not part of it. Our plans are biased for the community. We want them to prosper and become independent.

Much of this preparation, involved local government staff immersing themselves in the host communities to get a true feel for the residents' fears and aspirations. Equally important,

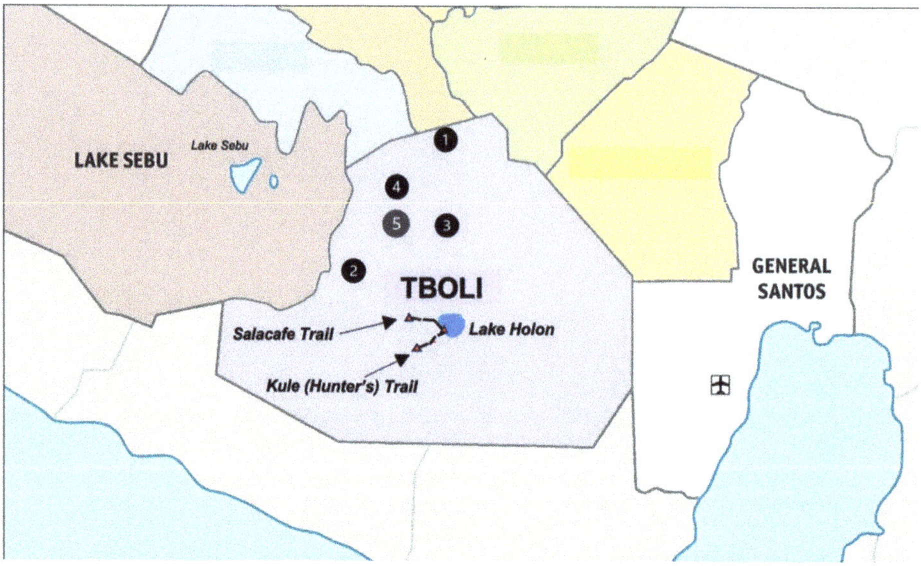

Fig. 11.5. Indicative locations of the tourist attraction sites of the Municipality of Tboli: Lake Holon (21 km from 5. Municipal Tourism Office); 1. Lamhaku El Kini Hot Springs (22.5 km); 2. Bakngeg River Cave (15.5 km); 3. Hidak Falls (6 km); 4. Lemsnolon Cultural Village (4 km).

the community organization and immersion work of the tourism and culture offices allowed them to better understand how to integrate Tboli culture, values, and traditions into the tourism activities. They acknowledged that most of the initial tourism activities would be nature-based. At the same time, sensitive and responsible harmonization of cultural heritage would create a unique, sustainable visitor experience. In this regard, experts from national government agencies, such as the National Commission for Culture and the Arts (NCCA) and the Department of Tourism (DOT), worked with the local government in cultural mapping, research, and related groundwork.

The pilot model for Tboli's community-based tourism program was *Sitio Kule*, the community from which the popular trail to Lake Holon started. In 2014, the residents built a *Guno Bong* to use for community gatherings and meetings. They had also carved out the beginnings of the Kule Hunters' Trail to Lake Holon. The local Tourism Office and DOT began providing basic hospitality and tourism training programs such that the *Guno Bong* evolved into a multi-purpose receiving center that served community members as well as tourists (Alfasain, 2021). Residents were also trained to incorporate authentic Tboli cultural practices and culinary traditions into the services of the Center, as well as in the homestays around it.

Recognizing that the tourism projects were businesses at their core, the local government assisted the residents in organizing themselves into People's Organizations (PO)[2] who would be responsible for managing each attraction. Local government community organizers also facilitated the official registration and incorporation of each PO in line with government laws and regulations for civil society organizations (CSOs).

Each PO appointed members to serve as managers and staff for the tourism businesses. In turn, the local government and DOT provided them with training workshops and toolkits on bookkeeping, financial management, and other essential business management skills, as well as modules on disaster risk reduction and management (DRRM) and environmental conservation. This became the model of community organizing and capacity development that the Tboli local government implemented in the other host villages.

The Tboli Tourism Office describes the role of local government as supportive rather than supervisory. One of its most important functions, in this regard, has been to establish a policy environment that will ensure the protection and conservation of Lake Holon and its environs.

The municipal government has implemented a strict limit of 200 visitors per day as the tourism carrying capacity for Lake Holon. They have also continued the practice of closing Lake Holon regularly to all visitors for, on average, two months at the start of each year (South Cotabato News, 2016; Philippine News Agency, 2020; Sarmiento, 2020). This enables the natural environment to rejuvenate. It also enables the Tourism Office to engage in what Hilado calls 'listening time' (R. Hilado, 2023, personal communication). The feedback from the host communities received during this time is an important input for the future development plans for the sites.

In 2016, the municipal government also began collecting a 50-peso environmental users' fee (equivalent to 1 US dollar) from all visitors to the natural sites. A total of 70% of this fee, which is separate from the direct revenues earned from tourist activities and services, goes to the community and funds their local development projects. At the end of each year, the community also receives 25% of entrance fees collected by the local government from the respective sites.

Outcomes

Official records of the Tourism Office show that registered visitors to Lake Holon rose from 5600 tourists in 2015 to almost 17,500 in 2018. In addition, local government revenue from tourism doubled in the period from 2016 to 2018, while total direct tourism employment grew from 275 to 396 workers in the same period.

In addition to the residents directly engaged in the operation of visitor centers, accommodation, and attraction sites, many other community members participate as tour guides, producers of cultural merchandise and souvenirs, or cultural artists and performers. The local government estimates that close to 1,700 people now work in tourism, all of whom now have health insurance – either as a benefit for the people directly employed to manage the sites and visitor centers, or as a benefit of membership in the various tourism associations organized for the tour guides, boat operators, and other small players.

According to the Municipal Tourism Office, all of the community-based tourism enterprises have been officially enrolled in the Philippine Government Electronic Procurement System or PHILGEPS. All are now qualified to supply goods and services needed by government agencies. For example, the Lemsnolon Cultural Village now hosts government meetings at its *Guno Bong*. Local and national government offices in the area now also procure Tboli souvenirs, workshop paraphernalia and other requirements from the different community enterprises.

For DOT Regional Director Dillera, the story of Tboli tourism goes beyond economics:

> And tourism really made a big difference on the Tboli people themselves. They seemed to have developed pride of place, pride of culture. [Before tourism] they would wear the Tboli clothing only during special occasions. But now, there are visitors almost every day. Wearing the traditional attire has become [a normal] part of their daily clothing [again].
>
> [Most especially] the stories of the people were recognized. Community members who offer *habal-habal*[3] services, the tour guides, etc. all have narratives. Each has their own story about their experiences and their lives. (N.N. Dillera, 2023, personal communication)

In 2016, Tboli's efforts were recognized by Green Destinations[4], an international tourism certification program, which listed Lake Holon as one of the Top 100 Sustainable Destinations in the World (South Cotabato News, 2016). The following year, Lake Holon made their Top 100 list again (Green Destinations, 2017).

This case study revolves around the theme of community-based tourism. Some writers prefer to use community-led tourism or community-driven tourism, to put a stronger emphasis on the active involvement and leadership of the local community in decision making and management of tourism. The common thread that ties them together is the principle of placing the community at the center of the destination planning and management process.

For additional insight into community-based tourism as an approach to sustainable and inclusive tourism development, readers may start with some the work of Ashley and Roe (1988); Murphy (1988); Okazaki (2008); and Zielinski *et al.* (2021).

Notes

[1] Philippine cities and municipalities are divided into villages called barangays, which are the smallest units of local governance in the country.

[2] People's Organization (PO) is the official term used by the Philippine government to describe a type of non-governmental organization (NGO) that is established by the members of a community to undertake collective action and address community concerns and needs.

[3] A mode of public transportation where individual passengers ride on the backseat of the motorcycle, holding on to the back of the driver during the ride. *Habal-habal* is used by many independent travelers to get from the town center to the Lake Holon visitor center.

[4] Available at: https://www.greendestinations.org/ (accessed 1 August 2023).

References

Alfasain, G.V. (2016) T'boli's community-based tourism. *Yadu Karu's Blog*. Available at: https://www.yadukaru.com/2016/02/tbolis-community-based-tourism.html (accessed 29 March 2023).

Alfasain, G.V. (2021) On leading Sitio Kule. *SunStar/Davao*. Available at: https://www.sunstar.com.ph/article/1897207/davao/lifestyle/on-leading-sitio-kule (accessed 29 March 2023).

Ashley, C. and Roe, D. (1988) Enhancing community involvement in wildlife tourism: Issues and challenges. *IIED Wildlife and Development Series* Series No.11. International Institute for Environment and Development. Available at: https://www.iied.org/7784iied (accessed 5 December 2023).

Bengwayan-Anongos, A. (2022) Philippines. Country report. In: Mamo, D. (ed.) *The Indigenous World 2022*. International Work Group for Indigenous Affairs, Copenhagen, pp. 279–292.

Green Destinations (2017) 2017 Sustainable Destinations Top 100. 101 Destinations Win Global Sustainability Competition. Available at: https://www.reykjanesbaer.is/static/files/Frettamyndir/skyringamyndir_gogn/2017-sd-top-100-press-release.pdf (accessed 1 August 2023).

Miranda, M. (2023) T'boli People: Homestay and Cultural Immersion at T'boli Town. *Project Gora*. Available at: https://projectgora.com/tboli-people/ (accessed 29 March 2023).

Municipal Government of Tboli (2021) Comprehensive Development Plan 2021–2026. The Municipality of Tboli. Available at: https://www.tboli.gov.ph/comprehensive-development-plan-2021-2026/ (accessed 18 July 2023).

Murphy, P.E. (1988) Community driven tourism. *Tourism Management* 9(2), 96–104. DOI: 10.1016/0261-5177(88)90019-2.

Okazaki, E. (2008) A community-based tourism model: Its conception and use. *Journal of Sustainable Tourism* 16(5), 511–529. DOI: 10.1080/09669580802159594.

Perez-Brito, C. (2021) *No Data, No Story: What the Absence of Indigenous Peoples-Specific Data Reveals*. The World Bank, Washington, DC. Available at: https://blogs.worldbank.org/opendata/no-data-no-story-what-absence-indigenous-peoples-specific-data-reveals (accessed 1 April 2023).

Philippine News Agency (2014) Development of ecotourism sites in South Cotabato's Lake Holon starts. *Philippines Today*. Available at: https://www.philippinestoday.net/archives/14427 (accessed 29 March 2023).

Philippine News Agency (2020) T'boli's famed Lake Holon shut down for rehab. BusinessMirror. Available at: https://businessmirror.com.ph/2020/01/13/tbolis-famed-lake-holon-shut-down-for-rehab/ (accessed 29 March 2023).

Sarmiento, B. (2020) South Cotabato's Lake Holon to welcome back tourists in January 2021. *MindaNews*. Available at: https://www.mindanews.com/top-stories/2020/12/south-cotabatos-lake-holon-to-welcome-back-tourists-in-january-2021/ (accessed 1 April 2023).

South Cotabato News (2016) Lake Sebu and Tboli among World's Top 100 Sustainable Destinations for 2016. Available at: https://www.southcotabatonews.com/2016/09/lake-sebu-and-tboli-sustainable-destinations-top100.html (accessed 12 July 2023).

Zielinski, S., Jeong, Y. and Milanés, C. (2021) Factors that influence community-based tourism (CBT) in developing and developed countries. *Tourism Geographies* 23(5–6), 1040–1072. DOI: 10.1080/14616688.2020.1786156.

12 The Guesthouse Phenomenon in the Maldives – Development and Issues

Mohamed Shenaan and Heike Schänzel
School of Hospitality & Tourism, Auckland University of Technology, Auckland, New Zealand

Origin: Shenaan, M. (2022) Exploring Guesthouse and Cuisine Experiences: The Strategic and Sustainable Development of the Guesthouse Phenomenon in the Maldives. PhD thesis, Auckland University of Technology.

Summary

The tourism industry in the Maldives has long been dominated by the 'one island, one resort' model, where luxury resorts on uninhabited islands kept tourists isolated from local communities. However, recent legislative changes have opened up new opportunities for community-based tourism (CBT) with the establishment of guesthouses on inhabited islands. This shift aims to increase local participation in the tourism sector and has led to greater interaction between tourists and the island communities. The guesthouse industry has witnessed rapid growth, surpassing the number of resorts, providing more affordable options for visitors, and offering the potential for economic and social benefits for local communities.

Despite the benefits, the lack of clear guidelines and varying sizes and facilities of guesthouses has caused confusion among stakeholders and tourists. The increased interaction between conservative local communities and tourists has also led to conflicts.

This case study is based on a research project conducted in September 2019 exploring the sustainable development opportunities and challenges of the guesthouse tourism industry in the Maldives. Extensive in person interviews were conducted with tourists staying at three local island islands in the country, as well as guesthouse operators and managers from the same islands.

The Value and Interest of the Case Study

Readers of this case study will gain an insight into the diversification of tourism in the Maldives, leading to increased interactions between local communities and the industry. The case examines the growth of community-based tourism in the Maldives, focusing on issues of sustainability, the challenges faced, and the impacts of the changes it has caused.

The Maldives and Tourism

The Republic of Maldives situated in the Indian Ocean, comprises of over 1,000 islands distributed across 26 natural atolls. Despite its vast area (approximately 90,000 km^2), the land area is relatively small (approximately 298 km^2), with the urbanized capital city, Malé, being one of the most densely populated in the world, with over 210,000 residents living on less than 2 km^2 of land. The islands are of varying sizes but are

Fig. 12.1. A group of low-lying islands in the Maldives. Source: Asad Photo Maldives (2018).

on average approximately 1 km^2, with most less than 1m above sea level (Thoha, 2020). Fig. 12.1 illustrates the small size and low elevation of the islands' geography.

The inception of tourism in the Maldives dates to 1972 when a small number of European tourists arrived. At its outset, the industry had modest beginnings, with just two resorts available for visitors in its inaugural year. Approximately 1000 tourists visited during that time (Ministry of Tourism, 2013; Kachroo-Levine, 2021). These resorts were specifically constructed on uninhabited islands. They were the only exclusive commercial enterprises on each island, and introduced the 'one island, one resort' concept or 'enclave tourism' to the Maldives (Scheyvens, 2011).

By the end of the 1970s, 25 resorts were operating under the enclave tourism model, and the central government had taken over the previously independent running of the industry. The 'one island, one resort' concept has flourished, and there are now 170 tourist resorts (as of May 2023) (Ministry of Tourism, 2023a). With the resorts being set on private islands, which are leased by the Maldivian government to private entrepreneurs and businesses, this model of tourism allows each island to be only accessed by paying guests and workers (Dell'Agnese, 2019). The resorts are often developed with attached water villas providing guests with panoramic ocean views and the ideal visual spectacle for promotional marketing materials (see Fig. 12.2). The country has become one of the leading luxury tourism destinations within the last half-century, and by the end of 2019, before the COVID-19 outbreak, the country welcomed more than 1.7 million international visitors (National Bureau of Statistics, 2020).

Known for white sandy beaches and palm-fringed coral islands surrounded by warm clear seas, the destination has traditionally focused on '3S' (sun, sea, and sand) tourism. The resorts have proven to be a popular attraction for wealthy foreign tourists, initially mainly from traditional European tourist markets (Ministry of Tourism, 2013; Zubair and Bouchon, 2014). Typically known for high-end tourism, attracting honeymooners and the wealthy, the country

Fig. 12.2. Water villa accommodation attached to a Maldivian resort island. Source: Asad Photo Maldives (2019).

became well established as a premium tourist destination.

The strict implementation of policies to restrict interactions between tourists and locals was driven by the desire to preserve socio-religious customs. This approach stemmed from the stark contrast between the religiously conservative local communities and the tourists engaging in perceived hedonistic leisure activities (Shakeela and Weaver, 2018). The concerns primarily revolved around potential negative socio-cultural influences of tourism, such as issues related to immodest clothing, alcohol, and drugs (Ahmed, 2018). As a result, the enclave tourism industry, focusing on isolation, was developed to address these concerns. Although this strategy offered luxurious privacy for tourists and appeared to limit negative socio-cultural impacts on locals, it inadvertently hindered most locals from fully benefiting from tourism (Shakeela and Weaver, 2018; Dell'Agnese, 2019).

Despite the initial success of the enclave tourism industry, the government of Maldives recognized the need for a more inclusive approach to tourism development. To increase local participation in tourism, a shift in government policies occurred in 2008. This marked a turning point for the Maldivian tourism industry, with the advent of guesthouse tourism.

Guesthouse Tourism

In 2008, a newly elected government made amendments to the Tourism Act, allowing for tourism development on all 200 local inhabited islands. This led to the rise of what locally became known as 'guesthouses' (Buckley *et al.*, 2017). The government aimed to democratize the tourism industry, giving local communities greater control over decision-making and ensuring more widespread economic benefits from tourism.

Fig. 12.3. A Guesthouse on Dhiffushi island. Source: Araamu Holidays and Spa (2022). Permission given by owner.

The burgeoning trend of guesthouse tourism is believed to have the potential to bring about significant transformations in local island economies. This is evidenced by the increase in employment and entrepreneurship opportunities for locals, either directly within guesthouses or through subsidiary businesses that have emerged to cater to tourists, such as marine recreation, restaurants, and souvenir shops (Zubair and Bouchon, 2014).

Guesthouse characteristics

Guesthouses in the Maldives offer an affordable alternative to luxury resorts, thus opening the country to new market segments (Asian Development Bank, 2019). Despite government regulations stipulating a minimum of three rooms and six beds per guesthouse, there is a lack of a clear definition of them. This has resulted in diverse accommodations and an absence of standardized guidelines for development and management.

Guesthouses come in varying sizes and designs, as illustrated in Figs. 12.3 and 12.4, and are often at owners' homes, which are renovated and remodelled to convert them into tourist accommodation facilities. Some are small boutique-style hotels, while others are repurposed homes with only a few rooms (Pappayah *et al.*, 2017). Bowen *et al.* (2017, p. 726) describe Maldivian guesthouses as 'small/medium-sized hotels and small/medium-sized holiday apartment complexes'. In 2020, a 12-storey, 253-bed complex with an infinity pool and spa facilities, Kaani Palm Beach guesthouse, opened on the island of Maafushi (Avas, 2020). This points to the varied nature of this type of accommodation in the country.

The official definition of a guesthouse by the Ministry of Tourism has evolved annually, from being described as 'Low-cost accommodation,

Fig. 12.4. A guesthouse on Maafushi island. Photograph by author.

on local inhabited islands, providing lodging, or board and lodging' in 2014 (Ministry of Tourism, 2014), to 'Local island accommodation for low to mid-market segments, offering bed and breakfast' in 2018 (Ministry of Tourism, 2018). A more suitable definition is 'Low-to-medium cost, local and externally owned, local inhabited island accommodation in the Maldives, of varying sizes, most offering lodging, board, and tourist activities' (Shenaan, 2022, p. 191). This definition covers the broad range of characteristics that make up Maldivian guesthouses. The introduction of this new category of accommodation has led to a remarkable surge in the number of guesthouses being established within local island communities.

The growth of guesthouse tourism

In recent years, the Maldives Ministry of Tourism has classified tourist lodgings in the country into four different categories of accommodation: resorts, hotels, safari vessels, and guesthouses (Ministry of Tourism, 2018). The enclave resorts continue to dominate, with almost 70% of the total bed capacity. 'Hotels' are the least common and are usually based on urbanized inhabited islands, and 'safari vessels' are popular with marine recreational tourists. By the end of 2013, tourist bed nights at guesthouses only accounted for 1% of total bed nights (resorts accounted for 95%). By the end of January 2023, this had risen to 15% of total tourist bed nights (resorts 82%) (Ministry of Tourism, 2023b).

At the same time, 881 guesthouses were in operation, with a combined bed capacity of more than 14,400, making up 24% of all tourist beds available in the country (Ministry of Tourism, 2023b). Fig. 12.5 illustrates the number of these accommodation facilities in operation between 2012 and the end of January 2023, especially highlighting the rapid growth of the guesthouse industry, with the number

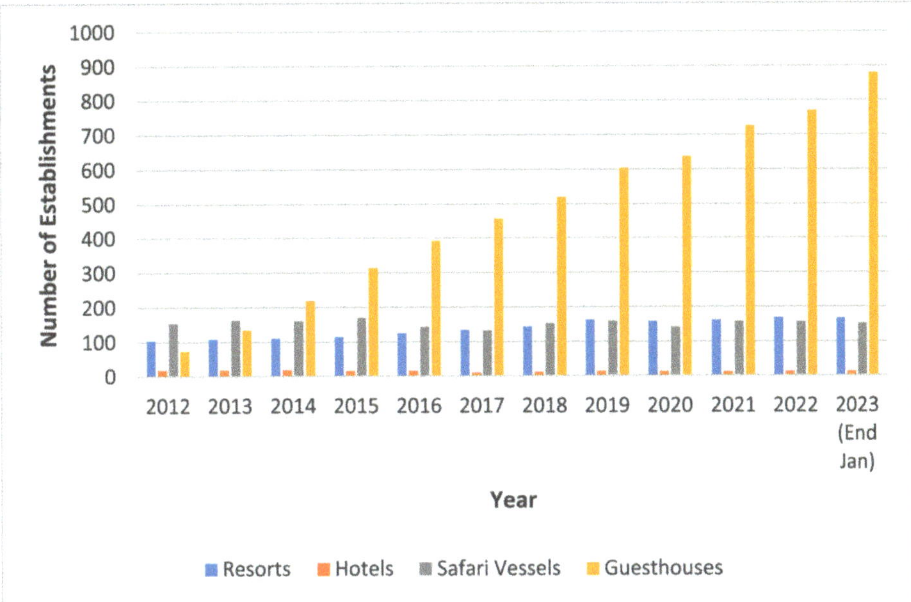

Fig. 12.5. Maldives tourist accommodation establishments 2012–January 2023. Source: National Bureau of Statistics (2023); Ministry of Tourism (2023b).

of guesthouses surpassing all other types of accommodation in 2014.

Challenges

The rapid expansion of the guesthouse industry has led to several emerging challenges, including unregulated growth, cultural and societal conflicts, limited local participation, and a potential lack of authentic experiences for tourists.

Uncontrolled growth and confused identity

The rapid growth of guesthouse tourism has highlighted the vulnerability of the islands. Due to their small size, many islands have limited space for tourism development, leading to an imbalance where the number of tourists exceeds the local population and puts a strain on essential resources such as natural areas like beaches. As a result, locals have been affected by the encroachment of their recreational spaces by tourists. On some islands, guesthouse owners have taken measures to segregate public areas, thereby restricting locals from accessing these spaces (Lama and Becker, 2019).

The lack of a clear definition or concept for guesthouses has resulted in a wide range of accommodation options being offered to tourists. This includes the emergence of multi-story hotel-like accommodation, which, given the limited size of the islands, has led to issues of overcrowding and sustainability concerns. The absence of clear guidelines for guesthouse development and management has allowed for this diversification. It requires careful consideration from both tourism authorities and operators to ensure that sustainable practices are implemented moving forward (Shenaan, 2022).

Research on tourism and development in various destinations has highlighted similar negative impacts alongside other undesirable social and environmental repercussions such as pollution and an increase in crime (e.g. Aquino *et al.*, 2018; Albu, 2019; Kongbuamai *et al.*, 2020). Waste management has emerged as a significant concern in the Maldives. Ironically, some tourists may complain about locals discarding rubbish in the sea and on the beach without realizing that the increase in tourist numbers has exacerbated these issues further.

Social issues and cultural contradictions

Uncontrolled development in the pursuit of meeting the demands of tourism has also increased the potential for conflict between tourists and locals in the Maldives. As the landscape has transformed to aid and accommodate the influx of visitors, clashes have arisen due to cultural differences, values, and practices between the two groups. The liberal lifestyle and behaviour of many tourists often contrast with the more traditional and conservative norms of the local communities, leading to misunderstandings and tensions. Issues surrounding dress code, public behaviour, and leisure activities can and have become flashpoints, fuelling social friction.

The Maldives, an Islamic country for centuries, has previously adeptly managed to strike a delicate balance between accommodating hedonistic tourists and respecting the conservative religious practices of the local population through the implementation of enclave tourism resort strategies (Henderson, 2008; Shakeela and Weaver, 2018). However, the rise of guesthouse tourism has brought tourists and locals closer, leading to cultural conflicts. Similar issues related to tourism development in other Muslim regions are not new, and the Maldives faces comparable challenges in its guesthouse tourism development. Despite the prohibition of alcohol on local islands, intoxicated tourists returning from resort trips to their guesthouses have caused incidents with locals.

On local islands, tourists are expected to adhere to modest dress codes when using beach facilities, as mandated by the Ministry of Foreign Affairs (Government of Maldives, 2020). Concerns have arisen regarding the immodest dress of some tourists. However, tourists may lack knowledge of what is expected of them in terms of cultural practices as much of the promotional material related to guesthouses include the typical bikini-clad tourist, sun, sea, and sand images. To accommodate tourists seeking such experiences, 'bikini beaches' have been established. Many of these beaches have been fenced off since the rise of guesthouse tourism. In some cases, signs explicitly prohibiting locals from accessing these areas have been displayed, leading to escalating conflicts between local islanders, guesthouse operators, and local councils that allow such segregation (Abbas, 2018; Lama and Becker, 2019).

Balancing the desire for economic growth and tourism revenue with the preservation of local traditions and cultural values is a complex challenge that requires careful consideration and dialogue between stakeholders to mitigate potential conflicts.

Lack of local participation and lack of authentic experiences

Despite the initial goal of guesthouse development to distribute tourism benefits among the broader population, issues such as those mentioned above, and others related to a lack of direct benefits for locals have hindered local community participation in the industry and exacerbated tensions.

Traditionally, guesthouses in the Maldives have been managed by locals and family members of owners (Zubair and Bouchon, 2014; Buckley *et al.*, 2017). However, there has been a noticeable shift towards guesthouse ownership by individuals from outside the local island community. Properties are now being sold to large businesses, primarily from the capital city of Male', with little or no connection to the island communities where the guesthouses are located (Shenaan, 2022, Ministry of Tourism, n.d.). Additionally, some properties are developed in partnership with foreign owners, with the local owner often taking a silent role in the business. Recent changes in the law have allowed foreigners to own up to 49% of local guesthouse businesses (Avas, 2021).

The traditional model of local ownership, where families from the island community own and operate guesthouses, helps to reduce economic leakages and strengthens ties with the local communities. However, the influx of external parties purchasing guesthouses has led to increased construction of guesthouses on the islands, with local land being sold to wealthy business owners from Male' and other external investors. These operators tend to prioritize economic profit over community benefits, resulting in an increase in foreign employees within the guesthouse sector. This can lead to economic leakages from the island communities, with the benefits of guesthouse development not fully reaching the locals.

Effective local participation in tourism is essential for communities to see economic benefits from CBT (Roxas *et al.*, 2020). For CBT to be successful, natural resources must be used sustainably (Giampiccoli *et al.*, 2020). Overuse of resources can lead to negative impacts outweighing the benefits, resulting in local apprehension towards tourism increasing (Postma and Schmuecker, 2017).

Tourism development can have negative sociocultural impacts, including overcrowding and conflicts over differing religious and moral values. Another concern is the standardization of identity and the loss of authentic practices (Frent, 2016). Many guesthouse owners and operators are unsure about the unique selling points of their offerings and often try to emulate the identity of traditional resort industries by providing foreign cuisine and resort-like recreational activities. While recreational activities are a major attraction for tourists, those visiting local communities often seek more authentic experiences. Many of these visitors seek genuine interactions with hosts, but these experiences are often hindered by the guesthouses' tendency towards a more impersonal, hotel-type experience (Shenaan, 2022). As a result, there is a disconnect between what guesthouses provide and what tourists expect, potentially leading to disappointment and dissatisfaction.

The deeper, more meaningful, and transformative experiences that make certain tourist encounters unforgettable are often lacking. This gap may be partly attributed to the absence of a clear identity and concept of what a guesthouse should embody. Guesthouses often fail to capitalize on local practices to promote the essence of local culture. This includes the limited availability of authentic local cuisine at these establishments. Instead, most guesthouses offer 'Western' style fast food or Southeast Asian cuisine (Shenaan *et al.*, 2021). This lack of genuine local activities and cuisine not only deprives tourists of richer experiences but also hinders local communities from fully participating in the industry and reaping its benefits.

Benefits and a Sustainable Pathway

The emergence of guesthouse tourism in the Maldives has yielded several notable positive impacts, benefiting both local communities and the nation. One such impact is the diversification of tourism offerings, making the Maldives more accessible to a wider range of tourists and market segments. Guesthouses provide affordable options for budget-conscious travellers, offering an alternative to traditional high-end and exclusive resort experiences (Chia and Muiz, 2021; Shenaan, 2022). This expansion in the tourism landscape has opened new avenues for economic growth and engagement with local communities.

The growth of guesthouses has proven instrumental in generating enhanced economic opportunities for residents with an increase in employment and entrepreneurship within these communities (Shakeela and Weaver, 2018; Chia and Anas, 2022; Shenaan, 2022). This shift has empowered the local youth, who previously faced limited prospects and often had to migrate to the capital city or seek employment in the resort sector. The establishment of guesthouses has spurred ancillary businesses to thrive on many islands, capitalizing on the tourist influx by offering services such as restaurants, souvenir shops, and marine recreational activities, an example of which is seen in Fig. 12.6 (Salgado, 2014).

Even with the lack of cultural offerings, guesthouse tourism has the potential to offer a unique experiential dimension to travellers, allowing them to delve into the rich cultural fabric of the Maldives through direct interaction with local communities. This includes engaging in local culinary experiences. Despite the limited selection of local food on guesthouse menus (Shenaan *et al.*, 2021), visitors staying on local islands expressed a desire for a greater selection of local cuisine, as well as other genuine cultural encounters

Fig. 12.6. A business providing recreation activities on Maafushi island. Photograph by author.

(Shenaan, 2022). Such authentic experiences often resonate deeply with budget travellers, backpackers, and those seeking to escape the confines of conventional mass tourism, fostering lasting memories and meaningful connections with the destination (Schänzel *et al.*, 2022).

The guesthouse industry does face complications, particularly in relation to sustainability, cultural preservation, and maintaining authentic experiences (Ahmed, 2018; Shakeela and Weaver, 2018). Striking a balance between catering to tourist demands while preserving local traditions and the environment requires thoughtful consideration and management. Considering these challenges, stakeholders in the industry need to remain committed to addressing these concerns to ensure the long-term viability and inclusivity of guesthouse tourism.

A sustainable pathway for guesthouse tourism to develop requires collaboration between various stakeholders. The central government plays a crucial role in developing policies and regulations, ensuring compliance, and providing support and training for guesthouse owners. The government has only recently (2019) actively started promoting guesthouses at international tourism fairs in partnership with operators, where previously they had concentrated solely on promoting the resort sector (Miadhu, 2019). Since then, this relationship has continued. Guesthouse owners and the government need to engage with local communities, leveraging local culture and knowledge to create authentic experiences for guests. There should also be clear guidance for tourists regarding expected behaviour when visiting. Marketing efforts can focus on promoting guesthouse tourism as a cultural experience, attracting tourists seeking authentic local interactions.

Local communities, if willing, should participate in consultations with local and central government, as well as guesthouse operators, regarding the viability and enthusiasm for offering personalized and culturally rich experiences to guests. There is the potential to benefit from intersectoral linkages, and employment opportunities while contributing to the development of the guesthouse industry. This includes industries such as the local agricultural sector which can benefit from the inclusion of local cuisine on guesthouse menus. Such links and benefits have already been noted in other small island nation (Laeis *et al.*, 2019). Guesthouse tourists need to follow guidelines in order to respect local culture and customs and allay local communities' fears of negative influences of visitors (Ahmed, 2018; Shakeela and Weaver, 2018), while experiencing culturally rich and immersive encounters.

Flexibility and adaptability are also essential for the future direction of guesthouse tourism, tailoring sustainable development approaches to different island characteristics, ensuring ongoing communication between stakeholders, and allowing for adjustments based on changing needs and challenges. Island size, population, facilities, as well as cultural behaviour, vary throughout the archipelago, and some locals may oppose the influx of tourists in their communities. To ensure the sustainable development of the sector, it is crucial to respect their perspectives.

Overall, a positive future direction for guesthouse tourism in the Maldives involves a collaborative effort between various stakeholders to ensure sustainable development that benefits both local communities and tourists. By addressing issues such as sustainability, cultural preservation, and authentic experiences, guesthouse tourism in the Maldives has the potential to thrive in a positive manner.

References

Abbas, N. (2018) Hulhumale. Available at: https://twitter.com/NishwanAbbas/status/1075657833939558400 (accessed 15 July 2023).

Ahmed, M. (2018) Guesthouses in the Maldives: Towards community-based tourism. *International Journal of Innovation and Economic Development, Inovatus Services Ltd* 4, 44–50. DOI: 10.18775/ijied.1849-7551-7020.2015.41.2004.

Albu, R. (2019) Study on the effects of tourism development on the local community of Brasov. *Bulletin of the Transilvania University of Brasov. Series V: Economic Sciences* 12, 37–42. DOI: 10.31926/but.es.2019.12.61.2.5.

Aquino, R.S., Lück, M. and Schänzel, H.A. (2018) A conceptual framework of tourism social entrepreneurship for sustainable community development. *Journal of Hospitality and Tourism Management* 37, 23–32. DOI: 10.1016/j.jhtm.2018.09.001.

Araamu Holidays and Spa (2022) Araamu Holidays and Spa Guesthouse, Dhiffushi, Maldives. Available at: https://araamuholidays.com/ (accessed 16 December 2023).

Asad Photo Maldives (2018) Bird's eye view photography of islands. Available at: https://www.pexels.com/photo/bird-s-eye-view-photography-of-islands-1268869/ (accessed 6 June 2023).

Asad Photo Maldives (2019) Aerial View of Outrigger island Resort at Konotta, Maldives. Available at: https://www.pexels.com/photo/aerial-view-of-outrigger-island-resort-at-konotta-maldives-9482139/ (accessed 3 August 2023).

Asian Development Bank (2019) Maldives: Economic Update 2019. Available at: https://www.adb.org/documents/maldives-economic-update-2019 (accessed 4 August 2023).

Avas (2020) New Guesthouse Opened in Maafushi, Largest to Date. Available at: https://avas.mv/en/90174 (accessed 1 August 2023).

Avas (2021) Foreigners to be Allowed to Invest in Guesthouse Business. Available at: https://avas.mv/en/97076 (accessed 16 December 2023).

Bowen, D., Zubair, S. and Altinay, L. (2017) Politics and tourism destination development: The evolution of power. *Journal of Travel Research* 56, 725–743. DOI: 10.1177/0047287516666719.

Buckley, R.C., Guitart, D. and Shakeela, A. (2017) Contested surf tourism resources in the Maldives. *Annals of Tourism Research* 64, 185–199. DOI: 10.1016/j.annals.2017.03.005.

Chia, K.W. and Anas, A. (2022) (Re)starting the tourism business during a crisis? Challenges and decisions to reopen of guest house entrepreneurs in the Maldives. *Current Issues in Tourism* 25, 1015–1020. DOI: 10.1080/13683500.2022.2030681.

Chia, K.W. and Muiz, A. (2021) The nature and challenges of guest house business in the Maldives: An exploratory study. *Tourism and Hospitality Research* 21, 3–14. DOI: 10.1177/1467358420926688.

Dell'Agnese, E. (2019) Islands within Islands?' The Maldivian resort, between segregation and integration. *Tourism Geographies* 21, 749–765. DOI: 10.1080/14616688.2018.1505941.

Frent, C. (2016) An overview on the negative impacts of tourism. *Revista de Turism – Studii Si Cercetari in Turism* 22, 32–37. Available at: http://www.revistadeturism.ro/rdt/article/view/344 (accessed 5 August 2023).

Giampiccoli, A., Mtapuri, O. and Dłużewska, A. (2020) Investigating the intersection between sustainable tourism and community-based tourism. *Tourism* 68, 415–433. DOI: 10.37741/t.68.4.4.

Government of Maldives (2020) Travelling to the Maldives. Available at: https://www.gov.mv/en/publications/travelling-to-the-maldives (accessed 2 June 2023).

Henderson, J.C. (2008) The politics of tourism: A perspective from the Maldives. *Tourismos* 3, 99–115. Available at: http://saruna.mnu.edu.mv/jspui/bitstream/123456789/13916/1/The%20Politics%20of%20Tourism%20-%20A%20Perspective%20from%20the%20Maldives.pdf (accessed 2 August 2023).

Kachroo-Levine, M. (2021) How the Maldives Transformed from a Fishing Archipelago to a Tropical Hot Spot in 50 Years. Available at: https://www.travelandleisure.com/trip-ideas/island-vacations/maldives-history (accessed 3 June 2023).

Kongbuamai, N., Bui, Q., Yousaf, H.M.A.U. and Liu, Y. (2020) The impact of tourism and natural resources on the ecological footprint: A case study of ASEAN countries. *Environmental Science and Pollution Research* 27, 19251–19264. DOI: 10.1007/s11356-020-08582-x.

Laeis, G.C.M., Scheyvens, R.A. and Morris, C. (2019) Cuisine: A new concept for analysing tourism-agriculture linkages? *Journal of Tourism and Cultural Change* 18, 1–16. DOI: 10.1080/14766825.2019.1624763.

Lama, P.D. and Becker, P. (2019) Conflicts in adaptation: Case studies from Nepal and the Maldives. *Disaster Prevention and Management: An International Journal* 28, 304–318. DOI: 10.1108/DPM-12-2018-0393.

Miadhu (2019) Guest House Association of Maldives Takes Part in the Largest Travel Fair of India. Available at: https://miadhu.mv/article/en/11540 (accessed 18 November 2023).

Ministry of Tourism (2013) Fourth Tourism Master Plan—Volume 1. Available at: https://www.tourism.gov.mv/en/downloads/master_plan (accessed 20 June 2023).

Ministry of Tourism (2014) Tourism Yearbook 2014. Available at: https://www.tourism.gov.mv/en/page/tourism_year_book (accessed 6 June 2023).

Ministry of Tourism (2018) Tourism Yearbook 2018. Available at: https://www.tourism.gov.mv/en/page/tourism_year_book (accessed 6 June 2023).

Ministry of Tourism (2023a) Tourism Statistics May 2023. Available at: https://www.tourism.gov.mv/en/statistics/publications (accessed 10 July 2023).

Ministry of Tourism (2023b) Tourism Statistics January 2023. Available at: https://www.tourism.gov.mv/en/statistics/publications (accessed 17 July 2023).

Ministry of Tourism (n.d.) Registered Facilities. Available at: https://www.tourism.gov.mv/en/registered/facilities/filter-t3 (accessed 29 July 2023).

National Bureau of Statistics (2020) Statistical Yearbook of Maldives 2020. Available at: http://statisticsmaldives.gov.mv/yearbook/2020/ (accessed 15 July 2023).

National Bureau of Statistics (2023) Statistical Yearbook of Maldives 2020. Available at: http://statisticsmaldives.gov.mv/yearbook/2023/ (accessed 25 July 2023).

Pappayah, P.P., Muiz, A. and Jinus, D.J. (2017) Increasing the visibility of guesthouse operation in the Maldives through e-marketing. *Journal of Tourism and Leisure Studies* 2, 11–20. DOI: 10.18848/2470-9336/CGP/v02i02/11-20.

Postma, A. and Schmuecker, D. (2017) Understanding and overcoming negative impacts of tourism in city destinations: Conceptual model and strategic framework. *Journal of Tourism Futures* 3, 144–156. DOI: 10.1108/JTF-04-2017-0022.

Roxas, F.M.Y., Rivera, J.P.R. and Gutierrez, E.L.M. (2020) Framework for creating sustainable tourism using systems thinking. *Current Issues in Tourism* 23, 280–296. DOI: 10.1080/13683500.2018.1534805.

Salgado, S. (2014) Maafushi Island Shows the Way for Inclusive Wealth Creation through Tourism. Available at: https://blogs.worldbank.org/endpovertyinsouthasia/maafushi-island-shows-way-inclusive-wealth-creation-through-tourism (accessed 10 December 2023).

Schänzel, H., Urie, G. and Lynch, P. (2022) Slow hospitality: Family perspectives of holiday experiences at beacg fale in Samoa. *Hospitality & Society* 12, 95–116. DOI: 10.1386/hosp_00051_1.

Scheyvens, R. (2011) The challenge of sustainable tourism development in the Maldives: Understanding the social and political dimensions of sustainability. *Asia Pacific Viewpoint* 52, 148–164. DOI: 10.1111/j.1467-8373.2011.01447.x.

Shakeela, A. and Weaver, D. (2018) 'Managed evils' of hedonistic tourism in the maldives: Islamic social representations and their mediation of local social exchange. *Annals of Tourism Research* 71, 13–24. DOI: 10.1016/j.annals.2018.04.003.

Shenaan, M. (2022) Exploring guesthouse and cuisine experiences: The strategic and sustainable development of the guesthouse phenomenon in the Maldives. PhD thesis. Auckland University of Technology, Auckland, New Zealand. Available at: https://hdl.handle.net/10292/15607 (accessed 1 August 2023).

Shenaan, M., Schänzel, H. and Berno, T. (2021) Investigating cuisine experiences in the Maldives: A novel research method utilizing tasting buffets. *Journal of Gastronomy and Tourism* 6, 63–80. DOI: 10.3727/216929722X16354101931998.

Thoha, N. (2020) Assessing Potential Climate Related Security Risks in the Maldives: Exploring the Future Climate-Induced Migration (Report 18/2020). Available at: https://docplayer.net/197543023-Assessing-potential-climaterelated-security-risks-in-the-maldives-exploring-the-future-climate-induced-migration.html (accessed 20 June 2023).

Zubair, F.N.I. and Bouchon, F. (2014) Maldives as a backpacker's destination: Supply and demand perspectives. *Procedia – Social and Behavioral Sciences* 144, 256–263. DOI: 10.1016/j.sbspro.2014.07.294.

13 Breaking Barriers, Building Futures: The Transformative Power of Tourism in Empowering Women in Munsiyari, India

Ranjana Tiwari
Department of Commerce & Management Studies, Jain (Deemed to be University), Bengaluru, Karnataka, India

Summary
In the picturesque village of Munsiyari, India, nestled in the Himalayan foothills, a group of local women found a way to turn tourism into a powerful tool for social change. Domestic violence was a pervasive issue in the community, with many women suffering in silence due to social stigma and lack of support for years until Malika Virdi, an avid mountaineer and social activist, came to their aid. The local women, with the guidance of Malika, took a bold step towards empowering themselves and ending domestic violence through the establishment of *Maati Sangathan*. Although the *Sangathan* was founded with the primary aim of promoting gender equality and eradicating the pervasive issue of domestic violence that had plagued the community for years, the women have successfully leveraged eco-tourism and handicrafts as ways to enhance their economic independence while also raising awareness about gender-based violence. This case study utilizes secondary published data to demonstrate the effectiveness of Sarmoli's Himalayan Ark, a community homestay program started under the aegis of *Maati Sangathan* that incentivizes the conservation of the Himalayan Ecosystem. The program has notably contributed to the restoration of high-altitude lakes, forests, and local communities, ultimately leading to its recognition as a thriving ecotourism model in Uttarakhand.

The Value and Interest of the Case Study

The case outlines how a community-led tourism initiative and its success in empowering women through tourism can provide valuable insights and a successful model for readers interested in social entrepreneurship, sustainable eco-tourism, and women's empowerment. By learning from the experiences of the people of Munsiyari, readers can better understand the challenges and opportunities involved in promoting social change through tourism and how to develop effective strategies for achieving long-term sustainability and empowerment.

The Rise of *Maati Sangathan*

We don't seek money. We seek love and respect. Money is like dirt on the hand; if you earn today, it will go away tomorrow. But not everyone gets love.

– Hirma, Service Provider, Himalayak Ark Homestay.

Uttarakhand, situated in the Northern Himalayas of India, unfortunately, ranks among the worst-performing states concerning domestic violence against women (Das, 2022). Even more troubling is the fact that the issue of domestic violence is deeply ingrained and widely accepted within the state, with both men and women justifying such behavior. Amidst this grim scenario, a glimmer of hope shines from the hills of the tiny, remote village of Sarmoli in Munsiyari, Uttarakhand. A quiet revolution began when Malika Virdi, an intrepid mountaineer and social activist, made a bold decision to leave Delhi and settle in Sarmoli with her partner and 2-year-old son in 1992.

The prevailing acceptance of domestic violence has created a significant barrier for women to seek help, resulting in a gross underreporting of cases. Several factors contribute to this reluctance, including social embarrassment, financial dependency, fear of retaliation, victim-shaming, and the complex bureaucratic procedures involved. However, the year 1994 marked a tragic turning point when two cases occurred within a month, involving husbands who, in a drunken state, burned their wives to death. Shockingly, a conspiracy of silence enveloped these appalling incidents, with no one willing to speak up and confront the truth (Society for Rural Urbal and Tribal Initiative, n.d.).

Mortified by the growing number of incidents, Malika decided to take the situation into her own hands. She and a few others came together, had meetings in several nearby areas, and rallied women by encouraging them to send postcards to local officials in communities all around requesting them to shut down the sale of alcohol. At a meeting in Munsiyari, thousands of women spoke up and requested that the selling of alcohol be prohibited in a note to the sub-divisional magistrate. This demonstration planted the '*Maati Sangathan*' (women's collective) sapling. Women began to gather in *Maati* to converse and share their pleasures and sufferings in a connection known as '*dukh-Sukh ka Rishta*' (relationship of sorrow-joy). Women gradually began to recognize their rights and voice their opposition to gender abuse.

Eventually, in 2003, Malika assumed the position of sarpanch (Head) of Sarpanch, Sarmoli-Jainti Van Panchayat. She endeavored to engender a sense of motivation among the populace to conserve their forest resources (Economic & Social Commission for Asia and the Pacific (United Nations), 2019). The community's way of life relies heavily on the forest, including their agriculture and overall lifestyle. Moreover, the forests have always been utilized and preserved as a shared resource. However, over time, people have reduced their reliance on the forest commons since commercial shopping markets have emerged. As they can now purchase gas cylinders and other daily necessities from the market, they are no longer dependent on the forest for agriculture and animal fodder. The prevailing attitude is, as stated by Malika, 'Why bother gathering firewood when we can simply cook with gas?' (Outlook Traveller, n.d).

Consequently, it became evident that any concerted effort towards conservation must address and incorporate the intricacies of livelihood concerns, as the two issues are inherently interwoven. Then, Malika embarked on a transformative journey to intertwine conservation with livelihoods.

The homestay program, initiated in 2004, has significantly bolstered the community's engagement in ecological conservation by linking it to their livelihoods (Nair and Desor, 2014). This realization prompted women with spare rooms in their houses to host tourists, though initially they were hesitant about welcoming strangers. After trying it out for a few months, they discovered that the program worked remarkably well, substantially increasing their family's income, with the added benefit of the money going directly to the women themselves. Moreover, this experience helped the women overcome their inhibitions and interact confidently with outsiders.

Throughout the years, *Maati* has consistently dedicated itself to addressing the issue of alcoholism, notably through the impactful Anti Liquor Protest in 2011 (Kunwar, 2011). They have also been committed to offering vital crisis intervention services. *Maati*'s renown stretches to about 50 adjacent villages, where individuals who have experienced rape, harassment, or domestic violence regularly turn to them for aid and encouragement. Moreover, the organization has actively organized demonstrations and campaigns, championing the cause of well-qualified staff and enhanced infrastructure in the region's educational institutions and healthcare facilities.

Further, recognizing the need to facilitate women's participation in collective activities, *Maati* has been actively exploring various avenues to create opportunities for women to earn a livelihood. One approach involves marketing assistance for traditional handwoven and knitted woolen products, such as carpets, rugs, blankets, and sweaters. In 2010, the organization initiated the *Maati* workshop to expand production and sales to a slightly larger, yet manageable, scale. Additionally, *Maati* orchestrates a biweekly *Mahila Haat* (women shopping street), where women can showcase and sell their agricultural produce, homemade snacks, and woolen items. This event serves as a platform for economic engagement and fosters connections between *Maati* and the broader community, particularly with women in the area (Nair and Desor, 2014).

Himalayan Ark – Community-Led Tourism Initiative

To blend the community's livelihoods and forest conservation, Malika proposed an alternative; a community-led homestay program. The genesis of Himalayan Ark was a nature-based tourism initiative to foster sustainable livelihoods for the local community. By enabling its members to engage in tourism-related activities, the initiative sought to generate income opportunities. In return, its members pledged active involvement in various conservation endeavors, including tree planting, land maintenance, and safeguarding measures (Nair and Desor, 2014).

The homestay initiative was conceptualized and implemented in 2004. Malika, being the head of the *Van Panchayat* (the village forest commons), opened up the opportunity for the entire community. She emphasized that individuals who hold rights in the *Van Panchayat* have the opportunity to volunteer and participate in the tourism enterprise known as 'Himalayan Ark.' However, there is a condition that they must dedicate time, provide voluntary labor, participate in *chauwkidari* (village watch duties), attend meetings, and actively engage in management tasks. Only those who proactively commit to conservation can join the enterprise and offer homestay services. Initially, in 2004, the venture started with the involvement of 13 families, chosen from a pool of 300 right holders in the *Van Panchayat*. The homestay program operates on a roster system and was initially supported by a network of 25 families. However, as of 2023, the number of families participating in the program has decreased to 18. Some families have opted out of the community homestay program and started individual homestays (Bansal, 2023).

This two-fold endeavor is multi-faceted. First, the homestay dimension enables visitors to immerse themselves in an authentic experience as they partake in shared meals, reside in the homes of local families, and engage in the exchange of skills and crafts (Walter, 2020). This unique interaction fosters a deeper connection between guests and their host families, enriching the cultural tapestry of the experience. The second facet revolves around the meticulous planning and execution of trekking escapades. The capable cadre of porters and guides, equipped with their acquired expertise, not only accompanies trekkers on their journeys but also adeptly manages the intricate logistics inherent in various types of treks (Nair and Desor, 2014).

Sustainable business model

The Himalayan Ark program follows a straightforward aggregator model. The Aggregator Business Model functions as a network in which a company collects information about specific service providers, establishes contracts with them, and promotes its services under its own brand (Duan, 2023). As a separate brand, the aggregator is able to provide consistent quality and pricing for the services, despite the fact that they are supplied by various partner providers. It is crucial to understand that the service providers remain independent and maintain ownership of their products and/or services. The aggregator serves solely to aid in mutually beneficial and unique marketing efforts. In Himalayan Ark's business model, the homeowners retain 80% of the generated revenue, while the remaining 20% is allocated to the management. The management team handles essential tasks such as marketing, managing bookings and payments, and providing guidance to rural families on enhancing their hospitality services (The Better Traveller, 2022).

Upon closer examination, it becomes evident that the Himalayan Ark's approach diverges significantly from typical aggregators. Their business model centers around the conservation of forest commons. In this unique setup, each participating family contributes 2% of their revenue to a *'sanrakshan kosh'* or conservation fund, dedicated to supporting the efforts of the *Van Panchayat* in preserving the environment. An additional 3% of their earnings is set aside as a savings instrument (Bansal, 2023). This fund is intended to be utilized for future homestay upgrades and to provide financial assistance to collective members during times of crisis or when loans are required. Through this approach, the Himalayan Ark demonstrates a genuine commitment to sustainable practices and the well-being of their community.

Embracing the power of rotating turns

In Malika's innovative approach, the allocation of homes is based on a rotating turns system, considering various factors such as the family's needs, including their other income sources, and their contribution to the community (Nair and Desor, 2014). The selection of hosts has no impact on the Himalayan Ark's financial performance since the company earns the same amount regardless of where the guest stays. The primary objective here is to encourage collaboration, not competition, and to prioritize collective decision making for the greater communal good. As of 2023, Sarmoli has 18 registered homestays under the Himalayan Ark homestay services, and several more can be found scattered across six neighboring villages (Bansal, 2023).

Further, the essence of community conservation lies in actively participating in the forest's upkeep. This involves dedicating labor to serve as guards, planting trees, and mending broken walls. Typically, these efforts occur approximately every 2 months. On certain occasions, the work is remunerated, while other times it takes the form of *shramdaan*, where individuals voluntarily contribute their labor to improve the community and the environment. Through these collective endeavors, communities unite to safeguard and nurture their natural heritage.

Unique selling proposition of the Himalayan Ark homestays

Himalayan Ark homestays stand out due to the authentic bond that visitors establish with both the location and its inhabitants. In contrast to conventional and impersonal hotels, these homestays offer a glimpse into the very soul of the village. Guests are warmly encouraged to embrace the hospitality and affection their host families provide while respecting the sanctity of their private spaces. It presents a unique opportunity to immerse oneself in the local way of life, participate in the family's daily activities, and nurture mutual respect.

At Himalayan Ark homestays, hospitality is thoughtfully curated to foster a harmonious bond between hosts and guests. Rather than placing meal orders, guests are treated to three wholesome meals that accommodate their dietary preferences (Das, 2021). Recognizing the responsibilities of the village women, breakfast and dinner are served hot, while lunch is conveniently packed to enable hosts to continue with their daily tasks. This arrangement ensures that the women of the household have the freedom to engage in their daily tasks, as it is not feasible for them to cater to guests throughout the entire day. This principle underscores the ethos of Himalayan Ark: 'She is the owner, and you are her guest; she should be at liberty to maintain her work rhythm' (Outlook Traveller, n.d). Upon returning in the evening, guests will be welcomed with tea and warm *pakodas*, followed by dinner. This approach fosters a distinct relationship rooted in mutual respect and shared experiences, quite different from the dynamic one would encounter in a typical hotel stay.

The visitors occasionally accompany their hosts in various village activities such as farming, and cooking. They also embark on culinary journeys, engage in activities like knitting and weaving, observe birds, and participate in local festivals. Notably, several knowledgeable women from the Sarmoli village, possessing deep insights into the local flora and fauna, have embraced roles as guides and naturalists. Also, within the framework of the Himalayan Ark initiative, the local inhabitants have curated diverse travel routes tailored for their visitors. These routes encompass a spectrum of experiences, spanning from elevated mountain excursions situated within altitudes ranging from 2,438 to 3,780 m, to more relaxed walks amidst the chestnut arboreal enclaves.

Another unique point of Himalayan Ark homestays is the kind of food they provide to their guests. Immersed in the local culture, the cuisine here truly reflects the Kumaon region. Guests are treated to a unique assortment of dishes not easily found elsewhere. The diet boasts a rich selection of whole grains, with staples like finger millet rotis and red rice gracing the table. The ingredients are predominantly organic and locally sourced, resulting in delectable and nourishing meals. Throughout their stay, guests are treated to an array of local delicacies encompassing the region's essence (Nair and Desor, 2014; Chitrakar *et al.*, 2020).

From the intriguing bhang ki chutney, a dip crafted from hemp seeds, to the rustic *madua ke roti* (a barley bread) deeply rooted in local tradition, and the flavorful *pahadi rajma*, (kidney beans) indigenous to the area – each dish offers a glimpse into the rich tapestry of Kumaoni cuisine. Incorporating unique tempering components such as *jambu* (a herb adorned with rosy flowers), *timur* (peppercorns), and *thoya* (resembling a black spice akin to cumin), which are predominantly unfamiliar beyond the expanse of the Himalayas, imparts an unparalleled essence to regular lentil preparations and vegetable curries (Bansal, 2023).

When women take on cooking responsibilities in a homestay setting, it can be a powerful avenue for empowerment. This role allows them to showcase their culinary skills and traditional knowledge, turning their expertise into a source of income and recognition (Chitrakar *et al.*, 2022). Moreover, it often leads to increased self-confidence and a stronger sense of independence as they engage with guests from diverse backgrounds, gaining a broader perspective and expanding their horizons. Through this engagement, women can break down gender stereotypes and contribute to their family's financial well-being. In many cases, this empowerment can extend beyond the kitchen, fostering a sense of agency and pride that positively impacts their lives within and outside the homestay environment.

Criticism and Challenges

The Himalayan Ark homestays have encountered their fair share of obstacles and critique. One prevailing challenge revolves around the notion that village homestays ought to provide budget accommodation, undermining the rich cultural immersion they deliver. The difficulty lies in accurately conveying the genuine essence of a homestay experience, which is why some guests might find the homestay overpriced (Outlook Traveller, n.d).

In addition, on a pragmatic level, collaborative work within a group can pose challenges. The competitive pressures of the market can be intense, leading some seven families to detach themselves from the community homestay framework. They question the necessity of waiting their turn to host and might opt to establish themselves as independent hosts to attract their own guests. However, it is evident that if an issue arises in one homestay, the entire village bears the repercussions and experiences adverse effects. The shift towards prioritizing personal gain at the expense of community well-being is also gradually influencing the villagers. In the presence of a collective, a sense of security and representation for the majority prevails. This shields the interests of community well-being from being overrun by market forces (Bansal, 2023).

Another persisting hurdle that remains is inadequate infrastructure, despite the Indian government's expressed intention to bolster homestays. Challenges encompass problems with connectivity and internet access, poorly maintained roads, malfunctioning ATMs, and the inability of guests to use credit cards in the area. The establishment and maintenance of these essential infrastructural components fall within the purview of the government, and their successful implementation is crucial for ensuring seamless last-mile connectivity. Only through establishing these facilities can rural homestays truly flourish and attain success (Nair and Desor, 2014).

Conclusion

This case study highlights the positive impact of community-led tourism initiatives on women's empowerment, showing how it can help address broader societal issues such as domestic abuse, economic issues, and gender inequality. It demonstrates how women can be empowered through tourism and contribute to the solution of larger societal problems (Foley *et al.*, 2018).

After witnessing domestic violence and abuse against women in her village, Sarmoli, Malika Virdi realized that women held a negligible position in the community. The men in the village were mostly alcoholics, leading to domestic violence against their wives and economic instability for families. All these issues stemmed from women's economic dependence on men in their families. Moreover, Malika believed that women who were financially strong would feel empowered (Dolezal and Novelli, 2022) and providing them with a small income could reduce family conflicts and enhance domestic and community cohesion (Foley *et al.*, 2018). She founded *Maati Sangathan*, a gathering for women to share their feelings, and started anti-alcohol campaigns in and around the village.

Tourism has the potential to be a powerful tool for empowering communities, and one way to achieve this is through entrepreneurship (Rogerson, 2002). Entrepreneurship, at its core, involves initiating change through innovation or creation. Supporting small-scale tourism businesses is more beneficial for local women than mass tourism (Vukovic *et al.*, 2023). Malika realized the potential of this innovation and came up with the idea of the 'Himalayan Ark' community-led homestay concept in Sarmoli, Uttarakhand, India, when she was elected as Sarpanch of the village.

Further, offering entrepreneurship opportunities through homestays can be a powerful tool in promoting positive change (Kulshreshtha and Kulshrestha, 2019). Studies have shown that economic development can significantly reduce gender inequality (Kabeer, 2016). Additionally, promoting gender empowerment, as demonstrated in this case study, benefits not only women's sustainable livelihoods but also the livelihoods of their entire families. Running a homestay business provided a valuable sense of psychological empowerment for women, ultimately boosting their self-esteem (Muslim *et al.*, 2017). This newfound confidence allowed women to broaden their horizons and break free from traditional gender roles that often relegated them to passive members of their community. By taking charge of their business, women could become active players in the community, gaining greater control over decision-making and access to education and training opportunities (Yaja *et al.*, 2023).

Further, *Maati Sangathan*, apart from the homestay program, assists women in earning a livelihood through marketing handwoven and knitted woolen products. They also organize a biweekly *Mahila Haat*, where women can showcase and sell their produce, snacks, and woolen items, fostering community connections. Nonetheless, the homestay program's collaborative approach yields greater rewards to the women of the families. They draw strength from the sense of fellowship and solidarity the organization and its community fosters. Beyond its economic benefits, the homestay program's true value lies in the social network it provides for women. It facilitates connection, exchange, and mutual assistance, resulting in a powerful sense of belonging.

Malika Virdi's Himalayan Ark exemplifies how entrepreneurial efforts in tourism can empower women and promote the principles of sustainable eco-tourism. With her extensive experience and expertise, she has developed a set of proven and effective community homestay program that have consistently yielded successful results. These methods have been recognized and valued by professionals in the field of eco tourism, responsible tourism and sustainable tourism as they provide valuable insights and lessons that can be applied in various contexts. The community, environment, and visitors are at the heart of her approach, aiming to balance tourism and the protection of the locals' rights. Paying attention to resource utilization and empowering the local population also play crucial roles in creating an authentically responsible and fulfilling travel experience.

References

Bansal, S. (2023) The Highs and Lows of Running an Eco-Tourism Collective in Munsiyari, Uttarakhand. *The Kodai Chronicles*. Available at: https://www.thekodaichronicle.com/community/boarding-the-himalayan-ark/ (accessed 21 August 2023).

Chitrakar, K., Carr, N. and Albrecht, J. (2020) Community-based homestay tourism and its influence on indigenous gastronomic heritage. *Journal of Gastronomy and Tourism* 4(2), 81–96.

Chitrakar, K., Carr, N. and Albrecht, J. (2022) Expanding the boundaries of local cuisine: The context of Himalayan community-based homestays. In: Nyaupane, G.P. and Timothy, D. (eds) *Tourism and Development in the Himalayas: Social, Environmental, and Economic Forces*. Routledge, Abingdon, UK, pp. 208–224.

Das, Y. (2021) Eco-tourism in Sarmoli: Here's how Uttarakhand's village nailed it. *Travel Leisure Asia*. Available at: https://www.travelandleisureasia.com/in/trips/the-conscious-traveller/eco-tourism-in-sarmoli-heres-how-uttarakhands-village-nailed-it/ (accessed 12 September 2023).

Das, K. (2022) Uttarakhand reports 36% rise in crime against women, 30% rise in rape cases. *Times of India*. Available at: http://timesofindia.indiatimes.com/articleshow/91012046.cms?utm_source=contentofinterest&utm_medium=text&utm_campaign=cppst (accessed 12 September 2023).

Dolezal, C. and Novelli, M. (2022) Power in community-based tourism: Empowerment and partnership in Bali. *Journal of Sustainable Tourism* 30(10), 2352–2370.

Duan, C. (2023) Thematic evolution, emerging trends for sharing economy business model research, and future research directions in the post-COVID-19 era. *R&D Management* 1–27. DOI: 10.1111/radm.12636.

Economic & Social Commission for Asia and the Pacific (United Nations) (2019) Life journeys of transformative leaders: Interview summaries for the report on 'pathways to influence: Promoting the role of women's transformative leadership to achieve the SDGs in Asia and the Pacific.'

Foley, C., Grabowski, S., Small, J. and Wearing, S. (2018) Women of the Kokoda: From poverty to empowerment in sustainable tourism development. *Tourism Culture & Communication* 18(1), 21–34.

Kabeer, N. (2016) Gender equality, economic growth, and women's agency: The 'endless variety' and 'monotonous similarity' of patriarchal constraints. *Feminist Economics* 22(1), 295–321.

Kulshreshtha, S. and Kulshrestha, R. (2019) The emerging importance of 'homestays' in the Indian hospitality sector. *Worldwide Hospitality and Tourism Themes* 11(4), 458–466.

Kunwar, D.S. (2011) A silent revolution has been brewing in Uttarakhand hills and plains. *Times of India*. Available at: http://timesofindia.indiatimes.com/articleshow/7989977.cms?utm_source=contentofinterest&utm_medium=text&utm_campaign=cppst (accessed 12 September 2023).

Muslim, H.F.M., Numata, S. and Yahya, N.A. (2017) Development of Malaysian homestay tourism: A review. *The International Journal of Tourism Science* 10, 65–74.

Nair, V. and Desor, S. (2014) *Maati*. Kalpavriksh, Pune.

Outlook Traveller (n.d) Responsible Tourism Works When Visitors Come but Don't Take Over the Landscape. Available at: https://www.responsibletourismindia.com/inspire-me/responsible-tourism-works-when-visitors-come-but-dont-take-over-the-landscape/414 (accessed 13 September 2023).

Rogerson, C.M. (2002) Tourism-led local economic development: The South African experience. *Urban Forum* 13, 95–119. DOI: 10.1007/s12132-002-0005-2.

Society for Rural Urbal and Tribal Initiative (n.d.) *Maati Sangathan*. Available at: https://www.sruti.org.in/maati-sangathan/ (accessed 13 September 2023).

The Better Traveller (2022) A Himalayan Homestay That's an Ark for Rural Empowerment. Available at: https://travel.ourbetterworld.org/story/journey/himalayan-homestay-is-an-ark-for-rural-empowerment (accessed 10 September 2023).

Vukovic, D.B., Petrovic, M., Maiti, M. and Vujko, A. (2023) Tourism development, entrepreneurship and women's empowerment–focus on serbian countryside. *Journal of Tourism Futures* 9(3), 417–437.

Walter, P. (2020) Community-based ecotourism projects as living museums. *Journal of Ecotourism* 19(3), 233–247.

Yaja, M., Kumar, A., Swamy, G.A. and Rout, H.B. (2023) Is homestay service an ideal business opportunity for women? *International Social Science Journal* 73(248), 463–477.

14 Community-Based Tourism as a Sustainable Approach to Tourism Development in Jibhi Valley, Himachal Pradesh, India

Parishmita Gogoi[1] and Alka Maheshwari[2]

[1]LEHO (Ladakh Environment & Health Organization), Leh, Ladakh, India; [2]Amity Institute of Travel & Tourism, Amity University, Noida, Uttar Pradesh, India

Summary

The Jibhi Valley of Himachal Pradesh, India, has vast potential for community-based tourism (CBT), offering unique village experiences with quality accommodation. CBT empowers locals to generate income and share their culture, showcasing handicrafts, local cuisines, and authentic experiences. Celebration of Jibhi's religious festivals and traditional rituals add a niche cultural perspective of the valley. The pristine environment mostly attracts nature enthusiasts, bird watchers, and hikers. Jibhi's prospects include cultural experiences, village stays, wildlife encounters, nature-based tourism, agro-tourism, wellness, art, and adventure.

The Value and Interest of the Case Study

Jibhi exemplifies the potential for economic growth and community development in rural areas through tourism (see Fig. 14.1 for a glimpse). Promoting responsible consumption of local resources and cultural heritage demonstrates how tourism can be a catalyst for improving livelihood and fostering a deeper appreciation for village lifestyles (see Fig. 14.2).

Background

Community-based tourism promotes all-inclusive participation and equity in the planning, decision-making, management, ownership, and distribution of benefits and costs within a community (Tasci *et al.*, 2013). It is a form of tourism that is owned and/or managed by local communities, with the primary goal of generating broader community benefits (Simmons, 1994; Goodwin and Santilli, 2009; Dodds *et al.*, 2018). It is distinguished by key elements such as community control and management, the preservation of cultural and natural heritage, empowerment, and fostering community development (Manyara and Jones, 2007; Okazaki, 2008). According to Mtapuri *et al.* (2022, p. 1),

> CBT can create commercial and social value to destinations, local businesses, as well as to residents. At the same time, it offers rich, immersive cultural experiences that can enhance the tourists' experiences when visiting different communities.

Fig. 14.1. Path to village serenity.

Implementing a strategic CBT model holds the potential to foster sustainable economic growth in a destination while preserving its natural environment and managing its carrying capacities effectively. Thus, it is 'an alternative to being able to provide community welfare and community empowerment towards sustainable tourism' (Bagus *et al.*, 2019, p. 1). It safeguards livelihoods, aids in poverty reduction, and promotes environmental sustainability. By transforming greedy tourism into green tourism, CBT emerges as a model for sustainable tourism (Asker *et al.*, 2010; Sawatsuk *et al.*, 2018; Giampiccoli *et al.*, 2020). Moreover, participatory planning lays the groundwork for enhancing tourism management skills within local communities. Improving skills in areas such as tour guiding, languages, communication, hygiene, and safety is vital to the success of CBT. Therefore, education and training play pivotal roles in empowering residents and fostering their capacity to actively engage in sustainable tourism practices (Novelli and Gebhardt, 2007; Tasci *et al.*, 2013; Bagus *et al.*, 2019).

Although CBT has gained popularity since the 1980s after the term was first used by Murphy (2013), up until now, insufficient studies have been conducted on comprehending the complexity of community livelihoods through tourism, especially in Jibhi Valley. According to Goodwin and Santilli (2009, p. 10), 'There are very few studies of its [CBT's] actual contribution to either conservation or community livelihood.' Although many studies (e.g. Gupta, 2015; Nag, 2018; Attri and Kaushal, 2019) have been undertaken on tourism development in different parts of Himachal Pradesh, Jibhi remains untouched in this aspect. Therefore, this empirical case study analyses the growth of Jibhi Valley as an exemplary tourism destination and highlights the dedication of the community and the Jibhi Valley Tourism Development

Fig. 14.2. Traditional Himachali house.

Association (JVTDA) towards the socio-cultural, economic, and environmental sustainability of the valley.

The geographical and tourism information about the valley was collected from the locals during the fieldwork. For the primary data, the author conducted her initial study in April 2021 by interviewing Mr Lalit Kumar, then the President of JVTDA, and the founder of Jibhi Adventures and owner of Wilderness Home Jibhi (homestay), along with more than 20 local stakeholders (*refer to Acknowledgement for details*) who shared their valuable insights highlighting the positive impacts of tourism development on the local economy and community. Following this, in August 2023, a telephone interview was conducted with Kumar to validate the previous study findings and update it with the latest developments in the valley. Kumar added that building a destination image by enhancing digital literacy is the latest approach to push for product standardization of Jibhi Valley as an ideal tourism destination. In March 2024, some more facts and data were validated through online communication with Mr Roshan Rana, an enthusiastic tourism professional, who is primarily a mountaineer and trek leader from the valley. All the participants named in this case study verbally agreed to the disclosure of their names and identities.

Brief about Jibhi Valley

Himachal Pradesh, India, is a scenic state in the western Himalayas that covers majestic snow-clad mountains up to a height of 7000m above sea level (NIC Himachal Pradesh, n.d.-a, n.d.-b). It encompasses 'deep gorges, thickly forested valleys, large lakes, terraced fields, and cascading streams' (Raghavan and Bhardwaj, 2024). Shimla, the state's summer capital, and Dharamshala, the winter capital, are nestled at an elevation of about 2,200m (Encyclopedia Britannica, 2024a) and 1,475m (Himachal Tourism, n.d.-a) above sea level, respectively.

The area of the case study (i.e. Jibhi Valley) is situated 1,600m above sea level, in the Kullu district in central Himachal Pradesh. It is currently gaining popularity as an experiential and ecotourism destination, associated with its serene environment and undiscovered charm (Himachal Tourism, n.d.-b; Ecotourism Society of Himachal Pradesh, n.d.; Encyclopedia Britannica, 2024b). The valley comprises of the villages of Shoja (*population:* 200), Ghiyagi

Fig. 14.3. Jibhi waterfall.

(700), Hirab (250), Sajwar (600), Sarthi (300), Sulhanu (300), Kuthachi (200), Rashala (200), Cheyunt (150), Soar (200), Kandhi (180), Bhalagaon (200), Tandi (300), Seri (400), Bahu (800), Binni (400), Cheni (700), and Cheth (600) with a total population of approximately 6,780 (R. Rana, Jibhi, 2024, personal communication). It is home to the famous Jibhi waterfall (see Figs. 14.3 and 14.4).

Although the construction of houses is gradually increasing, Jibhi still stands out for its evergreen forests, diverse landscape, rich vegetation, and culture. The forests of Jibhi primarily contain cider trees, silver oak, golden oak,

Fig. 14.4. Trail to Jibhi waterfall.

blue pines, spruce, and fir. Red Rhododendron (locally known as *Burans*), as shown in Fig. 14.5, is widely seen during its blooming season, and the locals prepare juice, jam, wine, and *chutney* out of its extracts. Some popular wild delicacies are yellow morel, commonly known as '*Gucchi*,' which is an expensive wild variety of mushroom (see Fig. 14.6), and fiddlehead ferns (local name: lingdi/lungdu/linger), as shown in Fig. 14.7.

Wheat and barley are the two main crops. Garlic, peas, and beans are some of the cash crops grown. Horse chestnut, walnut, peach, plum, apple, and persimmon are also part of the community's horticultural products.

Tourism in Jibhi Valley

Tourism in Jibhi Valley was introduced in 1992 by Mr Bhagwan Singh Rana when he pioneered his guest house named Dolly while hosting a small number of foreign travelers visiting the valley. He ran the business single-handedly for almost a decade. Later, a few more accommodation providers came into existence under the guidance of Mr Rana, and ever since then, tourism in Jibhi has witnessed immense growth. Following this, Mr Rana expanded his business to another accommodation unit named Rana Swiss Cottage. As of March 2024, a total of 150 registered tourist accommodation units are functional in the valley as homestays, cottages, and tree houses. At least 100 more such units are expected to register by April 2024 (L. Kumar and R. Rana, Jibhi, 2024, personal communication). Some members of the local community have engaged in collaborative efforts with individuals from neighboring regions, leasing land and jointly constructing high-quality tourist accommodation units. Through these partnerships, they mutually support and benefit each other in the promotion and development of tourism in the valley.

Fig. 14.5. Red rhododendron flowers (local name 'Burans').

Fig. 14.6. Yellow morels (local name 'Gucchi').

Fig. 14.7. Fiddlehead fern (local name 'Lingdi').

According to Kumar, the unique selling point (USP) of Jibhi is its peaceful nature and beautiful wildlife. Himalayan trout fishing remains a sought-after sport and a means of livelihood linked to the tourist inflow. The suitable climate supports brown and rainbow trout, promoting a type of responsible recreational sporting practice. The culinary part of Jibhi includes various local recipes, and siddu (stuffed and steamed wheat-based items) is one of the favorites among tourists. The valley has also introduced its locally produced variety of fruit wines such as apple, rhododendron, and apple cider for tourists.

Jibhi also offers adventures like waterfall rappelling, paragliding, river-crossing, trekking, hiking, trail running and mountain expeditions. Jibhi Adventures is one such organization which conducts all the above activities along with camping, yoga sessions, and recreational yet responsible trout fishing. Some of the popular treks in and around the valley are the Lumbhri trek – 5 days, Basleo Pass trek via Serolsar that reaches Kullu – 5 to 6 days, Great Himalayan National Park (GHNP) trek – 6 to 8 days, and day hikes to Raghupur Fort, Serolsar Lake, Balo temple, Chehni Fort, and GHNP entry gate, as well as to villages such as Sulhanu, Sarthi, Tandi, Seri, and other neighboring villages (R. Rana, Jibhi, 2024, personal communication).

The Serolsar (shown in Fig. 14.8) and Raghupur Fort day hikes are two popular trails, as one can be easily covered without any technical assistance. The hikes also hold religious and cultural significance. For example, the Serolsar Lake hike leads to a small temple of Budhi Nagin, as shown in Fig. 14.9. Thus, the valley's religious links offer pilgrimage as well as cultural tourism among both locals and tourists. The area's revered deities are Shesh Nag and Narayan Dev, and Skiran is a significant pilgrimage site that draws thousands of visitors. Chehni Kothi, a large historic fortified tower with a temple dedicated to Shringa Rishi, also attracts tourists. Some of the popular festivals are Faguli, a 4-day festival celebrated in February, as shown in Fig. 14.10, and Jibhi Mela in April after the week-long celebration of Baisakhi.

Role of the Jibhi Valley Tourism Development Association

The JVTDA was officially founded in 2004–2005 to bring development to the valley. In 2016, it was registered under the Societies

Fig. 14.8. The Jalori Pass to Serolsar Lake Trek.

Fig. 14.9. Serolsar Lake.

Fig. 14.10. Faguli celebration.

Registration Act, Government of India. Before its official foundation, the association was acting as an NGO called Envi-saviours (environment saviors). It has played a vital role in opposing the construction of some hydroelectric projects and other industrial activities that threatened the valley's ecosystem. Today, it runs based on a biannual election. During the initial fieldwork for this case study, the association had 194 members, comprising more than 85 active members and other secondary members. All the tourism activities conducted within the valley are mindfully looked after by the JVTDA. It has been actively organizing awareness campaigns about eco-tourism, CBT, and sustainability and has set many guidelines to preserve the serenity of the valley. It also acts as the rescue team of the valley in case of any emergency or injury.

During the COVID-19 lockdown, JVTDA was devoted to motivating all the local stakeholders and conducting training and awareness programs on the safety and hygiene guidelines for their homestays and tourism business units. Its latest achievement during the fieldwork was the implementation of a garbage collection system. It was successful in collecting waste and sending it to a garbage treatment plant in Manali. To implement a more cost-effective approach, the association has requested the government to support the valley in establishing a local treatment plant. There is a constant effort that goes on to create greater awareness among the community as well as tourists of the harmful effects on the environment caused by non-biodegradable waste materials.

Conservation Practices

While many visitors come to admire the valley's natural wonders, some also seek leisure activities like late-night gatherings, music, and alcohol consumption. Understanding the local culture and lifestyle is crucial before planning a trip to this rural destination. The valley's charm lies not only in its natural beauty but also in its rich traditions and peaceful ambience. Therefore, among the many preservation guidelines, JVTDA has a time allowance till only 08.00 pm to play loud music at any of the homestays or village cafes.

With the valley's increasing popularity, digital media influencers have begun to highlight its beauty on various platforms. However, there have been instances where the promotion of certain sites, like the depiction of the sacred 'Kulhi Katandi' water body (see Fig. 14.11) as 'Mini Thailand,' has caused concerns among the local community. Despite this, the locals

Fig. 14.11. Kulhi Katandi.

maintain a message of peace and harmony, urging tourists and entrepreneurs to respect their traditional beliefs and cultural heritage.

Another example of local conservation efforts is the annual open catch-and-release trout angling competition. Organized by the Trout Conservation and Angling Association of Kullu, this event fosters awareness across the district about the Himalayan trout species conservation. It encourages youth involvement in preserving the aquatic habitats and learning the necessary skills to generate livelihood out of it. The JVTDA is a proud member association, supporting the cause every year on behalf of the entire valley and underscoring the community's commitment towards nature and biodiversity. Additionally, the revival of traditional crafts like Pulla or Seraji Pulla, a type of footwear made from hemp, resonates with visitors seeking authentic cultural souvenirs. Also, the valley's youth actively engage in debates and discussions about nature conservation and cultural preservation, ensuring a sustainable future for their community.

While challenges such as limited funds for developmental activities persist, significant progress has been made in enhancing infrastructure and preserving the valley's natural resources. Through initiatives led by the JVTDA, the valley continues to thrive as a model of sustainable tourism. Efforts in requesting the government to establish road connectivity to remote villages and promote eco-friendly tourism practices are ongoing, ensuring that the valley remains a pristine sanctuary for generations to come.

Mitigation Practices

The JVTDA conducts value and impact analyses on village homestays, encouraging hosts to consider the potential negative impacts of mass influx on their resources. The approach centers on raising awareness of the valley's carrying capacity and the potential threats posed by unconstrained tourism. Therefore, the awareness programs target stakeholders beyond direct tourism practitioners, aiming to educate even the smallest business owners about the importance of sustainable tourism practices.

Some locals perceive tourism in the valley as merely mirroring the mainstream businesses seen in places like Manali and Kasol, two of the many mass tourism destinations of Himachal Pradesh, Kumar. Countering this, the JVTDA is dedicated to addressing concerns such as pollution, congestion, water scarcity, and unplanned construction resulting from uncontrolled tourism. L. Kumar, Jibhi (2023, personal communication) expressed, 'We do not want to replicate models from Manali, Kasol or any other mass tourism destinations from Himachal Pradesh. We want to maintain the essence of Jibhi alive and yet develop the place holistically.'

These efforts inspire responsible consumption and advocate for sustainable development practices within the valley.

Building a Destination Image and Way Towards Inclusivity

Despite challenges posed by promoters attracting an uncontrolled influx of tourists at low prices, Kumar emphasizes the importance of product standardization and fair pricing. 'There has to be a price tag to the services of the valley to maintain a standard' (L. Kumar, Jibhi, 2023, personal communication). According to Saayman and Giampiccoli (2016, p. 154),

> While challenges are present, community members should be directly involved and they should manage their marketing because even if challenges are present, marketing remains a crucial leverage point where community members can influence the balance between business and development goals.

Two years after the initial fieldwork, as a retired President of the JVTDA and a current chairperson of the association, Kumar focuses on building a destination image of Jibhi over the digital media platforms by raising the level of digital literacy among the local community. Through channels like a WhatsApp broadcast program with over 250 members, they have weekly engagements with the community stakeholders on topics and news updates related to tourism. The youth are encouraged to leverage social media platforms to mindfully promote the valley and its properties to attract responsible and nature-loving visitors.

Kumar also underscores the importance of understanding the potential impacts of tourism, especially among community members not directly involved in tourism-related activities. He aims to create a holistic environment in the valley that focuses on not only hosting guests in homestays but also introducing other economic activities to link the community through tourism. He supports establishing women-centric ventures, especially from the lower caste villages, to promote local craft and women's employment while combating the orthodox caste system in the valley. As a part of this initiative, 30 such interested women from two villages were chosen based on expertise by the JVTDA together with the local community. These women are encouraged to participate in cultural shows and practice their local handicrafts. L. Kumar, Jibhi (2023, personal communication) expressed

> More and more women are encouraged to practice and revive the local art forms, cultural folk dance and music as an economic activity. However, work is in progress to introduce a system to manage the activities sustainably. There will be no middle person interfering with these activities between the guests and women stakeholders so that the visitors buy or experience the products first-hand at the heart of the valley. All these practices are officially implemented and managed by JVTDA in mutual agreement with the village stakeholders.

Hence, the JVTDA has been proactive not only in guiding the stakeholders about the value and benefits of product standardization but also in promoting inclusive and harmonious sustainable growth within the community.

Recommendations

In addition to the current measures, implementing a few more initiatives can further uplift the tourism economy of Jibhi Valley. Firstly, conducting monthly awareness programs in all the local schools can educate children and their parents on crucial topics linked to tourism, including cultural preservation, sustainable living, hygiene, waste disposal, and resource management. Inviting volunteers and experts from diverse backgrounds can enrich these sessions. The program could also follow an environmental cleanliness and tree plantation drive to create a greater impact. Currently, a marathon is conducted annually in collaboration with the local senior secondary school. The objective behind this marathon is to spread awareness about the impacts of climate change caused by tourism and the responsibility of the local community, mainly the children and youth, to mitigate the issue. Kumar is eager to collaborate with youth from across the country and the world to organize interactive sessions between them and the local children.

Secondly, promoting local handicrafts and artisanal products more effectively in the form of souvenirs can motivate cultural preservation and generate income within the community. This can be initiated by collaborating with NGOs, travel companies, and academic institutions to support local communities by promoting indigenous art, crafts, experiences, and even culinary items. Linked to this, Kumar resonated with the author's idea of establishing a community museum in the valley. Additionally, 'with an appropriate approach to preserve the value of these age-old traditional practices and folk culture, research from an anthropological perspective can be conducted' (L. Kumar, Jibhi, 2023, personal communication). This means there is scope to study the impacts of tourism on local social complexities and provide insights into promoting sustainable livelihoods. This is where researchers can play their role.

Thirdly, organizing an annual cultural event can allow the locals to showcase their rich heritage to the visitors. In this case, apart from the trout catch-and-release angling competition, and traditional festivals, Jibhi can organize an independent annual event by merging a variety of adventure activities, cultural shows, and culinary programs into it. The event could also act as a platform to conduct informative discussions on sustainable livelihood development involving both tourists and locals. Kumar's vision is to organize an annual festival and create a roadmap for Jibhi's sustainable development for 7 to 10 years down the line. Seeking sponsorship or Corporate Social Responsibility funds from various private and public sector companies can provide stronger financial and promotional support in organizing such an event.

Conclusion

In today's global context, sustainable tourism is crucial. Jibhi Valley illustrates how tourism can be a force for good, benefiting tourists, locals, and the environment alike. Its model serves as an inspiration for other destinations seeking to achieve sustainable tourism practices. By blending traditional customs with modern practices, Jibhi offers visitors a unique experience while safeguarding its natural beauty. From embracing village life to providing comfortable accommodation and eco-friendly activities, Jibhi caters to diverse interests while prioritizing environmental protection.

Since its inception, the JVTDA has been at the forefront of ensuring that tourism respects local culture, preserves nature, and benefits the community. Through awareness campaigns and many conservation and preservation efforts, the association has fostered a sense of responsibility and stewardship towards the environment. Leveraging digital social media and engaging with stakeholders through digital platforms, the association has expanded its reach and strengthened community ties, ensuring that tourism development remains aligned with the values and aspirations of the local population.

By involving community stakeholders in decision-making processes and promoting fair pricing and product standardization, Jibhi Valley as a destination, with the support of the JVTDA, has created opportunities for economic empowerment and social inclusion, challenging traditional hierarchies and promoting equality within the community. Thus, Jibhi's remarkable transformation into a sustainable CBT destination stands as a unique exemplar. Not only does CBT significantly contribute to the local economy by generating job opportunities and boosting businesses but it also ensures the development of a liveable valley and sustainable prospects for future generations. This endeavor fosters a harmonious interplay between nature, culture, and community, promising enduring benefits through tourism for the years to come.

Acknowledgement

Apart from Mr Lalit Kumar, I would like to extend my heartfelt gratitude to the following individuals and establishments for their invaluable contribution to this case study:

1. Roshan Rana from Jibhi Adventures
2. Dinesh Kumar from Hope café
3. Rajnikant Rana, an aspirant
4. Rahul Kapula from The Hidden Burrow
5. Sanju Thakur from Wanderers Cottage
6. Bhagwan Singh Rana from Rana Swiss Cottage and Dolly Guest House
7. Surender Singh Chauhan from Jungle Book Cottage & Tree-house
8. Amit Vaidya from Jalori House
9. Mandeep Thakur from Acoustic Himalayas
10. Rahul Kumar and Mandeep from Mud House Jibhi
11. Ritesh from Jibhi House café & Cottages
12. Pooja Jain from Bunny's Cottage
13. Vijender Kumar from Eaglet Jibhi
14. Bhanchand Rishi from Pine Valley café
15. Parikshit Thakur from Doghri Chalets
16. Kanupriya and Anubhav from Maple Outhomes
17. Dalip Singh from Whispering Pine Cottage
18. Rahul Rana from Tandi White Hill Cottage
19. Yashwant from Lovely Tree House

Their support and cooperation were instrumental in making this case study possible. The names of a few other individuals are not mentioned in the list with respect to their personal comfort and preference of staying anonymous.

References

Asker, S.A., Boronyak, L.J., Carrard, N.R. and Paddon, M. (2010) *Effective Community Based Tourism: A Best Practice Manual*. Sustainable Tourism Cooperative Research Centre, Griffith University, Australia.

Attri, K. and Kaushal, V. (2019) Growth and development: A study of tourism industry of Himachal Pradesh. *Confluence Of Knowledge* 7(1), 46–54.

Bagus, S.I., Imade, S.U., Nyoman, S.I.A. and Putu, W.S.N. (2019) Community based tourism as sustainable tourism support. *Russian Journal of Agricultural and Socio-Economic Sciences* 94(10), 70–78.

Dodds, R., Ali, A. and Galaski, K. (2018) Mobilizing knowledge: Determining key elements for success and pitfalls in developing community-based tourism. *Current Issues In Tourism* 21(13), 1547–1568.

Ecotourism Society of Himachal Pradesh (n.d.) Jibhi and Tirthan Valley of Banjar in Kullu district. Ecotourism Society of Himachal Pradesh. Available at: https://himachalecotourism.hp.gov.in/banjar (accessed 15 March 2024).

Encyclopedia Britannica (2024a) Shimla | India, Map, Elevation, & History. *Encyclopedia Britannica*. Available at: https://www.britannica.com/place/Shimla (accessed 15 March 2024).

Encyclopedia Britannica (2024b) Kullu | India, Map, & Facts. *Encyclopedia Britannica*. Available at: https://www.britannica.com/place/Kullu (accessed 15 March 2024).

Giampiccoli, A., Mtapuri, O. and Dłużewska, A. (2020) Investigating the intersection between sustainable tourism and community-based tourism. *Tourism: An International Interdisciplinary Journal* 68(4), 415–433.

Goodwin, H. and Santilli, R. (2009) Community-based tourism: A success. *ICRT Occasional Paper* 11(1), 37.

Gupta, J. (2015) Tourism in Himachal Pradesh: A study of Kullu Manali. *International Journal of Engineering Research & Management* 2(3), 183–191.

Himachal Tourism (n.d.-a) Dharamshala Travel Guide – Dharamshala Tourism – How to reach Dharamshala. Available at: https://www.euttaranchal.com/tourism/dharamshala.php (accessed 15 March 2024).

Himachal Tourism (n.d.-b) Jibhi Valley in Himachal Pradesh. Available at: https://www.euttaranchal.com/tourism/jibhi.php (accessed 15 March 2024).

Manyara, G. and Jones, E. (2007) Community-based tourism enterprises development in Kenya: An exploration of their potential as avenues of poverty reduction. *Journal of Sustainable Tourism* 15(6), 628–644.

Mtapuri, O., Camilleri, M.A. and Dłużewska, A. (2022) Advancing community-based tourism approaches for the sustainable development of destinations. *Sustainable Development* 30(3), 423–432.

Murphy, P. (2013) Tourism: A community approach. *RLE Tourism*.

Nag, A. (2018) A study of tourism industry of Himachal Pradesh with special reference to ecotourism. *Sumedha Journal Of Management* 7(1), 85–106.

NIC Himachal Pradesh (n.d.-a) At a glance – Government of Himachal Pradesh, India. *NIC Himachal Pradesh*. Available at: https://himachal.nic.in/en-IN/at-a-glance.html

NIC Himachal Pradesh (n.d.-b) List of high altitude peaks in Himachal Pradesh. *NIC Himachal Pradesh*. Available at: https://himachal.nic.in/WriteReadData/l892s/22_l892s/himachal%20-%20Google%20Docs-63605469.pdf

Novelli, M. and Gebhardt, K. (2007) Community based tourism in Namibia: 'Reality show'or 'window dressing'? *Current Issues in Tourism* 10(5), 443–479.

Okazaki, E. (2008) A community-based tourism model: Its conception and use. *Journal of Sustainable Tourism* 16(5), 511–529.

Raghavan, C. and Bhardwaj, S.M. (2024) Himachal pradesh | History, Map, Capital, Government, & Languages. *Encyclopedia Britannica*. Available at: https://www.britannica.com/place/Himachal-Pradesh (accessed 15 March 2024).

Saayman, M. and Giampiccoli, A. (2016) Community-based and pro-poor tourism: Initial assessment of their relation to community development. *European Journal of Tourism Research* 12, 145–190.

Sawatsuk, B., Darmawijaya, I.G., Ratchusanti, S. and Phaokrueng, A. (2018) Factors determining the sustainable success of community-based tourism: Evidence of good corporate governance of Mae Kam Pong Homestay, Thailand. *International Journal of Business and Economic Affairs* 3(1), 13–40.

Simmons, D.G. (1994) Community participation in tourism planning. *Tourism Management* 15(2), 98–108.
Tasci, A.D.A., Semrad, K.J. and Yilmaz, S. (2013) COMMUNITY BASED TOURISM- finding the equilibrium in the COMCEC context, setting the pathway for the future.

15 Business Strategy and Development Towards Sustainability: A Story of Ban Bo Chet Luk Community Entrepreneurship, Thailand

Morakot Ditta-Apichai
College of Innovation and Management, Songkhla Rajabhat University, Songkhla Province, Thailand

Origin: This case study was written in collaboration with Mrs Nuraidah Mangkalu, a local guide at Ban Bo Chet Luk. It includes an interview with Mr Yu Na Longsman, Chairman of the Community-Based Tourism (CBT) Network, Satun Province, and Chairman of the Bo Chet Luk CBT Enterprise.

Summary

Community-based entrepreneurship has been acknowledged to be an essential instrument for the socio-economic development of rural communities and social upliftment, especially in developing countries such as Thailand. This case study covers community background, management history, business strategy, ecotourism, and sustainable development in the Ban Bo Chet Luk community. It analyses community-based tourism initiatives, business strategy, and entrepreneurial activities to share ecotourism experiences with visitors and raise awareness about preserving the natural resources and environment of the region.

The Value and Interest of the Case Study

Ban Bo Chet Luk Village, located in Pak Nam Sub-district, La-ngu District, Satun Province, has been a community-based tourism (CBT) community since 2004. However, the entrepreneurship of the local community is still studied in the village. This case study investigates community-based entrepreneurship by looking at cultural values, shared resources, linkages, and mutual trust nurtured through multi-stakeholder relations for functioning entrepreneurial activities in the community and sustainable aspects of the business in which community-based tourism is used to support sustainability. This case study should serve government policy (e.g. the need for entrepreneurship-related programs) and raise concerns (e.g. tourism standards) for the sustainability of community-based entrepreneurship.

Methods

This case study employed semi-structured in-person interviews with Mr Yuna Longsman, a community leader, and Mrs Nuraidah Mangkalu, a local tour guide of the Ban Bo Chet Luk village, both of whom agreed to be named in the case study, and on-site field notes by the author. Secondary data and information were sourced from both academic research and the community's own reports.

Background

A community-based tourism (CBT) initiative is 'a project, program, or collective action of people who belong to a community deciding to participate in or develop a small to medium-scale local tourism industry' (Spenceley, 2008, p. 288). This approach to tourism development aims to establish a mutually beneficial relationship between tourism activities and the host community, in accordance with the principles of sustainable development. Under the CBT model, the community makes the majority of tourist development decisions and benefits the most from them due to its incentives for balancing economic growth with the socio-cultural and environmental impacts of tourism operations (Rungchavalnont, 2023).

This case study highlights the importance of local empowerment in CBT business strategy and development towards sustainability by actively engaging the community in decision making processes to promote eco-friendly tourism tailored to the community's values and needs. The CBT provides opportunities for all members of the community, promotes economic diversity, contributes to cultural sustainability, builds social cohesion, lowers inequities, and ensures a more equitable and sustainable future.

The Ban Bo Chet Luk Community-Based Tourism Development and Business Strategy

Management history

Satun Province is a popular beach town destination in the Andaman Sea in southern Thailand, which has witnessed the expansion of tourism businesses and a rapid increase in the number of domestic and international tourists visiting the area. One of the tourist attractions is Ban Bo Chet Luk village. This community is part of the La-ngu district, a gateway for marine tourism and is under the responsibility of Mu Koh Phetra National Park in Satun Province. The expansion of tourism has had an environmental and socio-economic influence on the community. Although tourism has influenced the growth of the local economy, mainstream tourism activities cause detrimental effects on the local environment (e.g. tourists pick up beautiful stones as souvenirs and corals are destroyed) and the community (e.g. waste increases, disturbing noise, and improper dress) in the Ban Bo Chet Luk village and its neighborhood (Longsman, 2007).

Witnessing the drawbacks of mainstream tourism development, the local leaders of the Ban Bo Chet Luk community realized the necessity of the community-based tourism establishment. Therefore, Mr Yu Na Longsman, along with other community leaders and members of the Ban Bo Chet Luk natural resource protection group, established the CBT initiative in the early 1990s. It aims to conserve the community's natural resources, especially marine resources and mangrove forests, and to further economic development for the locals through tourism activities. There were initially 57 local members of the Ban Bo Chet Luk community's CBT enterprise, and each one owned shares (a share worth 100 Baht, or around US$3), with the mosque serving as its focal point (Phanwong *et al.*, 2021). The Ban Bo Chet Luk CBT enterprise provides tourists with marine tourism activities, the chance to learn about the local culture and the way of life of fishermen, and the ability to comprehend how people and the environment can coexist. The tourism activities are operated by the community members and the income, after deducting expenses, is divided among the group members. A portion of the profits is contributed to the community's social welfare for the underprivileged, according to eight Muslim principles (Fig. 15.1).

CBT development through a research-based approach

In the early stages, the local leaders and other members of the Ban Bo Chet Luk CBT enterprise faced challenges in planning and developing

Fig. 15.1. A meeting point for tourists at the Ban Bo Chet Luk village.

tourism programs and activities for their visitors due to a lack of knowledge about tourism development. Consequently, the community employed research as a tool to develop their community-based tourism. Using their knowledge and passion for community-based tourism, the community leaders successfully obtained their first research grant from the Thailand Research Fund (TRF) in 2004. This 3-year research project explored community history, local knowledge, social capital, and community resources for developing CBT, managing attractions, and establishing tourism programs and activities that are culturally acceptable (Longsman, 2007). The research project also involved a visit to Ban Mae Kampong, a CBT role model in Chiang Mai Province, to learn good practices and share knowledge within the community for the improvement of their CBT management system, services, and allocation of benefits (Longsman, 2007).

Following the success of the first study, community leaders continued to conduct research in an attempt to establish a CBT network in Satun Province in order to solve problems in knowledge transfer and business collaborations across CBT communities (Longsman, 2012). A project funded by the Thailand Science Research and Innovation (TSRI) served as a learning process for the ten CBT communities participating in it through the creation of a database of each community, the training of community leaders, the development of a support platform for the community members in its network, and the setting of the CBT standards in Satun Province (Longsman, 2012). This local CBT network in Satun Province operates as an independent organization.

Apart from conducting their own research, the community leaders have also participated in a series of studies carried out by education institutes (e.g. Boonmee and Tanasavate, 2019; Mongkolchaiarunya *et al.*, 2022). These research involvements continually provide opportunities for community development and knowledge enhancement. Additionally, the research-driven strategy enables the Ban Bo Chet Luk community to interact and establish connections with experts and other stakeholders, resulting in the extension of the community's network (Fig. 15.2).

Business strategy of the Ban Bo Chet Luk community enterprise

It must be emphasized that the Ban Bo Chet Luk residents' main source of income is fishing, with tourists acting as a secondary source. Aligning with the idea of 'community-based tourism (CBT)' as 'tourism by the local people for the local people' (Rungchavalnont, 2023), the Ban Bo Chet Luk community taps into opportunities from the rise of tourism to enable people from different generations to work together

Fig. 15.2. An interview.

for environmental conservation and cultural preservation, as well as local economic benefits.

According to the interview with Mr Yu Na Longsman (Chairman of the Bo Chet Luk CBT Enterprise), the business strategy emphasizes the authentic experiences of the way of life and local culture in harmony with nature. The core value of the CBT community enterprise is, therefore, focused on providing guests with a one-of-a-kind experience that immerses them in the local history, culture, and wisdom while also educating them about the local communities and their biodiversity. Through guided tours, storytelling, and educational experiences in relation to local traditions, tourists gain a deeper understanding of the community's way of life and the importance of sustainable practices. This understanding can lead to more responsible travel choices and increased support for environmentally friendly programs.

To deliver the value of CBT experiences, the members of the community enterprise planned and developed all tourism activities through their local coffeehouse forum (a roundtable discussion). Led by the Bo Chet Luk community leaders, the coffeehouse forum covers tasks relating to local development, the design of tourism activities, resource allocations, and problem solving. The forum is usually held every two or 3 months to allow members to update their tourist operations. However, if members have obstacles or problems, they can request a special forum to discuss problem-solving strategies with other members. Nonetheless, it is crucial to highlight that community leaders continue to play a significant role in deciding the direction of the Ban Bo Chet Luk CBT community.

The conversation with local guide Mrs Nuraidah Mangkalu showed how important the Internet has been in boosting the CBT community and raising awareness of natural resource and local culture preservation. The Ban Bo Chet Luk village has become a well-known CBT destination. Electronic word-of-mouth (eWOM) on social media (like Facebook and Instagram) has been pushed by tourists who visit the community due to its beautiful natural location, activities of nature and culture related preservation, and warm hospitality. Accordingly, the Ban Bo Chet Luk CBT business's primary marketing channel is now its self-generated, organic social media activity that encourages their visitors to share their travel experiences and photos in public on social media. Local guides and staff are encouraged to pick up photography skills in addition to tourism-related ones so they may assist with tourists' picture activities.

The CBT community also has Facebook pages to promote their tourism program and activities relating to local culture (e.g. fisherman's life) and events (e.g. Muslim festivals). Adding to that, the Thai government, both at the local (e.g. sub-district municipality and the provincial tourism and sports) and national (e.g. the Tourism Authority of Thailand) levels, plays an important role in marketing Ban Bo Chet Luk CBT village as one of the must-visit destinations in Satun Province on their official websites and tourism guides (e.g. Thailand Tourism and Sport, 2021). It is important to note that the Ban Bo Chet Luk community strives to be a good host. As a result, visitors to Ban Bo Chet Luk village enjoy more than simply a tourist experience. Instead, they also forge bonds with the local community. Consequently, the Ban Bo Chet Luk community enterprise garners a high number of revisits.

Tourism activities and hospitality at the Ban Bo Chet Luk village

Utilizing their local knowledge and passion for its marine resources, mangrove forest, and cultural traditions, the Ban Bo Chet Luk community-based enterprise has created a wide range of tourism activities for visitors. Mrs Nuraidah Mangkalu, a CBT local guide, provided an overview of some popular activities that visitors might partake in depending on the time they have to spend in the community:

- *A half-day tour (9.00 AM–2.00 PM)*: The program starts at the local pier, and the local guide takes the group to visit the 7-well historical landmark and back to the pier to take a local fishing boat, or 'Rue

Fig. 15.3. Prasat Hin Pun Yod.

Hau Tong', to the separated sea (sandbar) or Thale Waek (in Thai). After that tourists go kayaking to visit Prasat Hin Pun yod (the rock face looks like a castle with thousands of peaks) (see Fig. 15.3) and see Nautiloid fossils, 480 million years old, at the Fossil Bay. The program also includes local dishes with fresh seafood and seasonal fruits for lunch. The tourists listen to the folklore of Sampao Petra (two sacred stones that the local fishermen believe protect them during fishing) on the way back to the local pier.

- *A 2-day/1-night tour*: This program extends the half-day tour to the Bulon Islands, also known as Moo Koh Bulon, which is not included in the half-day program. Tourists can snorkel and see colorful corals and other marine life and stay overnight on the Bulon islands.
- *A tailor-made tour*: Aside from the typical tour programs, tourists can design a program that fits their time and needs. According to the Ban Bo Chet Luk local guide, the previous tailor-made program included planting mangrove at the community's mangrove forest conservation site, learning to fish like a local, and experiencing local cuisine with fresh and organic ingredients. These tours may also consist of tourist activities provided by other communities in the Satun CBT network.

The Ban Bo Chet Luk community's homestay

According to Mr Yu Na Longsman, in 2015 the Ban Bo Chet Luk community enterprise rented a plot of beachfront land from the residents to build the community's homestays for visitors who wanted to enjoy the peaceful natural surroundings and experience village life. Similar to CBT activities, the homestay was run entirely by locals, and the money earned was divided among the members according to their responsibilities. Due to the expiration of the rental agreement, the community homestays, which had been suspended during the COVID-19 pandemic, were later rebuilt not far from the local pier. The current homestays have five rooms and can serve up to 20 visitors. Each room costs 1000 Baht (about US$29) per night and includes basic amenities such as air conditioning, television, internet connection, clothing racks, towels, and toiletries. Additionally, the community enterprise's information center is located at the homestay (Fig. 15.4).

Economic Stability and Cultural Preservation Towards Sustainability

According to Mr Yu Na Longsman, the Ban Bo Chet Luk CBT enterprise works in partnership with key allies, including local tour guides, tour

Fig. 15.4. The community homestay.

agencies, homestays, long-tail boat owners, and the local community, to create community capacity in order to participate in the tourism chain and sustain their community-based business. All staff members and hosts for CBT programs are drawn entirely from the local community. The tourism-related activities are linked to the tourism value chain and represent sources of additional income for the communities involved. The CBT value chain benefits the Ban Bo Chet Luk community by retaining a large number of local inhabitants and youth in the village while providing a self-sufficient environment for the locals.

The CBT tour programs usually include a well-trained local guide, a long-tail boat, a Kayak set, and snorkeling equipment. The vehicles and the tour activities' equipment are owned by local residents who are members of the CBT community enterprise. The tourists participating in the program also have the opportunity to taste local foods, with the menu incorporating local ingredients. The CBT enterprise purchases local ingredients, including fish and seafood from local fishermen, as well as vegetables that their members and neighbors grow on their own farms. This approach ensures fresh, local, organic ingredients are supplied for tourists' local food experiences during their visit. These tourism-related activities have strengthened the local value chain over time, where culture is an essential component of the travel experience (Fig. 15.5).

Fig. 15.5. Local fish.

Satun UNESCO Global Geopark CBT Network

The Satun Geopark was named the first UNESCO Global Geopark in Thailand and one of eight in the Asia-Pacific in 2018. It is renowned for its tranquil, natural beauty, beaches, coastal islands, mountains, and foothills, as well as a diverse range of marine fossil species, including trilobites, brachiopods, stromatolites, conodonts, graptolites, tentaculites, and nautiloids (Global Geoparks Network, n.d.). The Satun UNESCO Global Geopark covers four districts, including Thungwa, La-ngu, Manang, and part of Mueang Satun. It also covers two national parks (Tarutao National Park and Mu Ko Phetra National Park) and one wildlife sanctuary.

The Ban Bo Chet Luk community is one of the top CBT enterprises in the Satun UNESCO Global Geopark due to its historical landmark, the 7-wells. According to the local history, sea gypsies arrived in the area nearly 200 years ago and dug seven wells for drinking water. Later, in the reign of King Rama VII, the name of the village was changed to 'Ban Bo Chet Luk' (the 7-well village) due to its identity (Thailand Tourism and Sport, 2021). The 7-wells is one of the top tourism attractions at the Geopark locations (Global Geoparks Network, n.d.). Other tourism-related enterprises available at the village include local food provision, homestays, souvenirs, and tour services. In addition to its socio-economic significance, the tourism community enterprise is crucial for the preservation of the geosites, particularly Prasat Hin Panyod, the most significant geosite in the Geopark, with its exceptional pinnacle karst landscape, sea cave, and secret lagoon (Phanwong *et al.*, 2021) (Fig. 15.6).

Challenges and Threats

Despite its successful history and local support, the Ban Bo Chet Luk CBT community also faces obstacles and risks that it is working to overcome in order to preserve stability in the long run. The four main challenges and threats are:

Education

Satun province is one of the top destinations in the Andaman Sea, hosting many international tourists. Although the Ban Bo Chet Luk CBT enterprise complies with Thailand's CBT standards, there is still a large gap in skill level and language, especially English. Only a few international tourists visit the Ban Bo Chet Luk community. Consequently, there is an opportunity for the CBT training to include communication skills in English that could help the

Fig. 15.6. 7-wells.

community expand their target group to include international visitors in the future.

Waste management

The management of waste in the area is a big concern for the Ban Bo Chet Luk community due to the isolated location. There is a canal and a large mangrove forest between the village and the Langu district, giving the area of Ban Bo Chet Luk the appearance of an island. As a result, the village only receives visits from the local government's waste trucks once or twice a week to pick up waste for disposal. The community minimizes trash to the greatest extent possible by utilizing few plastics and non-recyclable products.

Non-local investors

The Ban Bo Chet Luk community has been dealing with the persistent issue of external investors purchasing land from the local residents for the purpose of developing hotels and resorts as a result of the expansion of the tourism industry. This threat makes the community more vulnerable to changes in its way of life, local customs, and culture.

Local control constraints

The community was also concerned about having little authority over the tourist attractions in their area due to government restrictions. This lack of authority is a significant constraint to CBT programs because the community has little control over tourism planning, decision making, or destination management. This barrier makes the CBT community more vulnerable to tourism development due to policy changes.

Conclusions

The Ban Bo Chet Luk CBT enterprise is an example of how tourism can be used by locals to preserve natural resources, their culture and history, and provide an overall positive impact on the local economy. It is a community entrepreneurship model that has been operated for over 20 years by the locals in a remote area of Thailand where mass tourism is the mainstream, like in other developing countries. Today, the Ban Bo Chet Luk community works as a role model, offering good practices for the communities that share the same objective of cultural conservation, nature protection and preservation in the CBT network at the Andaman Sea. The Ban Bo Chet Luk community is a showcase to highlight the CBT model and its value chain can promote sustainability. Nonetheless, the community faces obstacles, such as language barriers, waste management, non-local capitalists, and a lack of authority to plan and manage their own natural resources, requiring assistance from external players, particularly the government and educational institutions. Tourism experts and organizations on a national, regional, and international level need to collaborate to make tourism sustainable and assist in cultural, economic, environmental, and social preservation in order to fulfill the tourism-related United Nations Sustainable Development Goals (SDG's) agenda.

References

Boonmee, J. and Tanasavate, K. (2019) Tourism management in Trang, Satun and Krabi provinces. *Rajamangala University of Technology Srivijaya Research Journal* 11(3), 491–503.

Global Geoparks Network (n.d.) *Satun UNESCO Global Geopark*. Global Geoparks Network. Available at: http://www.globalgeopark.org/GeoparkMap/geoparks/Thailand/12592.htm (accessed 29 August 2023).

Longsman, Y. (2007) *A Study of Community Potential for Tourism Management Ban Bo Jed Luk, Pak Nam Sub-district, La-ngu District, Satun Province* (Publication OCLC no.1281306222). Available at: http://digital.library.tu.ac.th/tu_dc/frontend/Info/item/dc:59354 (accessed 22 August 2023).

Longsman, Y. (2012) *A Project to Study Patterns and Tourism of Community Tourism Networks in Satun Province* (Publication OCLC no.1281305385). Available at: http://digital.library.tu.ac.th/tu_dc/frontend/Info/item/dc:57317 (accessed 22 August 2023).

Mongkolchaiarunya, J., Wisartsakul, W., Srakawee, K. and Buathong, S. (2022) *Value and Worth from Local Research*. Research Report, Puey Ungphakorn School of Development Studies, Thammasat University, Nonthaburi, Thailand.

Phanwong, P., Poomchamnong, N. and Stewart, T.N. (2021) *Knowledge Management of Creative Tourism by Communities in Satun Geopark*. Research Report, Faculty of Environment and Resource Studies, Mahidol University, Nakhon Pathom, Thailand.

Rungchavalnont, P. (2023) *Community-Based Tourism: Empowering Local Champions for Sustainable Tourism in Thailand*. United Nations Development Programme. Available at: https://www.undp.org/thailand/blog/community-based-tourism-empowering-local-champions-sustainable-tourism-thailand (accessed 29 August 2023).

Spenceley, A. (2008) Local impacts of community-based tourism in Southern Africa. In: Spenceley, A. (ed.) *Responsible Tourism: Critical Issues for Conservation and Development*. Routledge, London, pp. 285–303.

Thailand Tourism and Sport (2021) Ban Bo Chet Luk Community Tourism. Available at: https://thailandtourismdirectory.go.th/attraction/21604 (accessed 27 September 2023).

16 Baan Khoa Lak: A Case Study on Sustainable Rural Development Through Community-Based Tourism and Natural Resource Conservation

Krittawit Krittayaruangroj and Suparak Suriyankietkaew
College of Management, Mahidol University, Bangkok, Thailand

Origin: This case study is derived from data collected as part of a research project supported by Mahidol University (Basic Research Fund: fiscal year 2021), funded by Thailand Science Research and Innovation (TSRI). It features excerpts from interviews conducted with key stakeholders of Baan Khoa Lak, which have been edited for clarity and conciseness.

Summary

Baan Khoa Lak, a small village in Trang Province, Thailand, showcases rural sustainability. With around 183 families, it is a model for successful sustainable development. A standout initiative is the community-based tourism program, promoting sustainability while generating local income. Eco-tourism offerings like water rafting, Thai cooking classes, and homestays empower residents by involving them in decisions and nurturing a sense of ownership and belonging. A community forest diversifies tourism, safeguarding trees from harm, conserving resources, and upholding heritage. Baan Khoa Lak's achievements earned it honors, including Trang Province's cleanest village, Southern Region's happiness village winner, and Thailand's 2020 Best Responsible Tourism third prize. Community-based tourism uplifts living standards, spurs social connections, and fosters pride. Baan Khoa Lak epitomizes rural development's potential with limited means. Through environmental and cultural preservation, it attracts nature-loving tourists, enriching both economy and residents' lives.

The Value and Interest of the Case Study

The case study showcases a successful model of sustainable rural development through community-based tourism. It offers insights into how local communities can leverage their cultural heritage, natural resources, and collaborative efforts to achieve economic growth while preserving their environment and traditions. It also highlights the empowerment of the local community in driving their own development. The case demonstrates how community members can actively participate in decision making, resource management, and tourism activities, leading to improved livelihoods and a strengthened sense of ownership. This case serves as a compelling and instructive resource for academics, policy makers, and practitioners interested in community-based tourism, rural development, and sustainable practices.

Methodology

This research study utilizes a qualitative methodology, specifically employing a case study approach conducted in 2021. Data for this case were collected using a multi-method approach that embraced a multi-stakeholder perspective, facilitating the retrieval of information from a diverse array of stakeholders. Due to the varying situations of the COVID-19 pandemic in 2021 during the data collection period, and in adherence with the COVID-19 prevention measures mandated by the Thai government, the primary data source was online semi-structured interviews conducted with key informants. These people included the village head, the leader of the community-based social enterprise, members of the community enterprise, and residents.

Additionally, comprehensive online interviews were conducted with external stakeholders associated with the case. These stakeholders encompassed notably academics from Phuket Rajabhat University, Trang Campus, officials from the Numpud Subdistrict Administrative Organization and the three-panel judges of the Thailand Rural Tourism Award 2020, in which Baan Khoa Lak received two awards: Best village for company outing and second runner-up for best responsible tourism. All informants received a participant information sheet outlining the study objectives. Upon reviewing this information, participants were required to sign a consent form prior to the interview. Importantly, the evidence-based fieldwork research adhered to international ethical standards and was approved by the Mahidol University Central Institutional Review Board (MUCIRB). All interviewees quoted in this case study have agreed to disclose their names and positions. Supplementary data were sourced from publicly available outlets, including reputable newspapers and established institutes. Consequently, a data triangulation process was implemented to facilitate the following analysis.

Background: Geographical Setting, Demographics, and Historical Overview of Baan Khoa Lak

Baan Khoa Lak, a small village nestled in a valley surrounded by mountains, lies within Nam Phut sub-district, Muang District, Trang Province, in southern Thailand. Situated approximately 25 km from Trang city, this agricultural village initially comprised 3–4 main family clans. Presently, it is home to around 183 households, accommodating approximately 650 self-sufficient and harmonious residents within an area of 6,356 rai (1,017 ha) (Provincial Community Development Office of Trang, Community Development Department and Ministry of Interior, 2020). However, due to a growing population and expanding agricultural endeavors, the original Baan Khoa Lak area faced deforestation and encroachment challenges.

This small community has a significant history of fighting for self-sufficiency against communist groups since 1971. Between 1973 and 1975, the government became aware of the communist presence in the area and extended their support by providing weapons and military training to safeguard the well-being of the villagers. This event marked the beginning of the Baan Khoa Lak community coming together and strengthening their sense of belonging and ownership (S. Khunnui, Trang, 2021, personal communication).

In 1992, the government planned a large reservoir project in the village, which posed a significant threat to the watershed forest. Over 90% of the community members strongly opposed this project as it had the potential to disrupt the water resources and the upstream forest that the community relied on (S. Khunnui, Trang, 2021, personal communication). Moreover, the community determined that the large reservoir was not a genuine necessity. In response, they formed a group and submitted a petition to the late King Rama IX, seeking reconsideration and termination of the reservoir project. This became another milestone showcasing the unity and resilience of the Baan Khoa Lak community in their fight to protect their natural resources and environment.

Fig. 16.1. The tree ordination ceremony, a religious ritual involving Buddhist monks adorning trees with orange clerical robes. Photo: Tourism Authority, Thailand.

In 1997, to safeguard the abundant resources of the watershed forest in the region and counter the detrimental impacts of illegal deforestation on the local economy, food security, and biodiversity, the headman and committee members of Baan Khoa Lak implemented a unique practice known as the 'tree ordination ceremony.' Originally hailing from the Northern region of Thailand, they adapted and integrated this ceremony into their community's specific context in the South.

The tree ordination ceremony (Fig. 16.1) was introduced as a means to rescue local forests from logging activities and instill a sense of consciousness regarding the preservation of natural resources among the community members. This religious ritual involves Buddhist monks adorning the trees with orange clerical robes (Fig. 16.1), sanctifying them, and providing protection against damage, destruction, and deforestation. The act of wrapping the trees symbolizes their sacredness and serves as a powerful reminder of the importance of forest conservation in Baan Khoa Lak.

Since then, the members of Baan Khoa Lak have gathered and initiated various activities to foster cooperation and promote self-reliance within the community. In 2000, they established a group dedicated to conserving the watershed forests and expanded their network throughout Trang province. This network encompasses villages and communities across the upstream, midstream, and downstream forests, all working together to safeguard and preserve the natural resources and environment of the community.

The Development of Baan Khoa Lak Community Tourism Enterprise

The village's primary occupation revolves around rubber, palm, and fruit plantations. However, the rubber industry faced challenges due to price fluctuations caused by international regulations demanding traceability and forest preservation over the 5 year period from 2010 to 2015. Consequently, Thai rubber struggled

Fig. 16.2. Water rafting activities at Baan Khoa Lak, Trang Province. Photo: Tourism Authority, Thailand.

to command higher prices, impacting the community's income.

In response, Baan Khoa Lak established the Baan Khoa Lak rafting community enterprise in 2016, with the aim of enhancing the village's economic growth and overall well-being. The foundation of community tourism in Baan Khoa Lak lies in the preservation of watershed forests, cultural traditions, and heritage. The village employs a community management approach that involves active participation from its residents. This involves appointing the central volunteer development and self-defense committee as the core body for community management, collaborating with other community organizations to contribute to village development.

Within the realm of tourism, the Ban Khoa Lak rafting enterprise (Fig. 16.2) serves as a central component of the tourism operations. This enterprise collaborates with various professional groups within the community to welcome tourists. This involves offering community products such as souvenirs for sale and engaging visitors in cooking and craft workshops led by community members. Importantly, the ingredients and materials used for these activities are sourced locally within the community.

The initiative encompasses three primary objectives:

1. Effective management of tourism resources, facilitating well-organized activities that actively engage community members in the conservation of natural resources.
2. Economic advantages for the community through the sustainable utilization of local resources, thereby creating income-generating opportunities.
3. Production of high-quality community products linked to tourism, resulting in increased income, and amplified economic benefits.

Due to seasonal fluctuations, the village experiences periodic limited water availability, particularly during the dry season. Consequently, rafting activities cannot be provided, leading to a decline in tourist numbers. To address this, the community adopted a strategic approach to encourage tourism throughout the year. Through diversification of activities and attractions, Baan Khoa Lak aimed to appeal to tourists and establish a consistent income source. The village's dedication to sustainable development and resource management is geared towards creating a lasting impact on visitors, instilling a sense of environmental responsibility while enhancing community well-being.

Tourism Activities at Baan Khoa Lak

The community of Baan Khoa Lak places great emphasis on the preservation and nurturing of their natural resources, including the rivers and forests. All activities and products are designed to align with this principle, ensuring the sustainable utilization of these valuable resources. The products that are sold to visitors are carefully crafted to reflect the authentic culture and identity of the village, offering a genuine representation of their heritage. Furthermore, the community is committed to maintaining fair pricing for their products, ensuring that both the villagers and tourists receive equitable value from their transactions.

In terms of hospitality and customer service, strict guidelines are in place for the staff members to follow. It is explicitly forbidden for them to accept tips from tourists, as this ensures an unbiased and professional approach to their interactions. Moreover, the staff are expected to exhibit utmost politeness and respect towards all tourists, refraining from any signs or behavior that may be deemed impolite or offensive. The use of disrespectful language towards participants and tourists is strictly prohibited, fostering an atmosphere of mutual respect, and understanding.

The staff members are also encouraged to uphold principles of honesty and integrity in their duties, both towards the tourists and their colleagues. They are expected to support and assist each other in providing a safe and enjoyable experience for the visitors. By adhering to these guidelines and working together harmoniously, the staff aim to ensure that tourists feel welcomed, secure, and ultimately leave with happy memories of their time spent in Baan Khoa Lak.

Here are several tourism products and services offered by Baan Khoa Lak. Tourists can choose services according to their interests:

1. Water rafting in the Trang River. Rafting along the river for approximately 4 km. The water level is not deep and the current is not too strong, making this a suitable activity for people of all ages.
2. Traditional local lifestyle activities. Baan Khoa Lak provides visitors with the opportunity to immerse themselves in the authentic local way of life. They can actively participate in activities such as goat and black chicken breeding (Fig. 16.3), learn the art of southern-style cuisine cooking (Fig. 16.4), and engage in crafting workshops. These experiences offer a firsthand glimpse into the community's rich cultural heritage and traditional practices.
3. Community products. Baan Khoa Lak takes immense pride in its unique range of community products. Visitors have the chance to explore and purchase items like tie-dye fabrics, locally made Thai desserts such as crispy banana (Fig. 16.5), delectable chili paste, flavorful shrimp paste, and refreshing herbal teas. These products beautifully showcase the community's craftsmanship, traditional skills, and distinct flavors.

By offering such a diverse range of experiences, Baan Khoa Lak ensures that it caters to the various preferences of tourists. This approach guarantees a memorable and enriching visit for individuals seeking adventure, cultural immersion, and an authentic taste of the local traditions.

Outcomes of Baan Khoa Lak Community Tourism Enterprise

The enterprise's tourism activity serves as a significant source of income for the community. In addition to the primary revenue generated from the Ban Khoa Lak white-water rafting group,

Fig. 16.3. Goat breeding is a prevalent activity within the community of Baan Khoa Lak. Photo: Tourism Authority, Thailand.

which amounts to an average of 25,000 Baht (US$ 715) per month, there are various other income streams derived from tourists.

One such income source is the culinary skills of the local housewives. They utilize the abundant raw materials available in the community to prepare delicious meals for visitors, generating an average income of approximately 100,000 Baht (US$ 2,860) per month.

Furthermore, the sale of community products contributes to the overall revenue. These products include enticing treats like bananas chips and sweets, as well as refreshing herbal juices, exquisite wickerwork items, intricate woodwork, captivating souvenirs, and seasonal vegetables and fruits. The annual income from the sale of these community products amounts to around 300,000 Baht (US$ 8,580).

Community-based tourism brings not only economic benefits but also positively impacts Baan Khoa Lak's social cohesion. The collaborative nature of the enterprise cultivates unity, cooperation, and pride among villagers, united in offering memorable experiences for visitors. This fosters cultural exchange, promoting mutual understanding and appreciation. Baan Khoa Lak sustains its cultural heritage, adapting practices while maintaining its distinct identity. Their resilience safeguards heritage and passes it on to future generations. Simultaneously embracing modernity and heritage, the village shares its 'treasures' with visitors, nurturing a sense of belonging and enhancing appreciation. This revitalizes traditions, propelling community growth. Actively managing their cultural identity, Baan Khoa Lak welcomes cultural exchange, effectively preserving and enriching their heritage.

Baan Khoa Lak village has been recognized as a remarkable advocate for community engagement in forest conservation efforts, earning the distinguished accolade of 'Runner-Up' in the competition for 'The Best Village in Promoting Community Participation in Protected Forests,' awarded by the Department of National Parks, Wildlife, and Plant Conservation in 2014. The village's commitment to preserving the natural environment is commendable, as it actively promotes and encourages community participation in these crucial conservation initiatives.

Fig. 16.4. Baan Khoa Lak presents a variety of local southern-style dishes, inviting visitors to savor authentic flavors and even engage in cooking workshops. Photo: Tourism Authority, Thailand.

Overall, community-based tourism in Baan Khoa Lak generates not only substantial income but also nurtures social connections, environment conservation, cultural preservation, and a sense of shared identity within the community.

Success Factors of Baan Khoa Lak Community Tourism Enterprise

After conducting interviews with key stakeholders of the Baan Khoa Lak community-based tourism, several factors contributing to the success of this initiative were identified. These factors align with academic literature.

Leadership and management

Leadership is pivotal in community development, collaborating with members and stakeholders. A leader motivates participation, ownership, and accountability in community-based tourism decisions, enhancing capacity and equity (Suriyankietkaew *et al.*, 2022). Leadership negotiates shared goals and fosters experiential learning. Effective leaders influence and impact tourism planning (Uygun and Gupta, 2020). Without good leadership, community-based tourism can falter, with a short-term focus risking socio-cultural and environmental problems (Suriyankietkaew *et al.*, 2022). Without strategic planning, short-term tourism growth may undermine enterprises (Suriyankietkaew *et al.*, 2022), commodifying heritage (Cohen, 2008).

Emotional commitment thrives when leaders embody a shared vision, inspiring all stakeholders (Kantabutra, 2020). Effective leaders instill self-reliance for lasting community-based tourism success. For sustainability, leaders need broad support, planning, collaboration, and equitable benefit distribution (Suriyankietkaew *et al.*, 2022).

Baan Khoa Lak showcases strong local leadership in community-based tourism. The leader unites and motivates engagement, while the enterprise committee encourages involvement. The community retains decision control, forming

Fig. 16.5. Crafting crispy banana, which serves as a local souvenir and showcases the utilization of community-grown produce. Photo: Tourism Authority, Thailand.

a capable management team. The community leader undertakes measures to ensure the broad and fair dissemination of tourism benefits. The community's leadership approach facilitates the impartial allocation of advantages, exemplified through the creation of occupational groups. These groups unite residents engaged in similar occupations, fostering the exchange of ideas, collaborative planning, and coordinated endeavors. Activities such as goat and black chicken breeding, crispy banana production, homestays, cooking, and rafting are organized within these groups. The community leader's effectiveness is greatly enhanced by their organizational, managerial, and interpersonal competencies. The combination of these skills with the leader's authority and credibility results in the achievement of the prosperous community-based tourism model in Baan Khoa Lak. Our examination, carried out via a range of interviews, provides evidence to support the key role of leadership and management in the successful execution of community-based tourism in Baan Khoa Lak.

> I act as a community contact person and coordinator. As soon as I get inquiries and contact from visitors, I will delegate the job and assign it to each group based on the trip itinerary visitors' request. For example, if visitors request meals and catering during the visit, this request will be sent to a group of housewives in charge of the food. If visitors want to do the water rafting, this request will be sent to the rafting team. Therefore, the way we operate here is very systematic and empowers every community member. (Mr. Sawat Khunui, Head of Baan Khoa Lak community-based tourism enterprise)

> Baan Khoa Lak is a small community surrounded by abundant nature with watershed forests, rivers, and rich biodiversity. Under the existing community leadership, we were able to bring various resources or capital available here and used to benefit and generate income for people in the community and build the foundation of a strong community. (Mr. Somsak Noosang, Baan Khoa Lak community-based tourism enterprise's committee member)

Local ownership

Studies have consistently found that local community ownership fosters collective responsibility and emotional commitment towards community-based tourism (Stone and Stone, 2011; Saarinen, 2019). Increased awareness of tourism's societal and environmental impacts reinforces this commitment. A strong sense of belonging leads to full participation and collective responsibility driving sustainable tourism development (Thananusak and Suriyankietkaew, 2023)

This concept aligns with Organizational Citizenship Behavior (OCB) theory, where community members exert extra effort beyond formal roles. OCB's applicability is evident in communities without predefined roles, unlike business organizations (Fahimzadeh, 2017). OCB involves voluntary, supportive behavior that enhances work. Local ownership empowers Baan Khoa Lak's residents to manage enterprises and steer social-economic changes. They actively shape sustainable tourism projects, particularly leveraging natural attractions like watershed forests and rivers.

Our interviews underscore strong community ownership's pivotal role in Baan Khoa Lak's successful community-based tourism establishment.

> Baan Khoa Lak is a very small community. We are more united and cohesive because of the hardship and fighting communism influences we shared in the past. We know each other well and perceive everyone as our family members. Regarding the community-based tourism enterprise we set up, we take everyone's point of view, concern, and interest into account for every decision we make. We have our community's benefits, and people's well-being is our common goals. (Mr. Somsak Noosang, Baan Khoa Lak community-based tourism enterprise's committee member)

> We feel proud of our community. We have full ownership and shared responsibility to protect our community by conserving natural resources. This is for our children and the young generations. It is everyone's responsibility to maintain the watershed forest and clean water because it is not only for Baan Khoa Lak alone, but it also still connects with other neighborhood communities where we share water resources. (Mrs. Dara Thongpiang, Baan Khoa Lak community-based tourism enterprise's member)

> During the COVID-19 pandemic, everyone in the village is eager to get the vaccination. Once the vaccine was available, every community member quickly registered and vaccinated. It is not only for their health and safety but also because they consider that complete vaccination can create herd immunity as their shared responsibility to society. (Mr. Soontorn Changlek, Baan Khoa Lak community-based tourism enterprise's committee member)

Community participation

Community participation is pivotal for tourism development, especially for sustainable community-based tourism. In community-based tourism, the community owns and operates all tourism activities, making them central to planning and implementation (Sebele, 2010).

Effective community participation necessitates empowerment, encompassing the right to solve problems and control one's life. Collaborative efforts and equitable benefit-sharing mitigate resistance to community-based tourism (Jamal and Getz, 1995). Research indicates that holistic community involvement enhances community-based tourism performance and promotes sustainability (Kibicho, 2008). Social exchange theory explains varied community engagement levels (Nunkoo, 2016). Perceived benefits exceeding costs drive engagement. Favorable perceptions and acceptance of community-based tourism depend on value outweighing costs (Nunkoo, 2016).

In Baan Khoa Lak, community committees prioritize local involvement and overall development, allowing flexible membership. These committees function as active management teams, fostering regular participation and contribution through monthly meetings. Our interviews confirm the vital role of active community participation in the success of community-based tourism in Baan Khoa Lak.

> A village meeting is held every month. I encourage full participation from all members. Because the full participation will strengthen and unite our community. The regular meeting

> is also important because we need to update if any goes wrong or requires special attention to keep the neighborhood clean and safe at all times. (Mr. Sawat Khunui, Head of Baan Khoa Lak community-based tourism enterprise)

> Our village has clear rules and regulations that everyone should follow. Every household must follow the waste separation guideline and waste littering methods regarding waste management. Waste is sorted into organic, hazardous, and general waste. For the hazardous waste, our village will hand it over to the subdistrict Administrative Organization to collect and continue to decompose it. (Mr. Somsak Noosang, Baan Khoa Lak community-based tourism enterprise's committee member)

> We use the monthly village meeting forum to propose guidelines for developing tourist attractions, including the development of tourism services for tourists and visitors, by analyzing the community's situation. Professional development to villagers and members is also provided occasionally with support from academic universities in the area. (Mrs. Dara Thongpiang, Baan Khoa Lak community-based tourism enterprise's committee member)

Equitable sharing

Sustainable community-based tourism prioritizes equitable economic benefits for community members. Success relies on fair benefit distribution (Stone and Stone, 2011). This equitable sharing fosters positive attitudes and sustainable community-based tourism outcomes (Kibicho, 2008; Sebele, 2010). Sharing extends beyond monetary exchange to include knowledge. In the community-based tourism context, sharing can go beyond benefit sharing in the conventional view, but the community should internally and externally own knowledge sharing. According to Del Giudice and Maggioni (2014), innovation occurs through knowledge sharing with external stakeholders. Knowledge sharing is a prerequisite in the sustainable organization literature, leading to organizational innovation (Kantabutra, 2020).

Baan Khoa Lak community leader ensured widespread benefit distribution. He did this by making the decision, as a village headman and chair of the Tourism Committee, to use the structure of the existing water cooperative to manage the community-based tourism program. This decision meant that revenues generated from tourism activities would go into funds that benefit every community member.

The rafting enterprise allows up to two members per household to own a share, with each share costing 500 Baht (US$ 14)/share. Profits are allocated as follows: 40% for dividends, distributed annually among shareholders based on their shareholding. A total of 20% is earmarked for the rafting group to build savings for future investments, maintenance, or operational expenses. Another 20% is allocated to committee remuneration, while the remaining 20% is dedicated to member welfare, pooled in the community village fund. This fund supports various community welfare initiatives, including covering hospital fees, providing education scholarships, and offering funeral support. The board has the authority to convene a meeting to discuss the possibility of terminating the current situation. In the event that members fail to comply with the group's rules and regulations, decisions align with majority consensus. Our interviews affirm the crucial role of equitable benefit sharing in Baan Khoa Lak's community-based tourism success.

> Since Ban Khoa Lak Rafting Community Enterprise was registered, we have a clear benefit-sharing structure. Every year, members will receive their dividends. Currently, we have 149 members in the enterprise. We also set up the Village fund. It is our community savings. Currently, we have approximately 3 million Baht (US$ 85,800) in the village fund. The fund will be used on various occasions to improve community members' welfare, such as hospital fees, education scholarships, and funeral supports. (Mr. Sawat Khunui, Head of Baan Khoa Lak community-based tourism enterprise)

Conclusion

Baan Khoa Lak village is an example of successful community engagement in forest preservation. Additionally, it has developed an impressive community-based tourism model. The village takes pride in its abundant natural resources, including pristine watershed forests and water sources, and offers engaging rafting activities. Consequently, it has become an appealing destination for nature enthusiasts, attracting tourists throughout the year.

One of the challenges faced by the community is the insufficient income during the dry season. Due to the limited availability of water, the village is unable to offer rafting activities, which has led to a decrease in the number of tourists visiting the area. As a result, the community suffers from a lack of income during this period. To overcome this obstacle, the village has implemented strategic planning aimed at promoting tourism throughout the year. By organizing a variety of activities and attractions that can be enjoyed regardless of the season, the community aims to attract tourists and generate a steady income throughout the year.

The outbreak of the COVID-19 pandemic had a significant impact on Baan Khao Lak's rafting activities and community-based tourism. The implementation of travel restrictions made it challenging for tourists to visit the village, leading to a substantial decline in visitor numbers. Additionally, as the number of COVID-19 infections surged, the village was required to close its doors in adherence to government-imposed preventive measures. Consequently, the village experienced a sharp reduction in both tourist arrivals and the amount of money spent by visitors compared to the pre-pandemic period.

In response to these challenges, the village leader and committee devised a strategy to adapt to the circumstances by leveraging online marketplaces and social media platforms. Collaborating with the Tourism Authority of Thailand, Trang office, the sub-district municipality, and scholars from universities (Srivijaya University of Technology, Suan Dusit University), they received valuable guidance on effective online marketing techniques and establishing shops in e-marketplaces. Moreover, they sought advice on streamlining the operational process, from handling customer orders to product delivery, to ensure a seamless online shopping experience for customers.

The success of Baan Khoa Lak rafting enterprise relies on several crucial factors: strong leadership, local ownership, full community participation, and equitable sharing within the community. At Baan Khoa Lak rafting enterprise, the presence of capable local leaders and managers is clearly evident. These leaders play a pivotal role in uniting the community and motivating them to engage in various community initiatives. Additionally, the enterprise committee actively encourages voluntary participation from all members. The decision making processes in tourism development are controlled by the community through their appointed management team, allowing for a comprehensive and inclusive consultation process. The community leader takes the initiative to ensure that the benefits of tourism are fairly and widely distributed.

Baan Khoa Lak's community-based tourism advances multiple United Nations Sustainable Development Goals (SDGs), such as poverty alleviation, environmental sustainability, cultural preservation, and social inclusion. Tourism initiatives reduce poverty and encourage local economic growth, while also empowering women and senior residents. Community-based tourism's resource management aligns with SDGs 11, 12, and 15, while also supporting sustainable tourism practices for climate action (SDG 13) and strengthening local governance (SDG 16).

Baan Khoa Lak has received many awards, such as the first prize for most cleanliness village in Trang Province and the winner of the happiness village in Southern Region organized by the Bank of Agriculture and Agricultural Cooperatives. Notably, in 2020, it clinched the top honor for 'Best Village for Company Outing' and secured the third prize for 'Best Responsible Tourism' at the Thailand Rural Tourism Award, an event organized by the Tourism Authority of Thailand.

References

Cohen, E. (2008) The changing faces of contemporary tourism. *Society* 45(4), 330–333.

Del Giudice, M. and Maggioni, V. (2014) Managerial practices and operative directions of knowledge management within inter-firm networks: A global view. *Journal of Knowledge Management* 18(5), 841–846.

Fahimzadeh, H, *et al*. (2017) Measurement model of organizational citizenship behavior in tourism, case study: Travel agencies in Mazandaran and Guilan provinces. *Journal of Tourism Planning And Development* 6(21), 8–27.

Jamal, T.B. and Getz, D. (1995) Collaboration theory and community tourism planning. *Annals of Tourism Research* 22(1), 186–204.

Kantabutra, S. (2020) Toward an organizational theory of sustainability vision. *Sustainability* 12(3), 1125.

Kibicho, W. (2008) Community-based tourism: A factor-cluster segmentation approach. *Journal of Sustainable Tourism* 16(2), 211–231.

Nunkoo, R. (2016) Toward a more comprehensive use of social exchange theory to study residents' attitudes to tourism. *Procedia Economics and Finance* 39, 588–596.

Provincial Community Development Office of Trang, Community Development Department and Ministry of Interior (2020). Available at: https://trang.cdd.go.th/wp-content/uploads/sites/16/2020/12/%E0%B8%9A%E0%B9%89%E0%B8%B2%E0%B8%99%E0%B9%80%E0%B8%82%E0%B8%B2%E0%B8%AB%E0%B8%A5%E0%B8%B1%E0%B8%81.pdf (accessed 1 April 2023).

Saarinen, J. (2019) Communities and sustainable tourism development: Community impacts and local benefit creation in tourism. In: *A Research Agenda for Sustainable Tourism*. Edward Elgar Publishing, Cheltenham, UK, pp. 206–222.

Sebele, L.S. (2010) Community-based tourism ventures, benefits and challenges: Khama rhino sanctuary trust, central district, Botswana. *Tourism Management* 31(1), 136–146.

Stone, L.S. and Stone, T.M. (2011) Community-based tourism enterprises: Challenges and prospects for community participation; Khama Rhino sanctuary trust, Botswana. *Journal of Sustainable Tourism* 19(1), 97–114.

Suriyankietkaew, S., Krittayaruangroj, K. and Iamsawan, N. (2022) Sustainable leadership practices and competencies of SMEs for sustainability and resilience: A community-based social enterprise study. *Sustainability* 14(10), 5762.

Thananusak, T. and Suriyankietkaew, S. (2023) Unpacking key sustainability drivers for sustainable social enterprises: A community-based tourism perspective. *Sustainability* 15(4), 3401.

Uygun, S.V. and Gupta, S.K. (2020) Leadership and accountability for social development: Intellectual leadership and rectors. *International Journal of Indian Culture and Business Management* 21(2), 171–180.

Section 4: Resilience and Recovery Through Tourism

Addressing the resilience of tourist destinations in the face of adversity, this section explores how communities utilize tourism to navigate challenges like natural disasters and the COVID-19 pandemic. Cases from Malaysia, the Philippines, Vanuatu, Japan and Sri Lanka illustrate how tourism can aid in recovery and provide sustainable livelihoods.

17 Green Recovery and Regenerative Tourism: The Success Story of Miso Walai Homestay in Building a Resilient Rural Tourism Destination

Nadzirah Hosen[1], Isabel Paulino[2] and Amran Hamzah[1]
[1]Department of Urban and Regional Planning, Faculty of Built Environment and Surveying, Universiti Teknologi Malaysia, Johor Bahru, Johor, Malaysia;
[2]Department of Organisation, Business Management and Product Design, Faculty of Tourism, University of Girona, Girona, Spain

Summary

Although green recovery and regenerative tourism are well-documented in the literature, case studies at the destination level are lacking, and this is because the application of these concepts is still in its early stages. Therefore, this case study examines a rural tourism community in Lower Kinabatangan, Sabah (Malaysia). The authors used longitudinal analysis from 2007 to 2019, semi-structured interviews, and participant observations to illustrate how the area has managed to successfully recover from intense deforestation following years of exploitation by timber companies and also cushion the impact of COVID-19. The community pushed for a recovery agenda based on community-based ecotourism and green recovery, which resulted in the growth of regenerative tourism. This study examines how the community has seized new opportunities through ecotourism at Miso Walai Homestay, linked to a community forest-restoration programme, and how these opportunities were intensified during the COVID-19 pandemic to compensate for the lack of tourists. Consequently, this case study shows how Miso Walai Homestay has become a model of a regenerative approach that promotes a resilient rural tourism destination. Key success factors have been the agile community leadership, an early start in community forest restoration, partnerships with volunteer organizations, the establishment of trust and reputation, the promotion of citizen science, and the embrace of green recovery initiatives.

This case highlights the success of Miso Walai Homestay in leveraging green recovery initiatives to cushion the financial impact of two situations. First, after the withdrawal of the timber companies from the Kinabatangan region (Malaysia), a new model of community-based tourism was established through a community forest restoration programme. Second, during the lockdown period of the COVID-19 pandemic, income streams were consolidated from green recovery. The Miso Walai Homestay enjoyed a subsequent recovery stage in tandem with the gradual return of international tourists.

Background

In response to the gradual depopulation of rural areas, tourism has been earmarked as an economic activity that can bring about rural renaissance by creating new sources of wealth and employment (Cánoves *et al.*, 2006; Vergara-Muñoz and Martínez-Monedero, 2019), thus helping maintain the traditional values associated with rural life (Flores-Ruiz and Barroso-González, 2012). Although rural tourism can and should contribute to the revival of rural areas, many researchers and practitioners have been working on finding a balance between the various dimensions that influence economic, social, and environmental sustainability (Conti and Cravero-Igarza, 2010). The main aim was to grant benefits in terms of economic development and to also foster awareness of cultural history by preserving identity and recovering and protecting cultural and natural assets that have been lost.

In order to find this balance between the economic, social, and environmental dimensions, several community-based tourism and participatory governance initiatives emerged, enabling local communities to establish an intelligent management and governance regime (Randelli and Martellozzo, 2019; Purnomo and Rahmawati, 2021). In fact, the local community is an essential stakeholder in rural areas due to its moderating role in sustainable tourism development. When a consensus is reached on a development model, this helps to preserve the natural and cultural heritage, while at the same time promoting economic development (Monterrubio-Cordero, 2009; Cortés-Vázquez *et al.*, 2017). In this context, the concept of regenerative tourism has gained significant importance as an alternative to the traditional tourism models. Similar to sustainable tourism, it emphasizes achieving a harmonious equilibrium between human and natural ecosystems, while recognizing the significance of social, cultural, environmental, and economic well-being. However, regenerative tourism goes beyond the notion of simply 'not harming' the destination. In contrast to sustainable tourism, regenerative tourism perceives tourism as a catalyst for restoring both human and natural ecosystems. Thus, the objective of regenerative tourism is to engage travellers and the travel industry in actively creating a positive impact on the destination that fosters a cycle of beneficial effects on local communities and environments (Bellato *et al.*, 2023; OECD, 2023).

The outbreak of the COVID-19 pandemic resulted in a devastating drop in income for rural economies that had become dependent on tourism. Once the pandemic revealed the strong link between expanding cities, traditional tourist activity, deforestation, and public health, several recovery and regenerative strategies gained momentum, looking for solutions to a range of social, economic, and environmental challenges. These strategies highlighted the need to integrate local ecological knowledge and values into tourism governance, in an attempt to ensure that tourism activities contribute more value to communities than they take away. To do so, King (2022) identified five key principles for this shift, including the need to focus on community needs, improving ecosystem integrity, embracing diverse and inclusive business models, governing in a transparent and just manner, and enhancing conservation partnerships. Ultimately, from a socio-ecological perspective, green recovery contributes to building up the resilience of communities and economies against future shocks (Cote and Nightingale, 2012; Fabry and Zeghni, 2019; Heslinga, 2022), especially those pushed by 'fast variables' such as the COVID-19 pandemic (Cheer and Lew, 2017).

Methods

Utilizing a longitudinal approach at various intervals between 2007 and 2019, the methodology employed for this study encompassed the observation of changes and advancements in the local community over time (Ruspini, 1999; Caruana *et al.*, 2015), focusing on observation and analysis of the sequence of events, transformations, and strategies undertaken by the community. This enabled analysis of the community evolution concerning organization and leadership, shifts in community mentality, the growing acceptance of environmental conservation, and other aspects that collectively illuminate how the community's responses evolved in the face of changing circumstances. Furthermore, semi-structured interviews were conducted with key stakeholders, including government agencies, donors, and individuals within the tourism sector, who consented to the use of these data and being mentioned for research purposes. These interviews provide qualitative insights that delve into the decision making processes, challenges encountered, and adaptations executed by various stakeholders in response to

Fig. 17.1. Location of the Miso Walai Homestay.

the evolving situation. In addition, participant observations were employed to directly witness and understand the community's actions and reactions amidst the COVID-19 pandemic. This methodology offers a window into the community's resourcefulness, coping mechanisms, and shifts in behaviour arising from the sudden absence of tourists due to the pandemic.

About the Miso Walai Homestay

Miso Walai Homestay is a community-based tourism (CBT) destination situated in Mukim (sub-district) Batu Puteh, in the Kinabatangan district of Sabah, Malaysia (see Fig. 17.1). It is located along the Kinabatangan River, one of Malaysia's biodiversity hotspots. Over 250 species of birds inhabit the floodplains, including eight of ten species of hornbill, and over 90 species of mammals, such as *Orang Utan*, Pygmy elephants, and proboscis monkeys, making it one of the country's top wildlife tourism destinations (Majail and Webber, 2006).

The local *Orang Sungai* (River People) was primarily employed by timber companies in Sabah during the 1980s, when the sector was at its peak. It paid them well, but caused extensive damage to wildlife habitats and ecosystems in the Kinabatangan floodplains. It was not until the early 1990s that the Sabah government ceased large-scale uncontrolled logging activities, following intense international pressure. As a result, the Lower Kinabatangan Wildlife Sanctuary (LKWS) was declared in 1996. Consequently, those who had previously worked for timber companies and occasionally engaged in poaching, were in a tight spot, as they were losing their jobs and income. Cocoa farming was a temporary measure to ease the deteriorating economic situation, but it was abandoned after the cocoa price collapsed worldwide. Due to this, nearly 90% of the local community fell into a deep state of poverty, which forced many young people to leave their homes in search of employment in Sandakan, Kota Kinabalu, and Peninsular Malaysia.

In 1997, the government initiated a CBT project with the support of World Wild Fund Norway to address the income and livelihood loss caused by the official recognition of LKWS. The project was known as the Model for Ecologically Sustainable Community Conservation and Tourism (MESCOT). Its mission was to empower local communities in Lower Kinabatangan to run a homestay programme using ecotourism as a resource. As part of the initiative to strengthen the homestay organization, a tourism cooperative known *as Koperasi Pelancongan Mukim Batu Puteh* (KOPEL) was established in 2003, under the auspice of the Cooperative Commission of Malaysia (SKM). The membership of the tourism cooperative steadily increased from 72 members in 2003 to more than 353 members in 2022, demonstrating its success in eradicating poverty and promoting community-based tourism.

In this context, Miso Walai Homestay serves as a pivotal entry point for ecotourism exploration within the ecosystem of the Kinabatangan River. Tourists are afforded the choice to engage in guided river cruises along its waters, which provides a unique opportunity to witness the diverse assortment of flora and fauna that thrive within this locale. Simultaneously, the homestay partners with the village cooperative, KOPEL, to coordinate well-structured forest trekking expeditions. These expeditions offer visitors a platform for deep engagement with the local rainforest, fostering a profound understanding of the ecological significance that characterizes this region. Furthermore, tourists have the chance to reside among community members, offering them an authentic glimpse into the local way of life. This includes cultural programmes thoughtfully designed to cater to diverse groups, spanning from students to families and volunteers. By participating in activities like learning traditional cooking techniques and engaging in outdoor farming, tourists gain a first-hand understanding of the daily rhythms of the *Orang Sungai*.

To complement staying with the locals and experiencing their rural way of life, a tree-planting activity was added to the tourist experience. Over time, this activity expanded into a community forest restoration programme aimed at reviving the ecosystem. It became part of the Kinabatangan Corridor of Life initiative, which involves establishing an ecological passage to reconnect fragmented conservation areas with indigenous communities residing alongside the Kinabatangan River.

However, the travel restrictions during the COVID-19 pandemic had an adverse effect on KOPEL, since it relied almost entirely on international tourists. The COVID-19 pandemic highlighted the vulnerabilities of the tourism sector and presented a major opportunity to reassess the relationship of tourism with nature, climate, and the economy. During the pandemic, KOPEL worked alongside international universities to successfully cushion the impact of the lockdown and the absence of tourist arrivals. This included accelerating green recovery initiatives and other conservation practices, as well as fostering the use of citizen science among its members. Green regenerative tourism has emerged as a way to conduct response and recovery activities that tackle some of the underlying weaknesses of economies and societies. Thus, its aim was to help to 'build back better' with new forms of travel.

This case study reveals how the rural tourism community in the Miso Walai Homestay in Lower Kinabatangan, Sabah (Malaysia) has responded with resilient thinking to the gradual deterioration in wildlife, and the rapid changes brought about by the COVID-19 pandemic, by making the most of the opportunities presented by the green recovery agenda. The successful implementation of community-based tourism, coupled with the forest restoration initiative and tree planting activity, has played a pivotal role in bolstering the Kinabatangan Corridor of Life project, which has become the homestay's unique selling proposition (USP). This initiative has enabled the community to restore degraded areas, conserve the forest ecosystem, and provide a sustainable source of income through ecotourism activities. Furthermore, the community forest restoration activity has ensured the community's economic survival during lockdown, by providing employment opportunities and continuing to enhance the area's biodiversity, which further supports the development of ecotourism.

The results indicate that the community forest restoration programme was the key to sustaining the community economically during the lockdown and helped to promote sustainable tourism afterwards. The homestay exemplifies a remarkable regenerative approach that both mitigates the negative impact of tourism losses following COVID-19 and fosters a resilient rural tourism destination. This success story can inspire other rural destinations to adopt a green recovery approach and move beyond the conventional tourism offer, ultimately contributing to the sustainability and resilience of these destinations.

Critical Success Factors

The emergence of sustainable practices fostered regenerative tourism and cushioned the economic impact of the pandemic-induced crisis. In this section, we examine the key factors for effectively implementing green recovery which have been beneficial for economic, ecological, and social capital.

The first success factor came from the previous experience and contacts created from the CBT programme. From the inception of the CBT programme in 1996, Miso Walai Homestay had identified its target market and subsequently shaped the tourist experience accordingly, making contact with international tour operators and voluntourism companies who were looking for the niche market of ecotourism and wildlife tourism experiences. From those initial contacts, KOPEL fostered active partnerships with renowned voluntourism companies such as Raleigh International, GVI and Intrepid Travel to undertake the entire Tungog Rainforest Eco Camp. Although it took longer to develop the project using voluntourists to run the tree planting activity, as opposed to a private contractor, the partnership between the local community and voluntourists ensured a two-way learning process and nurtured a stronger community spirit and sense of ownership (see Fig. 17.2, which shows some pictures of voluntourists participating in the forest restoration projects).

The second success factor was the increased demand for voluntourism linked to community forest restoration activity after the track record of previous reforestation programmes. The experience in restoration programmes, which started in 1996, gave the community a head start in developing the capacity to implement a comprehensive forest restoration programme. Youths who used to work for logging companies cutting down trees found employment leading forest restoration projects. The local housewives oversaw the implementation of the community forest restoration projects commissioned by government agencies and donors. Furthermore, the community used its traditional ecological expertise to select the appropriate tree species and replanting methods to ensure a high degree of plant survival and diversity.

The third success factor was the investment in formal education. KOPEL's management team had limited formal education, but since 2010 some universities have approached them to conduct on-site training for local youth leaders in wildlife and water quality monitoring. Thus, local citizens were empowered with science skills to use the equipment for water quality monitoring, including analysing and interpreting the results. Furthermore, a group of youth leaders was helped through fully sponsored internship programmes in a Japanese university. Citizen science currently plays an essential role in enabling the community to succeed in managing the KOPEL conservation efforts. The community has also invested in future KOPEL leaders, ensuring a group of youth receive formal education in resource and environmental management. Recently, upon the completion of their tertiary studies, several of these youth have returned to the village to serve KOPEL instead of staying on to find jobs in the cities. This ensures that future generations gain valuable knowledge and training in the subject to further develop

Fig. 17.2. Pictures of the forest restoration activities.

KOPEL's methodology for habitat restoration, tree planting, and forest management, which ultimately helps to prevent out-migration by offering quality jobs to local young talent.

The fourth success factor was the agility of KOPEL's community leadership and the swiftness of its response in the face of the fast and unexpected nature of COVID-19. This enabled them to make prompt decisions to intensify conservation efforts and compensate for the loss of income from the decline in tourism. It is worth noting that KOPEL's swift response primarily aimed to consolidate income from green recovery initiatives during the lockdown and ongoing recovery phase. Subsequently, they effectively applied for community restoration projects, taking advantage of their previous contacts with conservation-related agencies and donors.

The fifth success factor was the previous experience in finding funding for forest restoration programmes. Prior to the pandemic, KOPEL managed to secure funding from international donors and public agencies, enabling them to successfully plant more than 400,000 trees, and also set and uphold a high standard in community forest restoration. This has allowed KOPEL to gain the trust of government agencies and international donors. The organization has also developed a reputation for being highly competent, committed, and trustworthy, which secured continued funding for community forest restoration projects. This support was the key to sustaining the community financially during the pandemic. Indeed, KOPEL increased the number of grants it received during the pandemic.

The sixth success factor represented the high engagement of the entire community with the project. Although many of the reasons for the success factors of actively involving the local community in conservation efforts have been extrinsic in nature, the 3 years of pandemic lockdown with hardly any tourists and only the forest restoration activity, gave the local community the opportunity to reflect about intrinsic reasons for continuing with the conservation programmes associated to ecotourism. Local guides, youth leaders, homestay operators, and even food and beverage staff interviewed during fieldwork in September 2022 agreed that conservation had become an important part of the Miso Walai Homestay's 'DNA'. This is reflected in the answer given by a 66-year-old woman in charge of managing one of the deeds for forest restoration when she was asked about her retirement:

> Why should I stop tending to the trees that I have planted? If I stay at home, I will only spend time watching TV and fiddling with my mobile phone... I will get sick. Here I'm active which makes me healthy... besides... I need to take care of my trees.

Moreover, KOPEL's community reforestation with voluntourists has set a successful precedent for new other forest restoration projects along the Kinabatangan River. Recently KOPEL has extended its forest restoration programme to the barren alluvial lands owned by big oil palm companies. Furthermore, new partners are joining, such as the NGO Danau Girang and ReGrow Borneo. Together with KOPEL, they have recently launched United Nations Development Project's in the Kinabatangan. As expressed by the current CEO, KOPEL's vision goes beyond just planting trees, as it also nurtures them in a regenerated landscape where wildlife can thrive:

> There are many forest restoration projects in the Kinabatangan. Some are limited to the planting of saplings without taking the necessary care required to ensure good growth. KOPEL's programme has avoided this pitfall... we have set a high standard.

Finally, the seventh success factor is KOPEL's effort to embrace new contemporary approaches to green recovery initiatives. This includes applying IoT technology (The Internet of Things) to adopt replanted trees or attracting sponsors through EcoMatcher to visit Miso Walai Homestay in person in the future.

All these actions and decisions have been KOPEL's success factors to effectively implement green recovery measures and mitigate the financial impact of COVID-19. Although tourism revenue has shown a gradual increase with the return of international tourists, it has yet to reach pre-pandemic levels. However, the KOPEL's success in leveraging green recovery during the lockdown has positioned it favourably to rebuild tourism through regenerative practices. Consequently, optimizing the connection between green recovery and regenerative tourism has become crucial.

Strategies for Developing Regenerative Tourism

This section proposes strategies for transforming green recovery initiatives into regenerative tourism experiences. These recommendations are inspired by the accomplishments and insights of KOPEL, firmly grounded in their key success factors that contribute to effective green recovery. The main objective is to create vacations that not only offer memorable experiences for visitors but also leave destinations in a better state than when the visitors initially arrived. To develop these strategies, we have utilized the five principles of regenerative tourism outlined by King (2022) as our guiding framework.

The first principle emphasizes prioritizing the needs of the community. KOPEL's journey showcases the resilience of a community that overcame adversity by restoring their natural surroundings. They rebuilt pride, fostered social cohesion, and enhanced community resilience around a shared vision of protecting their environment. Their success went beyond restoration, as they moved from the outskirts to the heart of the Kinabatangan Corridor of Life, forging strong relationships with various stakeholders. To achieve Principle No.1, recommended strategies include establishing a collective vision for the community's involvement in natural resource conservation, setting high standards in reforestation efforts, and enhancing CBT through interactive conservation-focused activities. This ensures visitors actively contribute to environmental preservation while enjoying immersive experiences.

The second principle focuses on enhancing ecosystem integrity and biodiversity. In this regard, KOPEL's reforestation programme stands out from the typical tree-planting ceremonies commonly seen in CBT initiatives in Malaysia. It goes beyond being a standalone community project by being fully integrated into the creation of a landscape-level ecological corridor known as the Kinabatangan Corridor of Life. Notably, KOPEL's forest restoration efforts have evolved to include the calculation and monitoring of carbon sequestration, involving educated youth and citizen scientists. Furthermore, the adoption of Internet of Things (IoT) technology enabled global communities to participate virtually in tree adoption and care. To achieve Principle No.2, recommended strategies include integrating community reforestation into mainstream forest restoration programmes to create ecological corridors, expanding the scope of community reforestation to encompass carbon sequestration calculations, and leveraging IoT technology to develop enticing virtual tourism packages that include the adoption and virtual care of planted trees.

The third principle centres on embracing diverse and inclusive business models. KOPEL's success lies in its robust financial and social capital, which played a pivotal role in fostering community resilience, particularly during unprecedented shocks like the COVID-19 pandemic. Unlike many CBT organizations, KOPEL wisely diversified its conservation efforts beyond traditional tourism activities. By incorporating initiatives such as lake restoration, biomass production, and solar energy generation, KOPEL achieved a delicate balance between profitability, equity, conservation, and social responsibility. As KOPEL embarks on the post-pandemic recovery phase, it has shifted its focus to becoming a training provider. Utilizing its dedicated training centre and quality chalets (Mandaa stay), KOPEL now offers courses on CBT, rural development, and conservation. Notably, their recognition as one of the winners of the UNWTO's Best Tourism Village initiative in 2021 has attracted the attention of Malaysian rural development agencies, positioning KOPEL as a preferred training provider with their own management serving as trainers. To achieve Principle No.3, the recommended strategies include diversifying the economic base to incorporate non-tourism income streams aligned with global priorities such as green recovery, optimizing the accumulated capacity, knowledge, and expertise to venture into training as a new avenue for employment and income, and leveraging awards and certifications to enhance branding as a potent marketing and promotional tool.

The fourth principle highlights the importance of transparent governance in the development of a tourism cooperative. KOPEL, formerly Miso Walai Homestay, ensures decision making through consensus and transparent financial transactions with annual audits. They distribute 80% of tourism income among members, allocate 20% for management costs and a community fund, and disburse conservation project income based on member services. During lockdown, KOPEL redistributed tasks and income to support affected members. They maintain positive relationships with government agencies, NGOs, and tour operators, collaborating on forest restoration and sustainable tourism. KOPEL also nurtures bonds with neighbouring villages, fostering a shared sense of place among the Orang Sungai communities. To achieve Principle No.4, establishing a formal cooperative and community fund after the maturity of a community-based tourism initiative led by local champions is recommended. Scaling up from niche to mainstream tourism involves strong connections with supply chain stakeholders. Building relationships with surrounding stakeholders and communities relies on creating a shared sense of place based on natural and cultural resources.

The final principle emphasizes the importance of enhancing regenerative partnerships. Government involvement is crucial for providing the policy framework for regenerative tourism to thrive, with a focus on understanding the symbiotic relationship between green recovery and regenerative tourism. International donors and local NGOs are now allocating funding for conservation projects aligned with the global focus on green recovery. However, community-based tourism organizations like KOPEL have a unique advantage in understanding the needs of different tourist segments within the context of regenerative tourism and aligning them with activities from green recovery initiatives. KOPEL's profound understanding enabled them to submit proposals to donors and partners to revitalize their community-based tourism affected by the pandemic. To achieve Principle No.5, recommended strategies include formulating a master plan for regenerative tourism that provides the policy framework, strategic direction, and stakeholder collaboration. Additionally, integrating interactive tourism elements such as IoT technology and virtual tours to complement green recovery initiatives, and enhancing the skills of nature and localized guides to include interpretation and storytelling related to regenerative tourism's peculiarities is recommended.

Conclusions

The case of Miso Walai Homestay highlights the significance of community engagement, proactive decision-making, and strategic partnerships in implementing green recovery and regenerative tourism. KOPEL's strategy to develop ecotourism including a community forest restoration programme after an intense period of deforestation, and its ability to adapt swiftly during the pandemic by intensifying conservation efforts, demonstrates the importance of resilience in uncertain times.

KOPEL's success stems from investing in local community capacity building and expertise, which has allowed them to lead conservation projects that benefit their ecotourism activity and ensure long-term sustainability and resilience in the face of changes. Collaborations with voluntourism companies, international donors, government agencies, and NGOs have played a vital role by providing financial support, knowledge exchange, and shared responsibility. Community empowerment through education and capacity building was instrumental in KOPEL's achievements. By providing scientific training and formal education opportunities, the community has acquired the skills needed for effective conservation and resource management. The community's intrinsic motivation, pride, ownership, and recognition of conservation as part of their identity have sustained their commitment even in challenging circumstances.

To promote regenerative tourism, it is essential for destinations to prioritize community needs, enhance ecosystem integrity, embrace diverse business models, promote transparent governance, and foster regenerative partnerships. By adopting these strategies, rural destinations can create tourism experiences that contribute to environmental preservation, while also providing economic benefits and empowering local communities. In conclusion, KOPEL's case exemplifies the importance of implementing green recovery and regenerative tourism. By following the outlined success factors and strategies, other rural destinations can work towards a more resilient and sustainable tourism industry.

Avenues for Study

This case study explores the successful implementation of community-based tourism and green recovery initiatives at Miso Walai Homestay and the KOPEL management. It emphasizes the critical success factors and strategies that contributed to their resilience during the COVID-19 pandemic, and their transition towards regenerative tourism practices after extensive deforestation activities. Several avenues for research and teaching are suggested to further explore the themes of the case.

One avenue is conducting a comparative analysis of similar community-based tourism initiatives in different regions or countries. This would provide insights into contextual factors that influence the effectiveness of green recovery and regenerative tourism, allowing for a deeper understanding of broader implications and transferability. By examining diverse case studies, common success factors, challenges, and strategies, it is possible to identify valuable insights for future initiatives.

Exploring stakeholder perspectives is crucial to understanding community-based tourism initiatives. Investigating the motivations, challenges, and perceptions of various stakeholders, including local community members, tour operators, NGOs, government agencies, and tourists, provides valuable insights into stakeholder dynamics and interactions. Analysing these perspectives can contribute to further understanding the factors influencing the success and sustainability of community-based tourism initiatives.

Policy analysis is another avenue to pursue to understand the role of government policies and regulations in facilitating or hindering community-based tourism and green recovery initiatives. By examining policy frameworks and regulatory environments, it is possible to gain insights into necessary policy support, incentives, and barriers that shape the implementation and scalability of these initiatives. Identifying policy recommendations supports the development of sustainable and regenerative tourism practices.

Expanding on the themes presented in the case study offers numerous research and teaching possibilities. By using comparative analysis, considering stakeholder perspectives, and conducting policy analysis, researchers can gain a deeper understanding of the critical success factors, challenges, and strategies that drive sustainability and resilience. This knowledge enhances academic scholarship while also informing the development and implementation of future community-based tourism projects and promoting more sustainable and regenerative practices in the tourism industry.

Acknowledgement

This publication has emerged from the project titled 'Rebuilding Rural Tourism in APEC Economies by Utilising Green Recovery Strategies' funded by APEC (TWG 05 2021A). This publication was also supported by the R+D+I ADAPTOUR project [contract number PID2020-112525RB-I00] funded by MCIN/AEI/10.13039/501100011033; and in collaboration with Generalitat de Catalunya, Departament d'empresa i treball. Finally, the co-author Isabel Paulino has been funded by the postdoctoral Margarita Salas fellowship program of postdoctoral Margarita Salas fellowship programme of the Spanish Ministry for Universities, through the European Union NextGenerationEU Funds [C21.I4.P1/AEI/10.13039/501100011033].

References

Bellato, L., Frantzeskaki, N. and Nygaard, C.A. (2023) Regenerative tourism: A conceptual framework leveraging theory and practice. *Tourism Geographies* 25(4), 1026–1046.

Cánoves, G., Villarino, M. and Herrera, L. (2006) Public policies, rural tourism and sustainability, a difficult balance. *Boletin De La Asociacion De Geografos Espanoles* 41, 199–217.

Caruana, E.J., Roman, M., Hernández-Sánchez, J. and Solli, P. (2015) Longitudinal studies. *Journal of Thoracic Disease* 7(11), E537–E540.

Cheer, J.M. and Lew, A.A. (2017) Understanding tourism resilience: Adapting to social, political, and economic change. In: *Tourism, Resilience and Sustainability*. Routledge, Oxon, UK, pp. 3–17.

Conti, A.L. and Cravero-Igarza, S. (2010) Patrimonio, comunidad local y turismo. *Notas En Turismo y Economía*.

Cortés-Vázquez, J., Jiménez-Esquinas, G. and Sánchez-Carretero, C. (2017) Heritage and participatory governance: An analysis of political strategies and social fractures in Spain. *Anthropology Today* 33(1), 15–18. DOI: 10.1111/1467-8322.12324.

Cote, M. and Nightingale, A.J. (2012) Resilience thinking meets social theory: Situating social change in socio-ecological systems (SES) research. *Progress in Human Geography* 36(4), 475–489.

Fabry, N. and Zeghni, S. (2019) Resilience, tourist destinations and governance: An analytical framework. In: *Tourismes et Adaptations*. Elya Editions, pp. 96–108.

Flores-Ruiz, D. and Barroso-González, M.O. (2012) El turismo como estrategia de desarrollo rural sostenible: Los parques andaluces. *Revista de Estudios Empresariales. Segunda Época* 1, 59–83.

Heslinga, J.H. (2022) Resilient destinations. In: *Encyclopedia of Tourism Management and Marketing*. Edward Elgar Publishing, Cheltenham, UK, pp. 706–708.

King, C. (2022) Beyond sustainability: A global study of nature-based solutions in regenerative tourism. *Travel and Tourism Research Association: Advancing Tourism Research Globally* 38, 219–227.

Majail, J. and Webber, D.A. (2006) Human dimension in conservation works in the lower kinabatangan: Sharing PFW's experience. In: *Fourth Sabah-Sarawak Environmental Convention*. p. 9.

Monterrubio-Cordero, J.C. (2009) Comunidad receptora: Elemento esencial en la gestión turística. *Gestión Turística* 11, 101–111.

OECD (2023) Focus on Green Recovery. Available at: https://www.oecd.org/coronavirus/en/themes/green-recovery (accessed 19 March 2023).

Purnomo, A. and Rahmawati, M. (2021) Community-based tourism: Capability and community participation in tourism development. In: *Community Empowerment through Research, Innovation and Open Access*. Routledge, London, pp. 139–144.

Randelli, F. and Martellozzo, F. (2019) Is rural tourism-induced built-up growth a threat for the sustainability of rural areas? The case study of Tuscany. *Land Use Policy* 86, 387–398.

Ruspini, E. (1999) Longitudinal research and the analysis of social change. *Quality & Quantity* 33(3), 1–22.

Vergara-Muñoz, J. and Martínez-Monedero, M. (2019) El despoblamiento rural y la conservación de su patrimonio. *REVISTA PH Instituto Andaluz Del Patrimonio Histórico* 98, 327–328.

18 The Realities of Peripheral Handicraft Livelihoods in the Face of Disasters in Vanuatu

Rachel Clissold[1,2], Ross Westoby[3], Karen E. McNamara[2] and Viviane Obed[4]
[1]International Centre for Environmental Management, Hanoi, Vietnam; [2]School of the Environment, The University of Queensland, St Lucia, Brisbane, Australia; [3]Griffith Institute for Tourism, Griffith University, Nathan, Brisbane, Australia; [4]Further Arts, Port Vila, Vanuatu

Summary
The South Pacific Islands region faces unique challenges and heightened impacts from disasters, which are influenced by structural characteristics of remoteness, smallness, insularity, peripherality, low elevations, and urban primacy. The combined catastrophic impacts of the COVID-19 pandemic, border closures, and multiple cyclones, have recently devastated livelihoods, especially those dependent on tourism, in Vanuatu. The voices of women and those working in the 'informal tourism sector' are rarely heard. This is an oversight considering marginalized groups need improved resilience-building prospects and well-targeted risk management and social protection initiatives. Drawing from 13 unstructured interviews with handicraft vendors in 2018 and 2021, this case study aims to investigate the lived experiences of female informal workers in a multi-hazard environment. Intersecting hazards causing significant loss included tropical cyclones, COVID-19, and the inequities and marginalizing forces of being women in the informal sector. Women's voices, knowledge, and needs are often sidelined by government, particularly through a lack of social protection policies and crisis responses. The gendered burdens thus accumulate, resulting in heavy workloads and emotional burdens. Evidence of handicraft vendors developing coping and response strategies without formal support include livelihood diversification, returning to traditional subsistence lifestyles, and collective financial risk-sharing, which have led to informal resilience-building outcomes.

The Value and Interest of the Case Study

The voices and needs of marginalized groups such as women and those working in the informal sector are often sidelined in disaster response, risk management, and social protection initiatives. This case study draws on 13 unstructured interviews with female, informal handicraft vendors in Vanuatu to share their lived experiences of the impacts and intersecting pressures from border closures, the COVID-19 pandemic, and multiple cyclones. In exploring the impacts faced and implemented coping and response strategies, this case study provides insights into how the Vanuatu Government and other organizations can better support the conditions and outcomes of targeted interventions for women (Fig. 18.1).

 Tourism Case Studies Asia-Pacific Region (eds Taufik Abdullah and Reni Polus)

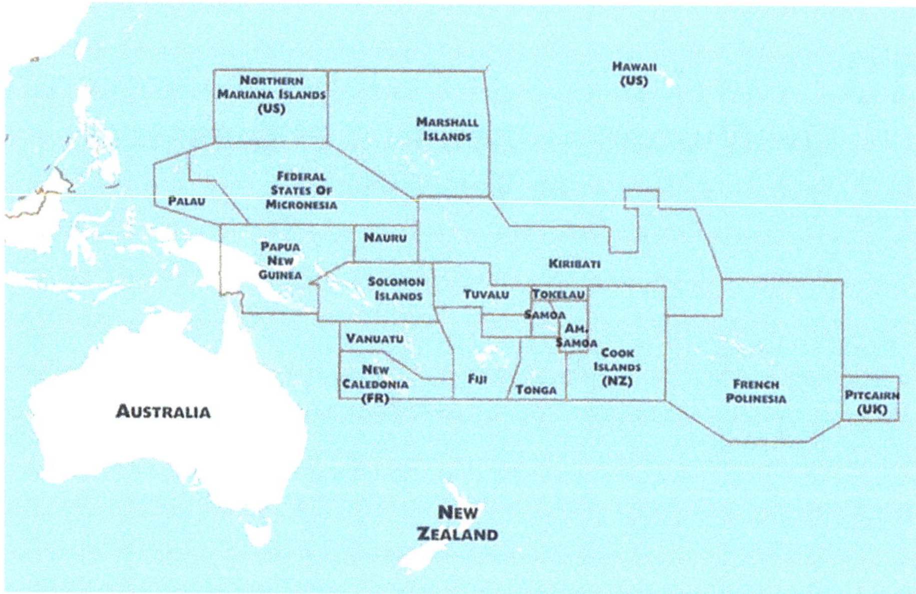

Fig. 18.1. South Pacific Islands region and Vanuatu (Craig *et al.*, 2020).

Background

Globally, women's entrepreneurship has been perceived as a driver of livelihood resilience and economic empowerment (ADB and TAF, 2018). It is increasingly recognized that working in tourism often serves as a first entry point for women to the world of work (Ferguson, 2010). Some studies emphasize that tourism benefits women beyond employment by supporting well-being, self-affirmation, and challenging gender barriers (Movono and Dahles, 2017; Kabil *et al.*, 2022), while other studies have demonstrated that tourism opportunities in the Global South tend to be low-skilled, casual work that can, in fact, conform to entrenched gender norms (Gentry, 2007; Movono and Dahles, 2017).

Vanuatu is an archipelago in the South Pacific that consists of 83 islands, covering 680,000 km^2 of the maritime zone and 12,336 km^2 of land (Republic of Vanuatu *et al.*, 2007). The population of Vanuatu sits at just over 300,000, with most people and infrastructure primarily located in low-lying coastal areas, making the nation vulnerable to cyclones and associated storm surges, sea level rise, inundation, and coastal erosion (Republic of Vanuatu *et al.*, 2007; GEF *et al.*, 2009). Before COVID-19, tourism in Vanuatu accounted for 34.7% of the country's national GDP and employed 36% of the population directly or indirectly (World Travel and Tourism Council, 2020).

Vanuatu ranks 108th out of 146 countries in terms of the Global Gender Gap Index, performing particularly poorly in terms of political empowerment, with very few women in parliament and ministerial positions. In terms of economic participation and opportunity, educational attainment, health, and survival, Vanuatu performed closer to global averages. The share of women workers in the informal sector working without formal work arrangements, social protection, and safety nets remains high at 72.4% (World Economic Forum, 2023). Marketplaces remain critical centres of economic and social exchange and have become a focus of the development sector in the Pacific, including as essential sites for women's empowerment (Underhill-Sem *et al.*, 2014; Busse and Sharp, 2019).

A range of interacting hazards and risks shape life in Vanuatu. Natural hazards are a core risk, with Vanuatu being the most at-risk in the World Risk Report (Aleksandrova *et al.*,

2021). A total of 76% of Vanuatu's population and 81% of its land is vulnerable to cyclones, earthquakes, volcanic eruptions, tsunamis, and landslides ((NACCC, 2007). Tropical Cyclone Harold (April 2020), for example, caused extensive damage to housing and infrastructure, affecting more than 158,000 people, approximately 58% of Vanuatu's population (Williams, 2020). The onset of the COVID-19 pandemic and associated border closures further impacted Vanuatu's highly tourist-dependent economy (GSTC, 2021). While borders are now open again, tourism numbers will take time to reach pre-pandemic levels, with airlines struggling to meet demands and two cyclones hitting many Vanuatu islands at the beginning of 2023.

In this case study, we examine the lived experiences, including responses, of women handicraft vendors in Vanuatu who have, in recent years, endured various disasters (hazards impacted by human vulnerability). We aim to understand how disasters affect 'invisible' members of the informal sector (i.e. the section of any economy that is neither taxed nor monitored by the government, despite being a viable form of employment, especially in developing and transition economies) whose livelihoods depend on the fringes of tourism. We draw on interviews with ni-Vanuatu women (i.e. women from Vanuatu) handicraft market vendors and document their experiences, including how they draw on endogenous resources (i.e. those locally existing strategies, knowledge, networks, institutions, and resources) and creative capacities to respond to risks.

Women's handicraft entrepreneurship is a unique niche market where women are central players. The handicraft industry across the Pacific region is synonymous with indigenous cultures, local materials, and skills passed intergenerationally among families. Handicrafts such as woven baskets, earrings, mats, bags, and clothing (Fig. 18.2) are often sold for domestic use to other ni-Vanuatu families but are also increasingly being adapted and sold to tourists.

The findings from this case study provide valuable insights for future policies to better acknowledge and address the realities, specific needs, and vulnerabilities of women handicraft vendors living in an environment with multiple interacting hazards and pressures.

Two sets of data were collected for this case study: (1) one group interview (with three participants) conducted in December 2018 (pre-pandemic) and (2) a series of unstructured/conversational interviews (n=12 with 13 participants) undertaken by one of the authors (a local ni-Vanuatu researcher) between January and May 2021. All participants were women with livelihoods tied in diverse ways (e.g. as vendors, intermediaries, or executive members of handicraft market associations) to the handicraft markets in Port Vila, Vanuatu's capital. The flow and structure of the interviews were flexible, thereby appropriately reflecting *Storian* – a Vanuatu-specific form of the Indigenous research method *Talanoa* – which means to 'chat, yarn, swap stories' (Crowley, 1995, p. 235 in Warrick, 2009). Latent content analysis was applied using NVivo to identify the key 'themes' and implied meanings that emerged from these storylines and participants' experiences. These methods allowed us to concentrate on deriving and sharing detailed and nuanced storylines and experiences of a few individuals and are not necessarily representative of all women's

(a) (b) (c)

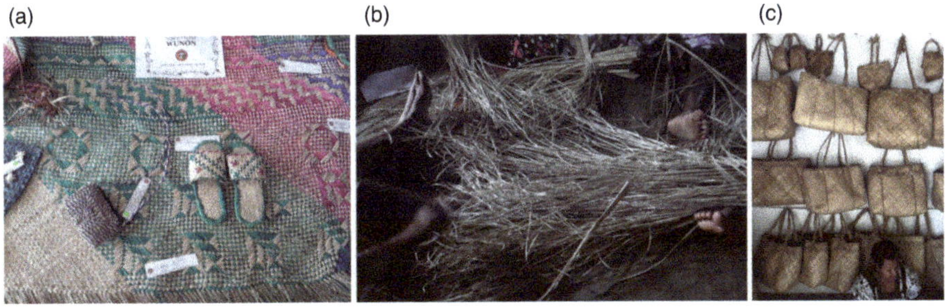

Fig. 18.2. Handicraft production and products sold at markets in Vanuatu.

experiences or perspectives in Vanuatu, especially with the small sample size. The results from this study do, however, have internal validity as it accurately portrays the multiple social realities of our participants, which future research can continue to build on and further refine.

Women in the Face of Risks, Pressure, and Hazards

In disaster contexts, the dominant narrative of women in the Global South has been that of disproportionate vulnerability or responsible caregiving – that is, the 'chief victim-and-caretaker' role (Resurrección, 2011). Women are depicted as more likely to be hurt or killed (Neumayer and Plümper, 2007); face violence (Fulu, 2007; Whittenbury, 2013); have compromised privacy, health, and safety (Fulu, 2007; Bradshaw and Fordham, 2013); and have low coping capacities due to socially embedded inequities that restrict mobility and access to relief (Scanlon, 1998; Sultana, 2011; Ayeb-Karlsson *et al.*, 2019). Existing studies have also documented some of the insecurities and issues that are inherent to marketplaces in the Pacific, and this includes women's security and safety, exploitation (e.g. unofficial fees), infrastructural and organizational obstacles and cultural beliefs about gender, among others (Dewey, 2011; Underhill-Sem *et al.*, 2014).

Narratives around women's agency, situated knowledge, and capabilities in disaster contexts are also critical to highlight (Arora-Jonsson, 2011). Often, where formal or governmental response measures are insufficient in the face of disasters, women have become critical agents of change, mobilizing their resources, adapting or acquiring the needed skills to ensure their families' survival and wellbeing (de Branco, 2009; Yonder *et al.*, 2009; Kusumasari, 2015). Women's social capital and networks, for example, have been identified as critical for the informal humanitarian distribution of critical resources (Drolet *et al.*, 2015), supporting income generation (de Branco, 2009), and providing a platform from which women can politically mobilize and become involved in local activism (de Branco, 2009; Yonder *et al.*, 2009; Enarson, 2012). At an individual level, women in high-pressure contexts have been documented as having critical capacities in using traditional knowledge to maintain their household's food and water security (Charan *et al.*, 2016), diversifying economic resources, and maintaining post-disaster income flows (Soares and Mullings, 2009; Islam *et al.*, 2014; Kusumasari, 2015).

Although women may display critical roles and capacities in disaster contexts, it is important to acknowledge another form of pressure that may be shaping these highly effective responses – that is, the gendered power configurations that impose various obligations and burdens on women to meet their expected roles as caregivers (Ariyabandu, 2003; Enarson *et al.*, 2007). As the central axis of family care, disasters can magnify women's responsibilities, burdening them with the need to secure assistance and resources in post-disaster settings (Wiest *et al.*, 1994; Khondker, 1996; Drolet *et al.*, 2015). The pressure and drive to meet these roles and responsibilities can be fed by shame, resulting in sacrificing or trade-offs in, their freedom and wellbeing (Béné *et al.*, 2014; Kusumasari, 2015). Studies that explore the capacities and response strategies of women must be aware of the kinds of power structures and expectations that may be constantly at play and the kinds of burdens this can produce for women (Arora-Jonsson, 2011; Resurrección, 2011; Tschakert and Machado, 2012).

Entrepreneurship during crises and disasters is an emerging research domain. Since COVID-19, literature on female entrepreneurialism in a time of crisis has burgeoned, with an improved understanding of the negative impacts on women-run business enterprises, learning experiences and coping strategies, and barriers to entrepreneurialism during crises (Grandy *et al.*, 2020; Hennekam and Shymko, 2020; Afshan *et al.*, 2021). A key finding is that crises such as the pandemic can occasion positive and negative effects, with devastating impacts on livelihoods but also the potential for motivating innovation and creativity, especially in the face of prolonged disasters (Linnenluecke and McKnight, 2017; Portuguez Castro and Gomez Zermeño, 2021). Few studies, however, examine the relationship between female entrepreneurs and crises in the context of tourism (Portuguez Castro and Gomez Zermeño, 2021;

Filimonau *et al.*, 2022). Filimonau *et al.* (2022) provided important lessons from Uzbekistan and Kazakhstan and called for more nuanced research into the cultural and societal drivers and barriers of female entrepreneurship in tourism, especially in disadvantaged communities. In terms of coping strategies, they emphasized the importance of social capital. Women entrepreneurs responded to crises through approaches that created mutual benefit, focusing on cooperation over competition, and the active learning undertaken by women tourism entrepreneurs (Filimonau *et al.*, 2022).

Impacts of a Multi-Hazard Environment on Women Handicraft Vendors in Vanuatu

The critical hazards that intersect and accumulate to affect ni-Vanuatu women handicraft vendors' livelihoods and wellbeing included the COVID-19 pandemic and associated border closures, multiple cyclones within a few years, and their status as women and informal workers (Fig. 18.3). The immediate and direct impacts of cyclones and COVID-19 were devastating losses to income and sustainability of business due to reduced sales (i.e. loss of customers, damaged marketplace infrastructure, damage to primary resources for handicrafts) and the reduced price of products. The handicraft women's capacities to adapt and recover were affected by their marginalization at the informal periphery of the tourism sector with little recognition and support from government, and other gendered pressures related to expectations from family. These intersecting impacts and pressures are exacerbated by the heavy workloads and emotional burdens that the women are left with to recover their livelihoods and businesses for themselves and their families.

The impacts of COVID-19, associated border closures, and the damage caused by cyclones were devastating to women's livelihoods and their businesses, primarily through significantly reducing sales but also the price of products. For example, as 'Vanuatu [is] highly dependent on tourism... [and] most of our products in Vanuatu are targeted to tourism', COVID-19 and border

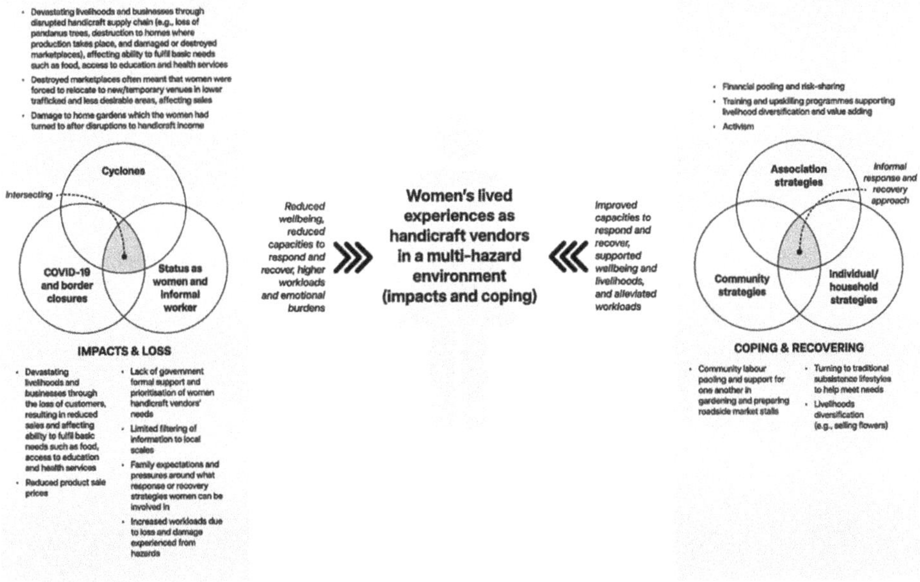

Fig. 18.3. Women's lived experiences as handicraft vendors in a multi-hazard environment.

closures had a direct impact on reduced sales due to the reduced number of customers, which the women considered as disproportionately affecting them: 'we are so affected compared to other businesses, we the mothers in the handicrafts are affected so much' (participant #6, 2021). For many women, this meant that they 'have not sold a thing', 'cannot maintain my sewing booth', and have 'not [been] able to start another business' (participants #6, #2, and #4, 2021).

Several women also discussed how, during COVID-19, revenue had reduced due to the decreasing sale price of their products: 'We understand that most people do not have jobs, so we must lower our prices to an amount where everyone can afford it. However, then again, this affects us because now we must receive less money than we used to'; 'Now we are lowering the price of things to 250vt [AU$3] but in the past, we were selling earrings at 1500vt or 2000vt [AU$18–24]' (participants #3 and #6, 2021). The women that worked as 'intermediaries' also explained how they had to 'make it clear to my producers to lower their prices because we cannot make as much as we used to... we all do not have enough money now... We are losing plenty of profits because we are selling on the prices we paid, so no profits are gained' (participant #7, 2021).

Tropical cyclones, one of which occurred during the COVID-19 pandemic in 2020 (Cyclone Harold), also reduced revenue for women but with more immediate and destructive impacts on the handicraft supply chain. Cyclones, for example, have interrupted handicraft production processes through the destruction of pandanus trees, which are the primary resource for many handicrafts: 'Most of the pandanus trees are blown down by the wind, and so it was tough to find leaves for quite a while, so production of mats and baskets was prolonged and even stop[ped]' (participant #9, 2021). Others who lost their homes also had interrupted production as they could no longer sew or make handicrafts until their homes were rebuilt: 'Our house got soaked with rainwater, and a coconut tree fell on our other house. So, I had to stop [sewing] for a while. After all the cleaning and putting the place back together, I started sewing again' (participant #5, 2021). The difficulties of lived experiences in a multi-hazard environment were also apparent when women explained that, due to the COVID-19 pandemic, they had become more dependent on their home gardens to supplement incomes and help meet the needs of the family (e.g. by growing crops or flowers to sell), only for cyclones to 'damage the trees and crops and spices' (participant #11, 2021). As a result, the 'community suffers long-term... they have to plant again, rehabilitate' (participant #11, 2021).

Another impact of cyclones on handicraft supply chains involves the destruction of and damage to sale points – that is, the marketplace infrastructure. This often means that handicraft vendors have had to move to new/temporary marketplace venues in lower trafficked and less desirable areas. After cyclone Pam in 2015, one participant explained: 'It [relocating] was a challenge for us because the convention centre [temporary venue] is back way and so the revenue was not good. Most of our mamas faced many difficulties' (participant #10, 2021).

Reduced revenue through decreased sales and reduced prices were emotionally distressing for the women, who explained how the pre-COVID handicraft business was thriving to the extent that many women 'bought land, built houses, and paid for their children's education through the money made by handicrafts' (participant #10, 2021). Several women explained that they had shifted from making 'enough money for about a month' or to 'keep me until the next cruise ship's arrival' to 'struggling now to make ends meet' (participants #2 and #1, 2021). Many highlighted how they could no longer fulfil basic needs such as food, school fees and bus fares, access to medical facilities, and contributions to community activities such as weddings and ceremonies (participant #1, #4, and #9, 2021). One participant summarized these cascading impacts and indicated that 'when it affects the financial part, it will contribute to other sectors' (participant #9, 2021). The stark contrast of financial security and general wellbeing pre-pandemic and during the pandemic was clear.

Beyond the intersecting impacts of COVID-19 and tropical cyclones, the handicraft women highlighted how their experiences are influenced by and exist within an inequitable and gendered system that creates specific, intersecting gendered barriers, burdens, expectations, and pressures that hinder their capacities to recover and adapt. One of the critical barriers several women experienced was the lack of recognition of their voice and needs: 'Vanuatu is indeed a male-dominated country. Therefore, it is a bit hard; they will not take you seriously' (participant #11, 2021). Working in the informal sector as a woman means 'voice' and 'representation' are sidelined from the various tiers of government: 'When the government wants to respond to our needs, people think it is not important, and they forget about us' (participant #2, 2021). This was evidenced in the lack of formal government support, with most women receiving nothing or a minimal amount they could not rely on. They explained that they 'did not receive any help from our leaders... None of us got any help from the village, government, or any association' and that 'some of us are asking the MP's and councillors to help us but they have not' (participants #4 and #6, 2021). For the few that may have business licenses, there was a 'stimulus package' from the national government that provided financial support, but overall, this package was perceived to be exclusive and had poor distribution among handicraft vendors, mainly due to poor leadership and limited filtering of information to local scales:

> They must take the responsibility to go out and provide information about us so we can have a piece in that too. But they failed to do so, and when we later heard, we went to get our money but found out that the government had stopped giving this money. We knew that our leaders also wanted this money, and we were upset because the money belonged to the small business holders. We followed up with each other, checking if any of us had received this money, but we still needed to do so. We waited until we knew it would never come, so we stopped asking. So, we knew that we were provided funds, but I need to understand how we were not entitled to any. (participant #2, 2021)

Another localized example of the lack of prioritization and recognition of informal sector female workers was in the aftermath of the 2015 tropical cyclone Pam, when the handicraft market infrastructure was destroyed. The women explained that the government did not prioritize the reconstruction of this infrastructure on which they depended and, even after it was finished, they were unable to move back as the marketplace was advertised and 'many other people came in instead' (participant #10, 2021). In this way, the government 'forget all about us. So shameful... it is so sad to see that, how they treated us' (group interview participant, 2018).

These gendered burdens can also filter to the household level, where some women were unable to participate in collective response or recovery strategies (e.g. as part of a women's group) 'because their partners are against the idea' (participant #12, 2021). This inability to participate further complicates individual women's capacities to adapt and recover.

Another noticeable gendered burden that was worsened as a result of income loss from COVID-19 and tropical cyclones, and further complicated the women's capacities to adapt and recover their tourism businesses was high workloads due to gendered responsibilities: 'Their responsibilities as wives, mothers, and community leaders have become more stressful now that most are out of work without pay' (participant #1, 2021). Some women's husbands also work overseas, heightening women's responsibilities to meet household needs. On top of being the primary income earners, the women's diverse responsibilities included, but were not limited to, cleaning, cooking, planting, maintaining, and harvesting crops, taking care of their children and extended family members, and attending to leadership roles in marketplace associations. One participant shared how intersections between COVID-19 restrictions and gendered roles and responsibilities affected capacities to recover their businesses: 'When my husband sees that I am at home, they expect a lot from us in the house, so I don't have enough time to complete all my work. But when I work in a booth, I see it is better because I focus more on my sewing. But here at my house, it's not balanced' (participant #2, 2021). Increased workloads have also emerged for women in association leadership positions as they try to find creative avenues to maintain the running of their associations with fewer resources as a result of COVID-19:

> before, we managed our time well because we receive an acceptable amount of money, and our finances are going well. However, now we must work harder to find ways to help these

> women. Sometimes we have sleepless nights given to work and plan to find a solution to how we can help all these 60 women under our control. It is not easy. And behind each woman, we must consider the families that depend on them...now with this situation [the pandemic], it gives us more pressure in our work because we need ideas as to how we can get money ... We work overtime, stay up late, make more plans. (participant #10, 2021)

High workloads have long been a challenge, with one group interview participant in 2018 sharing how heavy workloads are a core barrier to effective production and business growth: '... the women, they are doing lots of little jobs, but that is why it is one of the challenges where the flow of the [handicraft] products is coming, it is very slow...' (group interview participant, 2018). In this way, it is also a barrier to effective financial saving and resilience in the face of future risks.

Coping and Response Strategies: Collectivism, Livelihood Diversification, and Subsistence Livelihoods

Despite the lack of formal government support and effective coping and response strategies discussed at the collective scale, the handicraft women vendors highlighted five key avenues through which they used endogenous resources and were independently recovering their livelihoods and businesses: financial pooling and risk-sharing, training programmes which supported livelihood diversification and value adding, activism, community labour, and turning to traditional subsistence lifestyles. These were being implemented across the individual/household and community scales, with the support of marketplace or women's associations (Fig. 18.3).

The formal and informal networks of marketplace and women's associations were integral for financial pooling, risk-sharing, and the provision of training programmes. The president and secretary of a local handicraft association explained how the 66 female vendors in their association were intermittently running food markets as fundraisers, with funds then being used to support the women's needs and provide training (e.g. for livelihood diversification but also bookkeeping, leadership, and management training). Similarly, the chairwoman of a local women's group shared how they provided livelihood diversification and upskilling training (e.g. in dress painting, baking, and home-gardening, as chosen by local handicraft vendors themselves) and had their financial-risk sharing savings programme, whereby money saved by each member was shared and divided between the women.

Another important response and recovery strategy is that of activism, which was evident within some marketplace associations. For example, women involved in a Vanuatu arts and crafts association lobbied against undesirable government strategies (i.e. relocating women to a new area with few prospective customers). They successfully negotiated with local agencies for approval to share space at other existing marketplaces usually reserved for produce. Although this solution still had issues, the women could better secure post-cyclone income flows for their households: 'That would be a social problem if we were not served' (group interview participant, 2018).

An added benefit is that the associations themselves are more effectively cooperating and working together in the face of the COVID-19 pandemic: 'Before, associations hardly cooperate[d], but when COVID struck, we all began to cooperate well. The situation opened more ways for us to work together... So, our challenges are different because we have a stronger relationship these days' (participant #10, 2021). Others expressed the importance of informal responses at the community scale, beyond handicraft or women's associations, which allowed for inclusive coping and recovery processes, especially through the sharing of labour: 'To cope with the situation, we continue to work together as a community or church so we can achieve our needs together' (participant #9, 2021). The sharing of community labour and support revolved around gardening and preparing roadside marketplaces so vendors could continue to sell their food and products.

At an individual or household scale, livelihood diversification emerged as the critical response strategy for women to cope and respond to reduced tourist numbers and losses to income. Several of the women, for example, turned to growing and selling flowers at local marketplaces: 'A lot of the women here are also coming into selling flowers because of the drop in number of tourists coming into the country leading to less purchase[s]' (participant #1, 2021). Alongside selling flowers to locals, many were selling local food: 'Along with the other women in the community, we sell finger food for 20vt [AU$0.25], and I see that this helped a lot too' (participant #4, 2021). For the women interviewed in 2018, the strategy for securing post-cyclone income flows involved taking advantage of the arrival of international aid workers by selling locally made handicrafts to them: 'As they arrived at the airport, instead of seeing sad faces, they see smiling faces... so they [volunteers] would like to bring something back home as a souvenir' (group interview participant, 2018).

In response to these losses to income, with cascading impacts on food security, many women and their households were turning to their traditional subsistence lifestyles to survive: 'To cope with the pandemic, we only continue to plant our food in our garden so that we can save money for other essentials we may need' (participant #1, 2021). These women who were able to depend on their gardens considered themselves 'lucky', as 'when we are in this situation [the pandemic], we can always depend on our crops', and the crops can be used to '...sell local food... [and] get us money' as well as 'give us healthy food for ourselves and our households' (participants #3 and #7, 2021). The tensions of living in a multi-hazard environment are, however, evident here, as turning to subsistence lifestyles and crops may help in the face of the pandemic but may simultaneously create additional vulnerability in the face of cyclones: 'the impacts of 'natural' disasters [cyclones] are severe in communities, especially the agricultural sector' (participant #11, 2021).

Implications for Policy, Practice, and Research

The overlap between gender, tourism development, and disaster are little understood (Filimonau *et al.*, 2022). Stories of gender empowerment, beyond the neoliberal conceptualizations of economic improvement that formal or informal tourism livelihoods generate, miss the many structural inequalities that prevent true gender equality (Cole, 2018).

This case study has provided place-based insights into the kinds of lived experiences, responses, capacities, and roles of ni-Vanuatu handicraft vendors that must be better acknowledged, addressed, built on and utilized. First, it was clear that women handicraft vendors in Port Vila face various hazards and pressures (i.e. the COVID-19 pandemic, cyclones, gendered barriers and household responsibilities, and status as primary earners and informal workers) that interact and accumulate to severely impact livelihoods, and exacerbate heavy workloads and emotional burdens (i.e. stress). The pressure on handicraft vendor women and the tendency to have their needs and voices sidelined means that government support and social protection policies are desperately missing (due to poor leadership and communication channels, lack of recognition for the informal sector, and gendered inequities).

Heavy workloads and emotional burdens illustrate how these women bear the costs of adjusting to multiple shocks and pressures, acquiring the responsibility of finding additional income while maintaining traditional gendered responsibilities in the home and community (Feeny *et al.*, 2013). Heavy workloads can be a significant barrier to the capacity to grow small and micro enterprises (i.e. by lacking time and production remaining slow), to cope and respond to shocks effectively and recover businesses, but also for women's empowerment, agency, and wellbeing. The fact that women are overwhelmingly represented in the micro and informal tourism space means that these kinds of restrictions to women's empowerment are often overlooked or obscured from view (Boonabaana, 2014). With this in mind, future social protection policies and resilience-building activities need to be aware of and implement efforts to reduce women's heavy workloads and burdens in a multi-hazard environment. Integrating proactive and culturally appropriate efforts that redress power structures and broader systems of inequality and oppression that continue to burden women in these ways is essential (Resurrección, 2011). Challenging traditional gendered patterns of labour and responsibility

(Ganapati, 2012; Jost *et al.*, 2016), establishing necessary support services in the tourism sector (e.g. childcare or more effective transport) or subsidies (e.g. cost of being involved in the marketplace) and ensuring better recognition for informal workers in the tourism sector must be considered.

Women are not passive victims. Evidence of vendors strategies to cope and respond to hazards at the individual/household and collective scales show their active engagement. The women handicraft vendors at the individual and household level demonstrate critical capacities to diversify livelihoods (e.g. selling flowers and local food) and utilize existing resources to cope (e.g. turning to traditional subsistence lifestyles and gardens to save money on food). Such strategies help address financial insecurities and burdens imposed by the lack of tourists (Bolwig *et al.*, 2010). It is important to ascertain a holistic understanding of vulnerability and support these kinds of household-level responses and strategies that build resilience in the face of multiple hazards and pressures.

It is clear that women handicraft vendors often counterbalanced the lack of access to formal support and recognition from formal institutions through a high level of engagement in their social networks and informal institutions. Existing local institutions enabled financial risk-sharing and opportunities for fundraising, while providing critical training activities to build overall resilience. These collective strategies align closely with other studies which have found that social networks and collaboration beyond the immediate family are integral to the resilience of women entrepreneurs and the recovery of their small tourism business activities (Filimonau *et al.*, 2022). Not only were women cooperating for mutual benefit in recovering businesses and fulfilling basic needs, but women entrepreneur networks were also using disasters as an opportunity to learn and rethink business directions, and prepare themselves for future disastrous events by investing in livelihood diversification and value-adding training programmes (Filimonau *et al.*, 2022). Local institutions may, therefore, be important entry points for future resilience-building activities, social protection policy, and distributing information and social support at the local scale for the tourism industry (see also Chishakwe *et al.*, 2012; Reid, 2016).

Support and resources for existing groups and their collective action efforts could be an intervention that drives women to cope and meet immediate needs (e.g. grassroots financial risk-sharing and fundraising) while also supporting them in more transformational ways (e.g. through training and upskilling programmes and providing a platform from which to lobby needs and rights) (Fisher, 2009; Yonder *et al.*, 2009). Supporting networks and constructive relationships between national, regional, and local domains, government, and civil society is a great opportunity going forward (Maskrey, 2011; Chishakwe *et al.*, 2012). Future studies should build on and further unpack the engagement of handicraft women vendors with the various local institutions they value and identify the key barriers, challenges, strengths, and opportunities for social support.

Conclusions and Ways Forward

Women handicraft vendors working in the informal sector in Vanuatu have been sidelined in formal response and recovery strategies, and yet they are not passive victims of those processes. Diverse capacities to respond and cope at the individual (e.g. livelihood diversification, turning to subsistence lifestyles) and collective scale (e.g. financial risk-sharing, fundraising) to multiple disasters is evident. Local institutions play a critical role in supporting recovery and building resilience, meaning these are legitimate entry points for future government efforts and potential social policies.

Although our study provides insights for social policy that may support women, we recognize that the experiences and perspectives presented here are unlikely to be the same for all women. We focus on women's voices within marginalized informal women-led tourism enterprises because they are often excluded and ignored in decision making across scales. Future studies should further refine, unpack, and build on our findings by engaging men and other genders for a more holistic understanding of the experiences of peripheral market vendors in Vanuatu and elsewhere.

Acknowledgements

We are grateful to all participants in this study for sharing their stories; thank you for your time and insights. We also wish to acknowledge the generosity of those who attended a fundraiser event (organized by the first three authors) in Brisbane to raise money (AU$3,750) and awareness to support improvements to women's vendor infrastructure at the handicraft market in Port Vila, as well as training opportunities, which was delivered by Further Arts (represented by the fourth author). The awareness generated and sharing of handicraft images at the dinner event was also an opportunity for genuine empathy as Australian patrons heard stories of some of these Pacific women.

References

ADB and TAF (Asian Development Bank and The Asia Foundation) (2018) Emerging lessons on women's entrepreneurship in Asia and the Pacific: Case studies from the Asian development bank and the Asia foundation. Available at: http://dx.doi.org/10.22617/TCS189585-2 (accessed 23 2018).November

Afshan, G., Shadid, S. and Tunio, M.N. (2021) Learning experiences of women entrepreneurs amidst COVID-19. *International Journal of Gender and Entrepreneurship* 13(2), 162–186.

Aleksandrova, M., Balasko, S., Kaltenborn, M., Malerba, D., Mucke, P. *et al.* (2021) World Risk Report 2021. Bundnis Entwicklung Hilft and Ruhr University Bochum – Institute for International Law of Peace and Armed Conflict. Available at: https://reliefweb.int/sites/reliefweb.int/files/resources/WorldRiskReport-2018.pdf (accessed 6 December 2023).

Ariyabandu, M.M. (2003) *Women: The Risk Managers in Natural Disasters*. Social Science Research Institute.

Arora-Jonsson, S. (2011) Virtue and vulnerability: Discourses on women, gender and climate change. *Global Environmental Change* 21(2), 744–751.

Ayeb-Karlsson, S., Kniveton, D., Cannon, T., Van Der Geest, K., Ahmed, I. *et al.* (2019) I will not go, I cannot go: Cultural and social limitations of disaster preparedness in Asia, Africa, and oOceania. *Disasters* 43(4), 752–770.

Béné, C., Newsham, A., Davies, M., Ulrichs, M. and Godfrey-Wood, R. (2014) Review article: Resilience, poverty and development. *Journal of International Development* 26(5), 598–623.

Bolwig, S., Ponte, S., Du Toit, A., Riisgaard, L. and Halberg, N. (2010) Integrating poverty and environmental concerns into value-chain analysis: A conceptual framework. *Development Policy Review* 28(2), 173–194.

Boonabaana, B. (2014) Negotiating gender and tourism work: Women's lived experiences in Uganda. *Tourism and Hospitality Research* 14(1–2), 27–36.

Bradshaw, S. and Fordham, M. (2013) *Hazards, Risks and, Disasters in Society*. Elsevier Inc, London.

Busse, M. and Sharp, T.L.M. (2019) Marketplaces and morality in Papua New Guinea: Place, personhood and exchange. *Oceania* 89(2), 126–153.

Charan, D., Kaur, M. and Singh, P. (2016) Indigenous Fijian Women's role in disaster risk management and climate change adaptation. *Pacific Asia Inquiry* 7(1), 17.

Chishakwe, N., Murray, L., Chambers, M., and International Institute for Environment and Development (2012) Building Climate Change Adaptation on Community Experiences: Lessons from Community-Based Natural Resource Management in Southern Africa. Available at: http://pubs.iied.org/pdfs/10030IIED.pdf (accessed 6 December 2023).

Cole, S. (2018) *Gender Equality and Tourism Beyond Empowerment*. CAB International, Wallingford, UK.

Craig, A.T., Heywood, A.E. and Hall, J. (2020) Risk of COVID-19 importation to the Pacific islands through global air travel. *Epidemiology and Infection* 148, 1–16.

de Branco, A. de M. (2009) Women responding to drought in Brazil. In: Enarson, E. and Chakrabarti, P.G.D. (eds) *Women, Gender and Disaster: Global Issues and Initiatives*. SAGE Publications India Pvt Ltd, New Delhi, pp. 261–272.

Dewey, S. (2011) Markets and women's market trading in the Pacific Islands. *Asian Women* 27(3), 1–23.

Drolet, J., Dominelli, L., Alston, M., Ersing, R., Mathbor, G. *et al.* (2015) Women rebuilding lives post-disaster: Innovative community practices for building resilience and promoting sustainable development. *Gender & Development* 23(3), 433–448.

Enarson, E., Fothergill, A. and Peek, L. (2007) Gender and disaster: Foundations and directions. In: Rodríguez, H., Quarantelli, E.L. and Dynes, R.R. (eds) *Handbook of Disaster Research*. Springer, New York, pp. 130–146.

Enarson, E.P. (2012) *Women Confronting Natural Disaster: From Vulnerability to Resilience*. Lynne Rienner Publishers, Boulder, CO.

Feeny, S., McDonald, L., Miller-Dawkins, M., Donahue, J. and Possi, A. (2013) *Household Vulnerability and Resilience to Shocks: Findings from Solomon Islands and Vanuatu*. Australian National University, Canberra. Available at: https://openresearch-repository.anu.edu.au/handle/1885/10054

Ferguson, L. (2010) *Analysing the Gender Dimensions of Tourism as a Development Strategy*. Instituto Complutense de Estudios Internacionales, Madrid.

Filimonau, V., Matyakubov, U., Matniyozov, M., Shaken, A. and Miroslaw, M. (2022) Women entrepreneurs in tourism in a time of a life event crisis. *Journal of Sustainable Tourism* 2022, 1–23.

Fisher, S. (2009) Sri lankan women's organisations responding to post-tsunami violence. In: Enarson, E. and Chakrabarti, P.G.D. (eds) *Women, Gender and Disaster: Global Issues and Initiatives*. SAGE Publications India Pvt Ltd, New Delhi, pp. 233–249.

Fulu, E. (2007) Gender, vulnerability, and the experts: Responding to the Maldives tsunami. *Development and Change* 38(5), 843–864.

Ganapati, N.E. (2012) In good company: Why social capital matters for women during disaster recovery. *Public Administration Review* 72(3), 419–427.

GEF, UNDP, and SPREP (2009) *Pacific Adaptation to Climate Change—Vanuatu: Report of in-Country Consultations*. GEF, UNDP and SPREP, Samoa.

Gentry, K. (2007) Belizean women and tourism work: Opportunity or impediment? *Annals of Tourism Research* 34(2), 477–496.

Global Sustainable Tourism Council (GSTC) (2021) Vanuatu Tourism Gets a Reboot: Destination Stewardship Report. Available at: https://www.gstcouncil.org/vanuatu-tourism-gets-a-reboot/ (accessed 7 December 2023).

Grandy, G., Cukier, W. and Gagnon, S. (2020) (In)visibility in the margins: COVID-19, women entrepreneurs and the need for inclusive recovery. *Gender in Management: An International Journal* 35(7/8), 667–675.

Hennekam, S. and Shymko, Y. (2020) Coping with the COVID-19 crisis: Force majeure and gender performativity. *Gender, Work and Organization* 27(5), 788–803.

Islam, M., Sallu, S., Hubacek, K. and Paavola, J. (2014) Limits and barriers to adaptation to climate variability and change in Bangladeshi coastal fishing communities. *Marine Policy* 43, 208–216.

Jost, C., Kyazze, F., Naab, J., Neelormi, S., Kinyangi, J. *et al.* (2016) Understanding gender dimensions of agriculture and climate change in smallholder farming communities. *Climate and Development* 8(2), 1–12.

Kabil, M., Abouelhassan Ali, M., Marzouk, A. and Dávid, L.D. (2022) Gender perspectives in tourism studies: A comparative bibliometric analysis in the MENA region. *Tourism Planning and Development* 2022, 1–13.

Khondker, H.H. (1996) Women and floods in Bangladesh. *International Journal of Mass Emergencies and Disasters* 14(3), 281–292.

Kusumasari, B. (2015) Women's adaptive capacity in post-disaster recovery in Indonesia. *Asian Social Science* 11(12), 281–289.

Linnenluecke, M.K. and McKnight, B. (2017) Community resilience to natural disasters: The role of disaster entrepreneurship. *Journal of Enterprising Communities: People and Places in the Global Economy* 11(1), 166–185.

Maskrey, A. (2011) Revisiting community-based disaster risk management. *Environmental Hazards* 10(1), 42–52.

Movono, A. and Dahles, H. (2017) Female empowerment and tourism: A focus on businesses in a Fijian village. *Asia Pacific Journal of Tourism Research* 22(6), 681–692. DOI: 10.1080/10941665.2017.1308397.

National Advisory Committee on Climate Change (NACCC) (2007) National Adaptation Programme for Action (NAPA). Available at: unfccc.int/resource/docs/napa/vut01.pdf (accessed 6 December 2023).

Neumayer, E. and Plümper, T. (2007) The gendered nature of natural disasters: The impact of catastrophic events on the gender gap in life expectancy, 1981–2002. *Annals of the Association of American Geographers* 97(3), 551–566.

Portuguez Castro, M. and Gomez Zermeño, M.G. (2021) Being an entrepreneur post-COVID-19-resilience in times of crisis: A systematic literature review. *Journal of Entrepreneurship in Emerging Economies* 13(4), 721–746.

Reid, H. (2016) Ecosystem- and community-based adaptation: Learning from community-based natural resource management. *Climate and Development* 8(1), 4–9.

Republic of Vanuatu, GEF, UNDP, UNFCCC, and NACC (2007) *National Adaptation Programme for Action (NAPA).*

Resurrección, B. (2011) The Gender and Climate Change Debate: *More of the Same or New Pathways of Thinking and doing?* (Asia security initiative policy series working paper no.10). RSIS Centre for Non-Traditional Security Studies.

Scanlon, J. (1998) The perspective of gender: A missing element in disaster response. In: Enarson, E. and Morrow, B.H. (eds) *The Gendered Terrain of Disaster: Through Women's Eyes*. Greenwood, New York, pp. 45–54.

Soares, J. and Mullings, A.Y. (2009) 'A We Run Tings': Women rebuilding Montserrat. In: Enarson, E. and Chakrabarti, P.G.D. (eds) *Women, Gender and Disaster: Global Issues and Initiatives*. SAGE Publications India Pvt Ltd, New Delhi, pp. 250–260.

Sultana, F. (2011) Suffering for water, suffering from water: Emotional geographies of resource access, control and conflict. *Geoforum* 42(2), 163–172. DOI: 10.1016/j.geoforum.2010.12.002.

Tschakert, P. and Machado, M. (2012) Gender justice and rights in climate change adaptation: Opportunities and pitfalls. *Ethics and Social Welfare* 6(3), 275–289. DOI: 10.1080/17496535.2012.704929.

Underhill-Sem, Y., Cox, E., Lacey, A. and Szamier, M. (2014) Changing the market culture in the Pacific: Assembling a conceptual framework from diverse knowledge and experiences. *Asia Pacific Viewpoint* 55(3), 306–318. DOI: 10.1111/apv.12063.

Warrick, O. (2009) Ethics and methods in research for community-based adaptation: Reflections from rural Vanuatu. In: Ashley, H., Kenton, N. and Milligan, A. (eds) *Participatory Learning and Action 60: Community-Based Adaptation to Climate Change*. International Institute of Environment and Development (IIED), London, pp. 76–87.

Whittenbury, K. (2013) Climate change, women's health, wellbeing and experiences of gender-based violence in Australia. In: Alston, M. and Whittenbury, K. (eds) *Research, Action and Policy: Addressing the Gendered Impacts of Climate Change*. Springer, Netherlands, pp. 207–221.

Wiest, R.E., Mocellin, J.S.P. and Motsisi, D.T. (1994) *The Needs of Women in Disasters and Emergencies*. [Report prepared for the Disaster Management Training Programme of the United Nations Development Programme and the Office of the United Nations Disaster Relief Coordinator]. Disaster Research Institute, University of Manitoba. Available at: https://www.gdnonline.org/resources/women-in-disaster-emergency.pdf (accessed 6 December 2023).

Williams, M. (2020) *Rapid Gender Analysis COVID-19 Vanuatu*. CARE International. Available at: https://www.care.org.au/wp-content/uploads/2020/04/Vanuatu-Rapid-Gender-Analysis-COVID-FINAL-17.04.20.pdf (accessed 6 December 2023).

World Economic Forum (2023) *Global Gender Gap Report: Insight Report June 2023*. World Economic Forum. Available at: https://www3.weforum.org/docs/WEF_GGGR_2023.pdf

World Travel and Tourism Council (2020) *Travel and Tourism Recovery Scenarios 2020 and Economic Impact from Covid-19*. World Travel and Tourism Council, London.

Yonder, A., Akçar, S. and Gopalan, P. (2009) Women's participation in disaster relief and recovery. In: Enarson, E. and Chakrabarti, P.G.D. (eds) *Women, Gender and Disaster: Global Issues and Initiatives*. SAGE Publications India Pvt Ltd, New Delhi, pp. 189–211.

19 Resilience and Loyalty in the Face of Adversity: A Case of Local Employees in Kerala's Backwater Tourism

E. Joshu Ajoon and Stu Hayes
Department of Tourism, University of Otago, Dunedin, New Zealand

Summary
This case study reports on Kerala's backwater tourism and the associated experiences of houseboat employees in the sector. Kerala's backwater tourism is a significant facet of the state's tourism industry, drawing both foreign and domestic tourists. Houseboats, referred to as 'kettuvallam,' play a central role in backwater tourism, harmonizing tradition with modernity. Despite gaining global recognition, challenges such as escalating costs and external crises, including the COVID-19 pandemic, have impacted Kerala's backwater tourism sector, leading to severe hardship and job insecurity among employees. Despite these adversities, interviews conducted as part of the case study reveal remarkable levels of ongoing resilience and loyalty on the part of houseboat employees, most of whom are local residents. Their resilience and loyalty are rooted in affective commitment and intrinsic motivation, wherein houseboat employees see themselves as custodians of tradition and ambassadors of the backwater region. In turn, their unwavering dedication may be seen as being a vital factor that can contribute to the ongoing adaptability and resilience of the backwater tourism sector as a whole.

The Value and Interest of the Case Study

This case study illustrates how a combination of affective commitment and intrinsic motivation drives backwater tourism employees to keep doing what they do, in spite of various adversities. In so doing, this case study contributes to the existing literature on local community engagement and employment in tourism businesses while offering valuable insights for human resource management in tourism organizations. We suggest that employing local communities in tourism jobs can result in higher levels of resilience and loyalty among employees. This, in turn, may help tourism organizations to address high turnover rates and build more resilient and adaptable tourism businesses.

Kerala's Backwater Tourism

Kerala tourism

Kerala, a state with a land area of 38,863 km^2 in the south-western region of India, is a popular destination with a variety of natural attractions. The state, fondly referred to as 'God's own country', has received awards and recognition from various entities like the United Nations World Tourism Organization (UNWTO), Pacific Asia Travel Association (PATA) and the International Centre for Responsible Tourism (ICRT) (Awards and Recognition, n.d.). These honours show the prominence of the Kerala district in tourism. Among the many attractions of Kerala, backwater tourism is central (Kokkranikal and Morrison, 2002).

Backwater tourism in Kerala

A backwater is a part of a river where the water is not reached by the current (i.e. where a river does not flow), or a smaller body of water that is out of the main current of a larger body (Joseph *et al.*, 2017). Backwater tourism, using 'kettuvallam' as houseboats, began in the Alappuzha district of Kerala in 1991. According to Kokkranikal and Morrison (2002), the revitalization of kettuvallam as houseboats began when a Kerala-based tourism operator initiated the idea of cruising through the backwater for long durations. To do this, they bought a kettuvallam and, with the help of the then remaining kettuvallam-artisans, renovated it by adding additional features like a bedroom, a bathroom, a kitchen, and a sun deck. The renewed kettuvallam (houseboat) attracted a significant number of both foreign and domestic tourists.

The physical geography of Alappuzha also played a crucial role in enticing tourists to experience backwater tourism. Alappuzha is a coastal district in Kerala and is home to several areas that sit below the sea level, including Kuttanad which sits 2.2m below sea level. According to Mr Josekutty Jose, the president of the All Kerala House Boat Owners Association (AKHBOA), foreign tourists constituted the majority of the tourists to Alappuzha in the initial days because they were interested to experience this geographical phenomenon. Following an initial influx of foreign tourists, the district also became popular with domestic tourists. Subsequently, since 1991 the demand for backwater tourism has burgeoned, becoming a major revenue-earning sector and a unique selling product for Kerala's tourism (Kokkranikal and Morrison, 2002).

Since 1991, a multitude of collaborative efforts between governmental and non-governmental entities have been instrumental in elevating the houseboat experience within the realm of responsible tourism in Kerala's backwaters. A significant advance was achieved through the implementation of 'responsible tourism' practices, spearheaded by the Responsible Tourism Mission, an agency established by the Government of Kerala (Mathew *et al.*, 2017). Rooted in the ethos of minimizing adverse economic, environmental, and societal impacts, this initiative has fostered transformations within the sector. A core transformative tenet of the agenda has been to empower local communities and amplify their economic prosperity (Responsible Tourism Mission, n.d.). This is achieved, in part, by amalgamating traditional livelihood practices related to kettuvallam within backwater tourism.

The term 'kettuvallam' is a combination of 'Kettu', meaning 'tied with ropes', and 'vallam', meaning 'boat'. The construction of kettuvallam is truly exceptional. According to Mathen (2012), the kettuvallam construction work involves assembling the hull using long wooden planks of Jack wood or Anjili planks, securely bound together using coir ropes interwoven with coconut fibre. Not a single nail is used in its construction. To ensure durability and water resistance, a coating of black resin, derived from boiled cashew nut shells, is applied to the boat's surface. The upper portion of the kettuvallam is crafted by skilfully weaving together natural materials like coir, bamboo, and panambu.

In recent times, certain modifications have been incorporated into the kettuvallam construction process so as to allow for the integration of modern fixtures and appliances. Such changes have been aimed at enhancing the overall tourist experience (Mathen, 2012). However, amidst these modernizations, the fundamental concept of construction has remained rooted in traditional practices, which are diligently focused on maintaining the essence and authenticity of the original design. Many Kettuvallams still rely on man-operated bamboo poles for navigation, thereby preserving the pristine blue-green waters of Kerala's extensive canal network. This unwavering commitment to heritage practices perpetuates the charm and authenticity of backwater tourism, offering travellers an authentic glimpse into the region's rich cultural legacy while simultaneously preserving the ecological integrity of its waterways.

Relatedly, as part of efforts to enhance backwater tourism, the Department of Tourism, Government of Kerala, introduced a houseboat classification scheme. This scheme categorizes houseboats into three categories: Silver Vessel, Gold Vessel, and the

Diamond Vessel (Kerala Tourism, 2019). The classification hinges on strict criteria set by the Department, encompassing responsible tourism practices, efficient waste management, transparent sharing of information, robust safety measures, and top-notch services in various areas like the kitchen, dining, and bedrooms. Crew requirements include uniform adherence, essential swimming skills, and English fluency. Other aspects, like room size, construction materials, furniture quality, energy efficiency, and LED lighting, are also evaluated. This meticulous classification not only acknowledges excellence but also contributes to elevating the overall backwater tourism experience by promoting higher standards and quality throughout the industry, reinforcing the commitment of the Kerala government to providing tourists with exceptional experiences while preserving the natural and cultural heritage of the backwaters. Crucially, in the context of this case study, this classification scheme also demands substantial local employment (at least 50%). This has helped to ensure that backwater tourism has remained 'local'.

The collaborative efforts of various backwater tourism stakeholders have, in the past, resulted in the development of various projects. A notable and recent example is the STREET project (Sustainable, Tangible, Responsible, Experiential, Ethnic Tourism hubs) that was implemented in 2022 by the Department of Tourism, Government of Kerala, with active participation from the local residents (Kerala Tourism, 2022). The STREET project aims to develop thematic streets in both new or already established destinations and take tourism deep into the interior and rural areas to help tourists experience the local specialities of each destination while decreasing the tourist concentration at the already established destinations (Kerala Tourism, n.d.). As part of the STREET project, stakeholders in the backwater tourism sector cleaned and revived the rivers and canals to expand backwater tourism to more regions and developed destination-specific tourism products to attract the tourists (Outlook, 2022). Such projects have yielded many benefits for the local community by increasing local participation in tourism, empowering women through their active involvement in tourism activities, preserving traditional jobs, reducing the use of plastic, and encouraging the promotion of recyclable and reusable materials across the community, improving waste management and promoting organic farming (Hiran, 2020; Responsible Tourism Mission, n.d.). In addition, backwater tourism has facilitated the transmission of traditional arts and crafts to younger generations while also supporting and promoting local heritage, ritualistic art forms, and local festivals and cuisines (Chandran, 2020).

Additionally, because of the growth of backwater tourism, villages in the backwater region that were once primarily reliant on agriculture have since gained global recognition due to the overall charm of the backwater experience. Domestic and international visitors have begun flocking to these backwater tourism regions. This influx has inspired numerous entrepreneurs to generate innovative ideas within the region, diversifying the products offered to visitors (Radhakrishnan, 2020). Backwater tourism has also benefited the businesses in the region, playing a crucial role in increasing the revenue of the region's hotels and resorts (Kerala Tourism, 2023). Apart from these benefits, backwater tourism has significantly increased employment opportunities for the local community, driven largely by the law that mandates 50% of employees to be local residents. The employment facet is underscored by Ajoon and Dileep (2020), revealing that approximately 45,000 families are dependent on the backwater tourism sector and around 14,000 people are employed in the houseboats playing in the backwaters.

Yet, despite the importance of, and potential for, backwater tourism in Kerala, the sector has, especially in recent times, been impacted by a variety of challenges. These challenges are outlined in the next section.

Adversities in backwater tourism

Backwater tourism in Kerala has faced and continues to face, numerous challenges due to rising prices for essential commodities and increased expenses. In addition, external events such as

COVID-19 have impacted the sector (Paul-A, 2022).

Ajoon and Dileep (2020) highlight the financial strain that arose due to rising prices for essential commodities and increased expenses. According to them, in the 1990s, when houseboat services first set sail, a single room in a houseboat would cost around INR6500.00. This was roughly equivalent to US$185 to US$215 (based on an exchange rate of US$1 to INR30–35). Diesel was priced at roughly INR35 per litre (around US$1 to US$1.17), employee salaries averaged INR3,000 (around US$85 to US$100), with an additional payment of INR25 per trip (around US$0.71 to US$0.83) as bata, which is a form of compensation provided to the employee for every trip they complete. The cost of essential commodities was relatively low. However, in more recent times the exchange rate has doubled (US$1 = INR70–75) and the economic landscape has evolved significantly, with diesel prices more than doubling, employee salaries surging to INR12,500, and the additional payment per trip increasing to INR290. Essential commodity expenses have also significantly increased. Surprisingly, despite these economic shifts, the cost of a single room on a houseboat has either remained the same or, in some cases, even decreased compared to the earlier years. In today's terms, you can book a single room for between INR5,500 and6,500, roughly equivalent to US$75–US$88, depending on the exchange rate.

These economic challenges have coincided with external events that have cast a shadow on the sector's prospects. Notably, the sector has been grappling with a series of adversities since 2017, beginning with the devastating Ockhi cyclone that was later followed by two severe floods in 2018 and 2019. The situation was further complicated by the outbreak of the Nipah virus in the same year (Paul-A, 2022).

Dileep *et al.* (2022) highlight how such adversities have caused severe financial difficulties for the backwater tourism sector, leading to delayed salaries for employees. Nevertheless, thanks to the support of unions, employees have managed to maintain their salary levels. In some cases, the tips from tourists helped the employees endure delayed payments.

In late 2019, just as the sector was navigating through these challenges, the COVID-19 pandemic struck, further prolonging issues. The state of Kerala went into lockdown starting from March 2020 and this persisted for several months. It was not until October 2020 that backwater tourism operations were allowed to operate, albeit with strict restrictions. According to Dileep *et al.* (2022), during the COVID-19 pandemic, the backwater tourism sector had to suspend operations, resulting in zero-earnings for the sector (Fig. 19.1). At the same time, the sector was still required to make significant payments. For example, operators were burdened with a monthly parking fee of INR4,000. Moreover, due to inactivity and collisions, while anchored during the pandemic, many boats suffered damage that was costly to repair (Figs 19.2 and 19.3). In addition, the sector had to bear the costs of providing COVID-19 protocol kits for the tourists in the reopening time and the sanitization of boats after each service. These costs had to be absorbed by the backwater tourism operators themselves, contributing to their financial distress.

According to Dileep *et al.* (2022), the cessation of operations brought about by COVID-19 resulted in job losses in the industry. While some employees were retained for cleaning jobs – predominantly those residing in proximity to the backwater region where the boats were anchored – the majority had to exit the sector completely. Many employees managed to secure alternative jobs in other sectors, while some endured the lockdown period without seeking other work opportunities. As the backwater tourism operation resumed, only 500 employees were called back (in comparison, there were around 14,000 employees prior to the COVID-19 pandemic), with most on a heavily reduced salary due to decreased tourist traffic.

The challenges faced by the backwater tourism employees during COVID-19 were studied by Dileep *et al.* (2022). They revealed seven psychological stages experienced by those employees during COVID-19. Initially, there was a sense of apathy (*Stage 1*) and overconfidence that stemmed from the region's history of overcoming crises. As the adversity grew, a shift to premonitions of danger occurred (*Stage 2*), associated with dwindling visitor arrivals, increasing frustrations, and blame being directed at local authorities. Panic (*Stage 3*) ensued as national lockdown measures paralyzed daily

Fig. 19.1. Houseboats anchored in Alleppey backwaters during COVID-19 lockdown.

life and left employees grappling with financial instability. With the lockdown's extension, inertia, irritation, and redundancy (*Stage 4*) took hold, corroding both entrepreneurial and employee morale, as properties deteriorated. As the lockdown persisted, hope (*Stage 5*) waned, prompting employees to seek alternative income sources and pivot to new fields, particularly among the younger workforce. The recovery phase (*Stage 6*) saw destinations cautiously reopening, albeit with limited tourist demand and high repair costs hampering a full resurgence. Ultimately, the industry adapted to the changed landscape (*Stage 7*), introducing protocols and flexible work arrangements, as employees embraced altered roles amidst reduced salaries but newfound contentment.

It is surprising to know that despite the havoc experienced, most local resident employees expressed a preference to continue working in the backwater tourism sector, as revealed in the study conducted by Ajoon and Dileep (2020). This paints a picture of commitment and emotional attachment, reflecting employee's resilience and loyalty towards the backwater tourism sector. Considering the potential importance of backwater tourism and the recent challenges being faced by it, this case study focuses on the backwater tourism employee's affective commitment and intrinsic motivation and its mediating role in resilience and loyalty. In other words, the case study examines what is it that drives backwater tourism employees to keep doing what they do, in spite of various adversities.

Method

A qualitative approach, based on semi-structured interviews, was used to collect rich descriptive responses from the employees. A total of 15 local resident backwater tourism employees from the Alappuzha district were interviewed shortly after the backwater tourism operations resumed, post-COVID, in late 2020. Purposive sampling was used to identify local resident employees who had experienced some

Fig. 19.2. Houseboat damaged due to prolonged inactivity during COVID-19 lockdown.

or all of the adversities previously described while working in the backwater tourism sector.

Findings and Discussion

The interviews provided valuable insights into the resilience and loyalty of backwater tourism employees when forced to navigate through uncertainties, so much so that they emerged stronger. In this section, we report on the difficulties faced by backwater tourism employees when challenged by external events that have impacted the sector. We also report on the responses of the employees to the difficulties and challenges. We found that despite myriad difficulties, backwater tourism employees chose to continue working in the sector. Their persistence is elucidated through comments that underscore the employee's emotional connection with the kettuvallam and backwater tourism. Such emotional connections represent a manifestation of affective commitment and intrinsic motivations. We draw on existing literature to substantiate this connection and its relationship with loyalty and resilience.

As outlined in an earlier section ('Adversities in backwater tourism'), the backwater tourism sector in Kerala has faced numerous adverse events that have disrupted normal operations. The rising costs of essential commodities and the management of employee's salaries and expenses have led to significant financial strain for the backwater tourism sector. In addition, the backwater tourism sector's sensitivity to various adverse external events (e.g. natural disasters, COVID-19) caused severe difficulties for the employees. As one employee mentioned:

> We have gone through various crises and have recovered from it successfully. Even if the crisis is at a different part of Kerala, it effects our business, the tourist arrival decreases. For example, the flood in 2019, though it occurred

Fig. 19.3. Renovation and repair of a damaged houseboat, preparing to resume services after the COVID-19 lockdown.

> in the northern part of Kerala it decreased the tourist arrival to Alappuzha [district is in the southern part of Kerala]. Similarly, the Nipah virus, it happened in Calicut district in the north, but we lost many reservations. Now the COVID-19, it has caused big difficulties to our business.

This emphasizes the sector's high sensitivity to external events. However, what is particularly surprising is the employee's continued interest in their work, even during extreme instances such as the complete standstill caused by COVID-19. Originating from the local community, these employees take pride in supporting the backwater tourism sector and find it challenging to consider leaving their jobs despite the sector's sensitivity to adversities.

One employee shared,

> This boat helped me and my family for years. I earned money for my family by working on this boat for all these years. It is a heartache for me to leave my boat.

Another employee said,

> One who knows this job would never leave this. This is a decent job and we are able to meet people from different parts of the world who come to see our place. We are showing the tourists our place and how we live. It is interesting for us. This nature of the job makes anyone fall in love with it.

Such comments highlight employees' emotional connection to their work and, indeed, the kettuvallam. This sentiment is further supported by another employee's statement, indicating a preference for engaging in cleaning tasks on the houseboat even during the lockdown period when the boats were anchored.

> *When the lockdown was going on, all our boats were anchored here* [Punnamada finishing point, Alapuzha], *I live nearby, so after few days of lockdown I used to come here in the evening to sit in the boat or do some cleaning works.*

The reason for developing such a strong connection with the job is primarily because they find the work decent, allowing them to interact with people from various places. The employees spoke about taking pleasure in sharing information about their hometown and lifestyle with the guests and often make efforts to maintain good relationships with their guests. In India, there is a long-standing philosophy that has been followed: 'Atithi Devo Bhava,' which means 'the guest is equivalent to God.' This ancient philosophy indicates the importance of hospitality and has an obvious connection with the employee's commitment to the backwater tourism sector. One employee mentioned this when asked about the guest-host relationships.

> As per our hospitality, we have to treat guests as God.

These comments illustrate the ways in which some houseboat employees have built an emotional connection with the kettuvallam and the work they engage in within the backwater tourism sector.

According to Meyer and Allen (1997), the emotional connection an employee has with their organization refers to 'affective commitment'. The employees expressed a strong sense of belonging, often describing their work as an integral part of their lives. The fact that some of them have family members who were also involved in the houseboat sector further deepens their emotional connection, creating a generational bond with the industry.

The backwater tourism employees we spoke with also demonstrated intrinsic motivation factors. Intrinsic motivation is another driving force that keeps employees dedicated to work, and stems from finding personal satisfaction, enjoyment, and fulfilment in the work itself, rather than relying solely on external rewards (Tremblay *et al.*, 2009). Based on our interviews, the houseboat employees appeared to find great pleasure in sharing their culture and home and connecting with people from different parts of the world. This sense of purpose and the joy they derive from making others happy and relaxed fuels their commitment to their job:

> I find immense joy in being a part of the houseboat industry. This isn't just a means to earn a livelihood; this houseboat holds a deep place in my heart and life. My forefathers have dedicated their efforts to this very boat, and now, I proudly carry on their legacy by contributing to its journey.
>
> Working on these houseboats is awesome. We take tourists on our boat to sail through this backwater. While we're sailing, we show them how we live, our houses, and what we do every day. We cook our traditional food for them and let them experience our way of living. The tourists enjoy seeing the stunning backwaters, the green fields, the rice paddies, and the fishing scenes. Seeing how excited and happy they get really makes us feel happy too.
>
> We make our guests happy and relaxed while on the sail. I show them around, explain to them everything we see while on the sail. Some are interested to enjoy the view and know it while some are interested to just see around and rest in the boat. For guests who are interested to know and experience, we take them through different routes and shows them many things. But if the guests are here for some quite time and relax, we provide them their space to relax and give them the best hospitality services.

The sentiments delineating the affective commitment and intrinsic motivation of the backwater tourism employees we spoke with, stand as a profound testament to their resilience and loyalty. According to Rusbult *et al.* (1988, p. 601), employee loyalty refers to 'passively but optimistically waiting for conditions to improve – giving public and private support to the organization, waiting and hoping for improvement, or practicing good citizenship'. The backwater tourism employees, despite the adversities and sensitive nature of the backwater tourism sector, waited for the conditions to improve so that they could resume their jobs. This demonstrates the loyalty of local employees in the backwater tourism sector.

This loyalty played a crucial role in enhancing resilience within the backwater tourism sector. Such findings echo those from other studies (e.g. Paul *et al.*, 2020) that established a relationship between resilience, commitment, and motivation. 'Resilience' is a positive change that happens in the face of an adverse event (Britt *et al.*, 2013). In the case of backwater tourism, despite facing several adverse events consecutively, the sector experienced resilience with the support of loyal employees. More than this, the sector thrived, as evidenced by the sector's ability to return to business and the subsequent achievements that followed the various adverse events. Therefore, the interplay between the local employee's loyalty, affective commitment, and intrinsic motivation emerged as a key determinant in the backwater tourism sector's ability to navigate and overcome adversities, thus underscoring the relationship between these factors in fostering resilience. Therefore, this case study establishes a connection between affective commitment, intrinsic motivation, resilience, and loyalty, among the local residents working in tourism jobs.

Conclusions

The main focus of this case study was to underscore the resilience and loyalty displayed by backwater tourism employees when confronted with adversity. Our analysis has established a direct link between these findings and the employees' affective commitment and intrinsic motivation, offering a deeper comprehension of the factors driving their dedication to the sector, even in the face of external challenges and consecutive adversities.

The narratives of backwater tourism employees, predominantly local residents, transcend surface-level engagement, fostering emotional connections and a genuine ardour for their roles in the sector. Importantly, these experiences epitomize the concepts of affective commitment and intrinsic motivation, as employees align themselves with the sector's meaningful activities and values, treating their work as an integral facet of their identity.

The assertion that their dedication springs from internal fulfilment, not external pressures, is substantiated by specific instances. For instance, employees expressed joy in guiding visitors through cultural exchanges, extending their commitment beyond the boats to embody custodianship of tradition and heritage. The joy derived from sharing their culture and home underscores their intrinsic motivation, reinforcing their commitment.

Remarkably, amid adversities, backwater tourism employees chose to continue working on the houseboats even when alternative employment options were available. This is an integral aspect of the employee's loyalty, contributing to the sector's resilience. Concrete examples, such as employees expressing an emotional difficulty associated with leaving the sector and voluntarily working on the houseboats during the lockdown period illustrate their loyalty to the sector.

In conclusion, the journey of backwater tourism employees stands as a testament to their loyalty and resilience. By delving into the specifics of affective commitment and intrinsic motivation, our case study offers a nuanced understanding of the factors driving the loyalty and resilience observed in these local residents.

Lessons Learned

Building upon the insights gained from our case study, we underscore the significance of local community engagement in the tourism sector. Our case study shows that local community engagement can be a catalyst for reducing employee turnover. As is demonstrated in the backwater tourism sector in Kerala, an engaged local community is loyal, committed, and motivated. This is highlighted by the backwater tourism employees in our study, for whom loyalty and resilience are intertwined.

Our case study demonstrates a clear link between affective commitment, intrinsic motivation, resilience, and loyalty among local resident employees in the tourism sector. This connection has not been explicitly studied in existing literature. However, it is important to note that our current case study is based on a specific location where people adhere to an ancient philosophy wherein 'the guest is equivalent to God.'

Identifying the mediating role of intrinsic motivation, affective commitment, loyalty, and resilience can help tourism businesses and human resource management to address high turnover rates and build more resilient and adaptable tourism businesses by engaging local residents in their operations.

References

Ajoon, E.J. and Dileep, M.R. (2020) Impact of global pandemic on smes employees: Building the scenario of backwater tourism in Kerala. *Asian Tourism Research* 362–371. Available at: https://www.researchgate.net/profile/Nimi-Markose/publication/369385130_Resident's_attitude_towards_responsible_tourism_in_the_Post_Covid-19_pandemic_scenario/links/64199de092cfd54f84193e63/Residents-attitude-towards-responsible-tourism-in-the-Post-Covid-19-pandemic-scenario.pdf#page=369 (accessed 19 July 2023).

Awards and Recognition (n.d.) Kerala Tourism. Available at: https://www.keralatourism.org/awards/ (accessed 8 July 2023).

Britt, T.W., Sinclair, R.R. and McFadden, A.C. (2013) Introduction: The meaning and importance of military resilience. In: Sinclair, R.R. and Britt, T.W. (eds) *Building Psychological Resilience in Military Personnel: Theory and Practice*. American Psychological Association, pp. 3–17. DOI: 10.1037/14190-001.

Chandran, A. (2020) *Tourism Sector Sets Eyes on Village, Folklore, Local Heritage for Recovery from Covid Crisis*. The New Indian Express, 23 May. Available at: https://www.newindianexpress.com/states/kerala/2020/may/23/tourism-sector-sets-eyes-on-villages-folklore-local-heritage-for-recovery-from-covid-crisis-2146832.html (accessed 22 November 2023).

Dileep, M.R., Ajoon, J. and Nair, B.B. (2022) COVID-19 and tourism stakeholders: Experience, behaviour and transformation. *Journal of Organizational Ethnography* 11(3), 275–293. DOI: 10.1108/JOE-07-2021-0043.

Hiran, U. (2020) Stakeholders Hopeful of Tourism Revival. The Hindu, 30 August. Available at: https://www.thehindu.com/news/national/kerala/stakeholders-hopeful-of-tourism-revival/article32482094.ece (accessed 22 November 2023).

Joseph, E.K., Kallarackal, T.K. and Varghese, B. (2017) Community-based waste management: Backwater tourism as a case example. *Atna Journal of Tourism Studies* 12(2), 85–94. DOI: 10.12727/ajts.18.5.

Kerala Tourism (2019) Houseboat Classification Scheme. Available at: http://keralatourism.gov.in/classificationofhouseboats/Houseboat.htm (accessed 19 July 2023).

Kerala Tourism (2022) Kerala Tourism launches the First Water STREET in Maravanthuruthu, Kottayam. Available at: https://www.keralatourism.org/newsletter/news/2022/water-street-project/2078 (accessed 23 November 2023).

Kerala Tourism (2023) Scenic Kumarakom Generates the Highest Revenue per Available Rooms in the Country. Available at: https://www.keralatourism.org/newsletter/news/2023/kumarakom-earned-highest-revenue/2188 (accessed 23 November 2023).

Kerala Tourism (n.d.) Street Project. Available at: https://www.keralatourism.org/responsible-tourism/street-project (accessed 28 November 2023).

Kokkranikal, J. and Morrison, A. (2002) Entrepreneurship and sustainable tourism: The houseboats of Kerala. *Tourism and Hospitality Research* 4(1), 7–20. DOI: 10.1177/146735840200400102.

Mathen, R. (2012) Houseboats in Kerala-constructional features and environmental issues. *IOSR Journal of Environmental Science, Toxicology and Food Technology* 1(6), 31–43.

Mathew, R., Chandran, A. and Swain, S.K. (2017) Backwater tourism: RT initiatives and socio-environmental dynamics. *SCMS Journal of Indian Management* 14(2), 113–127. Available at: https://www.scms.edu.in/uploads/journal/articles/article_10.pdf (accessed 18 October 2020).

Meyer, J.P. and Allen, N.J. (1997) *Commitment in the Workplace: Theory, Research, and Application*. Sage Publications, Thousand Oaks, CA.

Outlook (2022) Maravanthuruthu Water Street of Kerala Awaits Travellers. Available at: https://www.outlookindia.com/travel/kerala-tourism-launches-maravanthuruthu-water-street-news-194659 (accessed 2 December 2023).

Paul-A, S. (2022) *Houseboat industry in turbulent waters: Rising prices of fuels and essentials rock COVID-hit sector*. The Hindu. Available at: https://www.thehindu.com/news/national/kerala/houseboat-industry-in-turbulent-waters/article65411119.ece (accessed 28 August 2023).

Paul, H., Budhwar, P. and Bamel, U. (2020) Linking resilience and organizational commitment: Does happiness matter? *Journal of Organizational Effectiveness: People and Performance* 7(1), 21–37. DOI: 10.1108/JOEPP-11-2018-0087.

Radhakrishnan, S.A. (2020) *A blend of chess and tourism on Vembanad lake*. The Hindu. Available at: https://www.thehindu.com/news/national/kerala/a-blend-of-chess-and-tourism-on-vembanad-lake/article30646817.ece (accessed 22 November 2023).

Responsible Tourism Mission (n.d.) Kerala Tourism. Available at: https://www.keralatourism.org/responsible-tourism/ (accessed 23 July 2023).

Rusbult, C.E., Farrell, D., Rogers, G. and Mainous III, A.G. (1988) Impact of exchange variables on exit, voice, loyalty, and neglect: An integrative model of responses to declining job satisfaction. *Academy of Management Journal* 31(3), 599–627.

Tremblay, M.A., Blanchard, C.M., Taylor, S., Pelletier, L.G. and Villeneuve, M. (2009) Work extrinsic and intrinsic motivation scale: Its value for organizational psychology research. *Canadian Journal of Behavioural Science/Revue Canadienne Des Sciences Du Comportement* 41(4), 213. DOI: 10.1037/a0015167.

20 Responses from Businesses in Rural Communities Recovering from the 2011 Tohoku Disasters During the 2020 Tokyo Olympics

David N. Nguyen[1] and Gabriel Fuentes Cordoba[2]

[1]International Research Institute for Disaster Science, Tohoku University, Sendai, Japan; [2]Faculty of Liberal Arts, Sophia University, Tokyo, Japan

Summary

The 2011 Great East Japan Earthquake (GEJE) significantly impacted the Tohoku Region in north-eastern Japan, leading to a large number of human casualties and environmental devastation. It exacerbated pre-existing economic and demographic decline. Tourism development was promoted as part of economic recovery and revitalization efforts. With Japan securing the winning bid for the 2020 Tokyo Olympics, it was hoped that having some of the Olympic activities hosted in Tohoku would contribute to recovery efforts. This case study examines the perspectives of businesses in areas located near the Olympic venues in Tohoku, on what kind of effects the Olympics had on their operations.

The Value and Interest of the Case Study

This case study provides an insight into how mega-events, such as the Olympics, are used as a policy tool to revitalize the local economies recovering from disasters, especially in light of the COVID-19 pandemic. This case study explores two areas in the Tohoku Region, Miyagi and Fukushima, which adopted different approaches towards spectator policies during the time the games were held.

Background

On March 11, 2011, a magnitude 9.1 earthquake occurred at 2:46pm Japan Standard Time, with an epicenter located approximately 130 km east of the coast of Sendai, Japan, the largest city in the area. Following the earthquake were hundreds of aftershocks, and large tsunamis, reaching up to 40m in some areas. These all caused additional damage to many communities on top of the initial earthquake. The combination of earthquakes and tsunamis also led to three hydrogen explosions at the Fukushima Daiichi Nuclear Powerplant, located in Okuma in the eastern coast of Fukushima Prefecture. It is estimated that the disasters resulted in over 20,000 deaths and 2,500 missing people. In the aftermath of these disasters, it is estimated that US$286 billion was spent on reconstruction (Saito, 2021). The areas that experienced the heaviest damage were located in the Tohoku Region, an area located in the north-east of Japan, which consists of six prefectures: Akita, Aomori, Iwate, Yamagata, Miyagi, and Fukushima. Within the Tohoku Region, Iwate, Miyagi, and Fukushima Prefectures experienced the brunt of the disaster, with Miyagi Prefecture recording over 10,000 casualties and 1000 missing people. Fukushima Prefecture recorded over 2500 casualties, 200 missing people, and 60,000 residents relocated due to radiation fears (FDMA, 2020). Due to its significant impact on eastern Japan, the disasters are known locally as the 2011 Great East Japan Earthquake (GEJE).

The Socio-Economic Obstacles in the Tohoku Region

Prior to the 2011 GEJE, the Tohoku Region had already been experiencing decades of demographic and economic decline. All six prefectures had passed their peak population, with Akita, Aomori, and Iwate Prefectures reaching peak population in the 1980s, Yamagata and Fukushima Prefectures during the early 1990s, and Miyagi Prefecture in the early 2000s (Fukushima Prefecture, 2023a). The total fertility rate had already begun decreasing across the country by the 1970s and has remained on a consistent downward trend (Coulmas, 2007; Smil, 2007). The population decline also coincides with the declining economic development in the region. Fukushima Prefecture, which was once known as an industrial powerhouse, had already begun declining by the late 1990s (Fukushima Prefecture, 2023b). While the depopulation of rural Japan is not unique to the Tohoku Region, as nearly every region outside of Japan's three largest urban areas (Tokyo, Osaka, and Nagoya) has experienced similar issues, the scale of Tohoku's demographic and economic decline has been among the highest in the country.

It is estimated that the 2011 GEJE caused US$211 billion in direct damages. This makes it the costliest natural disaster in Japan to date, exceeding the costs of the 1995 Great Hanshin-Awaji (Kobe) earthquake. Based on the Japanese Cabinet Office's Gross Prefectural Product (GPP) numbers, the economies of both Miyagi and Fukushima Prefectures had already been declining since 2008. However, the 2011 GEJE led to an approximately 3% drop in GPP in Miyagi Prefecture, and a 8.5% drop in Fukushima Prefecture (Fig. 20.1). Although the economies in Miyagi and Fukushima Prefectures rebounded after 2011, the GPP in both prefectures began declining again in 2019, a year before the COVID-19 pandemic (Japanese Cabinet Office, 2023).

As the Tohoku Region experienced population decline, Tokyo and surrounding urban areas continued to experience population growth due to agglomeration (Ishikawa, 2020). Tohoku's young population increasingly chooses to migrate to Tokyo for educational or employment opportunities. Research also suggests that the opening of the high-speed rail line, known locally as the Shinkansen, between Tokyo and the Tohoku Region has increased migration out of the region. The departure of much of the working-age population in the region has led to a phenomenon known in Japanese as Shatta-Machi, or 'shutter streets' in English, where much of the commercial districts in Tohoku's cities consist of streets full of shops that are closed and/or vacant (Hirano, 2013). What remains in many of these cities are a population of children and the elderly (Fielding, 2017).

The disasters of 2011 had exacerbated existing socio-economic trends in Tohoku, resulting in the area experiencing a larger rate of out-migration. This was particularly evident in Fukushima Prefecture, where many residents evacuated their municipalities due to radiation concerns (Do, 2019; Nagamatsu *et al.*, 2020). Elsewhere, destroyed homes, lost livelihoods, and traumatic experiences have motivated residents to move to other locations within the country. It is estimated that 470,000 residents have evacuated and relocated since 2011 (Reconstruction Agency, 2017).

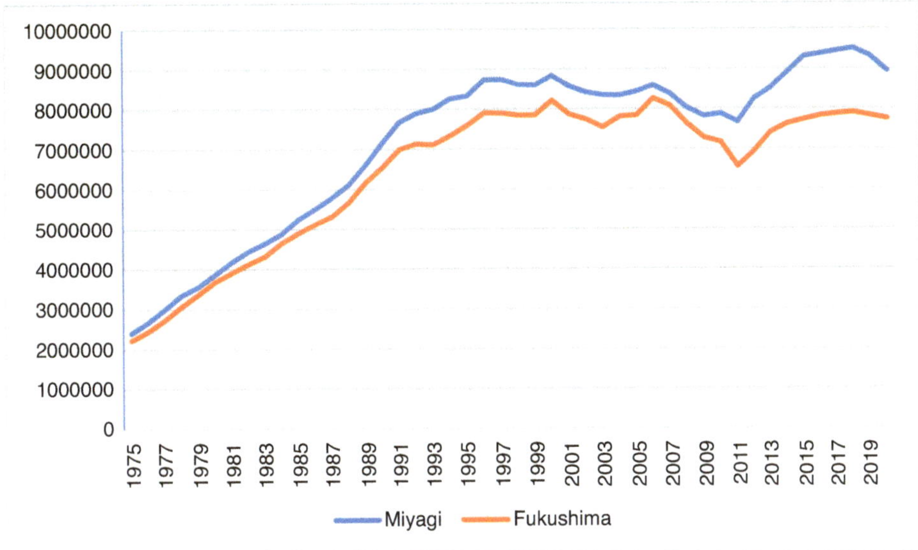

Fig. 20.1. Gross prefectural product (GPP) of Miyagi and Fukushima Prefecture by Million Japanese Yen. Note that from 2019, GPP calculation methods changed, so 2019 and 2020 numbers are estimates based on the percentage of growth observed between the years and applied to the numbers based on the older calculation (Japanese Cabinet Office, 2023).

Recovery Through the Olympics

In order to promote the recovery of the Japanese economy, the Japanese government enacted a series of travel development plans. These sought to facilitate an increase in inbound travel through the relaxation of visa entry policies, changes to airport infrastructure, and promotional activities. Inbound tourism to Japan had increased from 8.6 million visitors in 2010, the year prior to the disaster, to 31.8 million visitors in 2019, the year prior to the COVID-19 pandemic (JNTO, 2022). As inbound tourism to Japan began to surge, in 2013 the International Olympic Committee (IOC) awarded the 2020 Summer Olympics to Tokyo. While the official slogan of the Olympics was 'Discover Tomorrow', the games were also promoted as the 'Recovery and Reconstruction Games' by the Japanese government (IOC, 2021), as it was hoped that the event would increase inbound tourism during the games and that tourists would visit areas affected by major disasters.

However, at the beginning of 2020, the COVID-19 virus began spreading across the world, leading the World Health Organization to declare the virus as a pandemic on March 11, coincidentally the same day as the 2011 GEJE (WHO, 2020). Japan and the International Olympic Committee announced on March 24 that the Olympics would be delayed to the summer of 2021. The decision to delay the games drew mixed reactions within Japan as some groups had wanted to either further delay the games or cancel it due to fears of overwhelming the local medical system, while others had wanted the games to open in order to improve business, which was already affected by a downturn in consumption due to various COVID-related restrictions. Interviews among hotels and restaurants in Tokyo indicated that business operators were disappointed with the lack of economic benefits as a result of the postponement (Ueno *et al.*, 2021). Economic losses caused by the delay of the games and absence of spectators in most venues when the Olympics did occur were between US$1.3 billion, according to the Nomura Research Institute, and US$5 billion, according to Goldman Sachs estimates (Imahashi and Regalado, 2020; Ueno *et al.*, 2021).

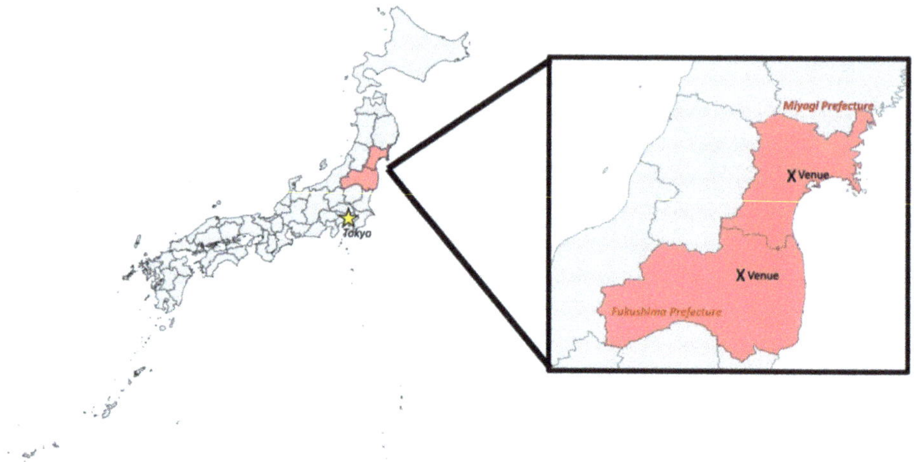

Fig. 20.2. Location of Miyagi and Fukushima Prefectures and the Olympic Venues in their respective prefectures.

This case study examines how businesses in Miyagi and Fukushima Prefecture fared during the 2020 Tokyo Olympics (Fig. 20.2). Specifically, it seeks to examine how much sales grew during the Olympics, if the businesses hired new employees, and whether they felt that they had benefitted from the games.

Case Study Area

For this study, our research focuses on Miyagi and Fukushima Prefecture, the only two prefectures in the region that were chosen to host Olympic venues. Miyagi was the site for the soccer events at Miyagi Stadium in Rifu City, and Fukushima hosted the baseball, softball, and other events near Tsuchiyu Onsen in Fukushima City (Tokyo Metropolitan Government, 2020). Survey collection was done at the following locations within Miyagi: Sendai station area, Matsushima Kaigan area, Naruko onsen, Sakunami, and Akiu. Sendai was chosen as it is the largest city in Miyagi, with Sendai Station serving as the primary transportation hub for spectators. Naruko Onsen, Matsushima Kaigan, Akiu, and Sakunami were chosen as they are the most popular tourist destinations in Miyagi. In the case of Fukushima, the cities of Fukushima City and Koriyama were chosen. Koriyama and Fukushima City are the largest cities in the prefecture and serve as the primary transportation hubs in the area. Within Fukushima City and Koriyama, businesses around Fukushima Station and Koriyama Station were chosen as visitors would have to transit there to reach the venue. Surveys were also conducted in Iizaka Onsen and Tsuchiyu Onsen, both popular tourist destinations in Fukushima City.

These two prefectures provide an interesting case study on the effects of the Olympics on local businesses, as the prefectural government adopted different approaches towards spectators. Miyagi Prefecture opted to continue holding the event, albeit at half capacity, allowing only local prefectural residents the chance to watch at the venue. In contrast, Fukushima Prefecture opted to close the venue to the general public, allowing only athletes and related staff inside.

Methodology

A survey was conducted in October 2021, 2 months after the conclusion of the 2020 Olympic games which were delayed until the summer of 2021. Survey collection was outsourced to a Japanese company called Rakuten Insights which specializes in survey collection at a national scale. Rakuten Insights conducted the surveys by utilizing its massive database of members throughout the country and contacted Rakuten members whose addresses were located in the survey areas. These members were then offered Rakuten points, which are used for online shopping on their website, for completing the surveys. For the purpose of this survey, Rakuten Insights randomly contacted members from its database whose addresses fell in the case study areas. In total 190 responses were collected from cities in Miyagi and Fukushima Prefectures. From the survey, we constructed dependent, independent, and control variables. We used three questions from the questionnaire to create the main dependent variables to use in the analysis. All of them are binary variables. Positive statements were assigned a value of one, while neutral and negative values were zero. The dependent variables are:

- Sales grew during the Olympics: We asked the question 'Did sales increase during the Olympics?' The variable takes the value of one if respondents answered either 'a lot more' or 'more', the variable is zero, otherwise.
- Hired new employees during the Olympics: We asked 'Did you hire new employees during the Olympics?' The variable takes the value of one, if respondents hired new employees, the variable is zero, otherwise.
- Business benefited from Olympics: We asked 'Did your business benefit from the Olympics?' The variable takes the value of one if respondents agreed or strongly agreed with the statement. The variable is zero, otherwise.

We analyzed the data by comparing sample means and implementing multivariate regression analysis. In our regression analysis, we included business-specific independent variables such as the number of employees, business age, and business type. We also included respondent-specific control variables such as age, gender, and hometown location. Finally, in the baseline specification, we included a location binary variable entitled 'Miyagi' that takes the value of one if the business is located in Miyagi and zero if the business is in Fukushima. This binary variable allowed us to examine whether there were differences in the perceived impacts of the Olympic games between these two prefectures, in terms of benefits to local businesses and the community.

We estimated the regressions using linear probability models with heteroskedasticity-robust standard errors. We regressed each dependent variable on the location of the business, respondent and business observable characteristics, and industry-specific fixed-effects. Industry-specific fixed-effects are binary variables that let us control for industry-specific characteristics. For example, some businesses are more likely to benefit from the Olympics than others given the nature of their activities. After introducing the fixed effects, we are certain that the differences are not driven by the type of business activity.

Results

Fig. 20.3 shows the mean of the main outcome variables by prefecture. In total, only 9% of respondents stated that their sales grew during the Olympics 8.7 and 9.2% of businesses in Miyagi and Fukushima, respectively, reported that their sales grew during the Olympics. With respect to hiring new employees, 30% of businesses in Miyagi reported that they hired new employees during the Olympic games, whereas only 19% in Fukushima did so. This simple difference is not statistically significant at the 5% significance threshold. Finally, approximately 13.6% of businesses located in Miyagi reported that they benefited during the Olympics, compared to 6.9% in Fukushima. As mentioned above, Miyagi Prefecture opted to hold the games with spectators at a limited capacity while Fukushima Prefecture held the games without. However, Fig. 20.3 suggests that the gains from the Olympics were minor in both locations. Our pooled sample includes businesses of different age and size. For example, 25% of businesses have 10 or less employees, 12% have between 11 and 20 employees, 10% between 21 and 50, 14% between 51 and 100, and 38% have more than 100 employees. The businesses surveyed in Fukushima were more likely to have more than 100 employees than those in Miyagi. Businesses surveyed in Fukushima are also on average, older than those surveyed in Miyagi. A total of

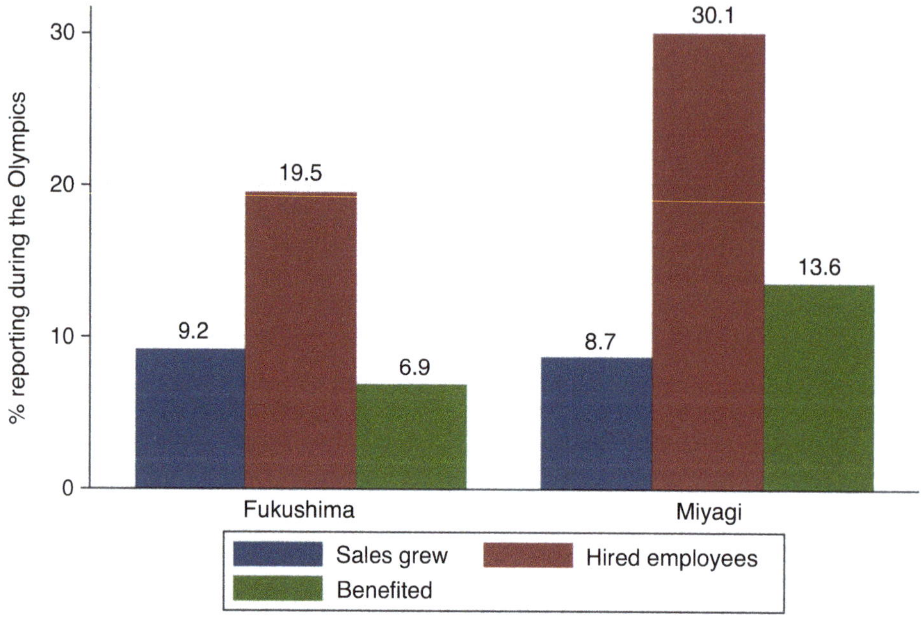

Fig. 20.3. Percentage of businesses that reported experiencing beneficial impacts during the Olympics in Fukushima and Miyagi.

87% of businesses in Fukushima were established more than 10 years ago, whereas 69% of businesses in Miyagi were at least 10 years old. Also, 3% of the businesses surveyed in Miyagi were less than a year old, while none of the businesses in Fukushima were less than a year old. These differences indicate that Fig. 20.3 has to be interpreted with extreme caution. To increase the validity of the findings presented in Fig. 20.3 and observe different trends in the data we implemented regression analyses, controlling for observable characteristics.

Fig. 20.4 replicates Fig. 20.3 and presents very similar results. Fig. 20.4 shows the differences in the self-reported benefits during the Olympics between Miyagi and Fukushima after controlling for the number of employees, business age, business type, gender, age, and hometown of the respondent. The differences in reporting that sales grew during the Olympics is not significant across businesses in Miyagi and Fukushima. The coefficient of Miyagi indicates that businesses surveyed there were 10 percentage points more likely to hire employees during the Olympics than businesses in Fukushima. However, this difference was not statistically significant at the 5% threshold. Similarly, businesses in Miyagi were more likely to report benefits from the Olympics than those in Fukushima, but again the estimated coefficient has a large standard error. In sum, businesses in Miyagi have more positive responses than those in Fukushima, but the difference are moderate in terms of magnitude and statistical significance.

Miyagi's policy of allowing spectators to visit the venues did not demonstrate a considerable impact on businesses when comparing their responses to those collected from Fukushima. However, small and large businesses may have different responses to this policy. Consequently, we explored if the differences in business size led to differing responses. To do this, we compared companies with more than 100 employees to those with less than 100 employees across Miyagi and Fukushima. When analyzing the responses by business size, regardless of their location in Fukushima or Miyagi, larger companies were more likely to self-report benefits from the Olympics than smaller companies. A total of 18% of the companies with more than 100

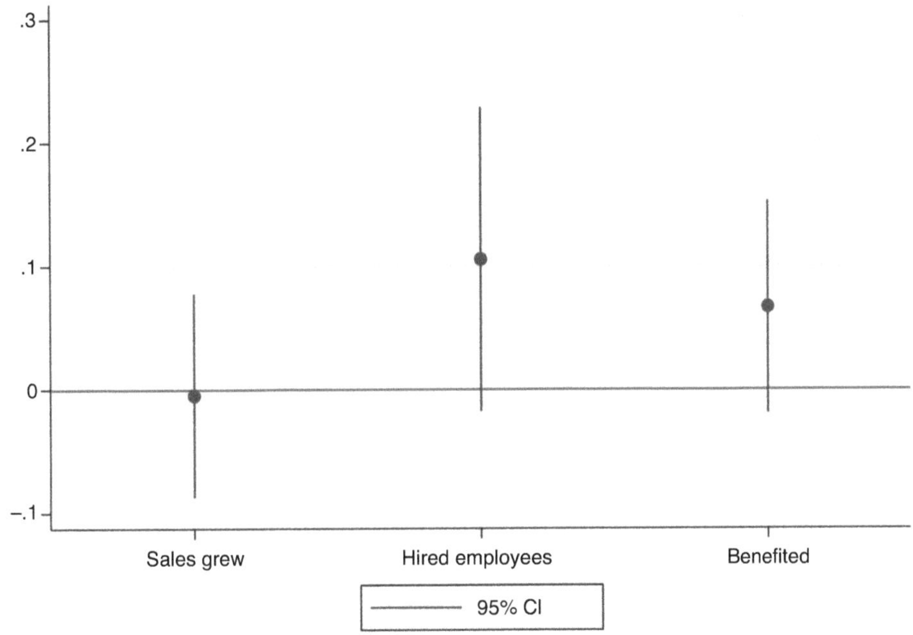

Fig. 20.4. Differences in the impact of the Olympics on reported sales, hiring, and benefits in Fukushima and Miyagi.

employees reported that they benefited during the Olympics whereas only 6% of companies with less than 100 employees reported benefiting during the Olympics. These differences are statistically significant when using both simple differences in means and multivariate regression analysis with industry-fixed effects. Large companies were also more likely to report hiring new employees and experiencing higher sales during the Olympics than small companies. But for these two cases, the differences are not statistically significant by the standard thresholds.

We also considered the possibility that older businesses may respond differently than newer businesses in regard to the perceived impacts of the Olympics. For ease of comparison, we divided the businesses into two groups: those established for 10 years or less and those established for more than 10 years. In Fig. 20.5, we show the percentage of businesses that reported higher sales, increased hiring, and experienced perceived benefits during the Olympics between the two groups. The figure on the left shows the pooled data and the right shows only Miyagi. When using the pooled data, younger and older businesses do not differ in their responses to the sales and benefits questions. But younger businesses reported that they were more likely to hire new employees during the Olympics compared to older businesses. When we restricted the data to Miyagi, we observed that the differences are large in magnitude with respect to hiring. About 44% of younger businesses in Miyagi hired new employees while only 24% of older businesses in Miyagi hired employees. Interestingly, the older businesses in Miyagi were more likely to report increasing sales and benefits during the Olympics. However, these differences are not statistically significant.

Conclusions

Following the triple disasters of 2011, tourism development was pushed as a policy tool to revitalize the economy of Japan, especially in areas affected by the disasters. It was hoped that by the government and local communities that in winning the bid to host the 2020

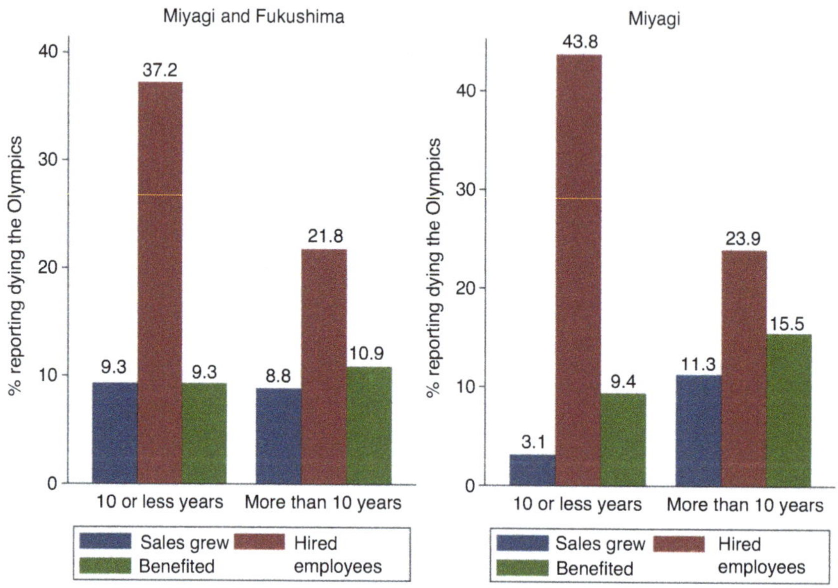

Fig. 20.5. Impact of the Olympics on reported sales, hiring, and benefits in by business age.

Tokyo Olympics, a surge of tourism would help propel the economic recovery process. However, the emergence of the COVID-19 pandemic led to a virtual standstill in global tourism. Communities hosting the event in the Tohoku region were caught at an impasse on whether to continue holding the games with spectators or not. Ultimately, Miyagi Prefecture opted to hold the games with spectators at half capacity while Fukushima Prefecture chose to hold the games without spectators.

Our survey examined how businesses in these two prefectures fared during the event. Although businesses in Miyagi reported more positive responses during the period of the Olympics compared to Fukushima, overall the perceived benefits of these games were minor. As with other case studies on the influence of the mega-events on local businesses, smaller businesses perceived that they benefitted the least during the event, as seen in London during the 2012 Summer Olympics and the 2006 World Cup in Berlin (Vlachos, 2015). The accommodation industry, a sector which usually benefits the most during a major tourism event such as the Olympics, reported the least amount of sales, which was clearly due to the halt in inbound tourism. The few spectators that could visit the venue in Miyagi were local residents, who did not need to stay at a hotel during the event.

While the circumstances that affected the Tokyo Olympics could be considered a rare event, due to the exceptional proliferation of the COVID-19 pandemic, the results highlight the vulnerabilities to external shocks of the tourism industry. Other external shocks, which have occurred in the past but on a lower scale than the COVID-19 pandemic, include instances such as the H1N1 pandemic and the Avian Influenza which caused significant impacts to global tourism. As governments around the world continue to bid for the rights to host major sporting-events in their country, destinations may need to re-consider the potential cost and benefits for local businesses, and develop contingency plans at the destination level, should a major crisis occur during the event. Solely relying on these mega-events as a panacea to solve socio-economic issues, without planning for a worst case scenario, could potentially lead to further strain on the destination's community and economy as well as reputational damage.

References

Coulmas, F. (2007) *Population Decline and Ageing in Japan – The Social Consequences*, Vol. 16. Routledge, London.

Do, X.B. (2019) Fukushima nuclear disaster displacement: How far people moved and determinants of evacuation destinations. *International Journal of Disaster Risk Reduction* 33, 235–252. DOI: 10.1016/j.ijdrr.2018.10.009.

FDMA (2020) Chapter 3 Overview of the Disaster. Available at: https://www.fdma.go.jp/disaster/higashinihon/item/higashinihon001_12_03-01_02.pdf

Fielding, T. (2017) Japan: Internal migration trends and processes since the 1950s. In: *Internal Migration in the Developed World*. Routledge, London, pp. 173–202.

Fukushima Prefecture (2023a) *Fukushima Prefecture Population Statistics*.

Fukushima Prefecture (2023b) *Long Term Economic Indicators Since 1975*.

Hirano, K. (2013) Difficulties in post-tsunami recnstruction plan following Japan's 3.11 mega disaster: Dilemma between protection and sustainability. *Journal of Japan Society of Civil Engineers* 1, 1–11.

Imahashi, R. and Regalado, F. (2020) Olympics delay to cost Japan $6bn in economic losses. *Nikkei Asia*. Available at: https://asia.nikkei.com/Spotlight/Tokyo-2020-Olympics/Olympics-delay-to-cost-Japan-6bn-in-economic-losses#:~:text=TOKYO%20%2D%2D%20Japan%20is%20faced,due%20to%20the%20coronavirus%20pandemic.

IOC (2021) *Tokyo 2020 'Recovery Olympics' Give Hope to Regions Affected by the Great East Japan Earthquake*. International Olympic Committee. Available at: https://olympics.com/ioc/news/tokyo-2020-recovery-olympics-give-hope-to-regions-affected-by-the-great-east-japan-earthquake

Ishikawa, Y. Bell, M., Bernard, A., Charles-Edwards, E. and Zhu, Y. (eds). (2020) *Internal Migration in Japan BT – Internal Migration in the Countries of Asia: A Cross-national Comparison*. Springer International Publishing, pp. 113–136. DOI: 10.1007/978-3-030-44010-7_7.

Japanese Cabinet Office (2023) *Prefectural Resident Statistical Forecast*.

JNTO (2022) *Japan Tourism Statistics*.

Nagamatsu, S., Rose, A. and Eyer, J. (2020) Return migration and decontamination after the 2011 fukushima nuclear power plant disaster. *Risk Analysis* 40(4), 800–817. DOI: 10.1111/risa.13432.

Reconstruction Agency (2017) *The Road to Recovery: Recovery and Reconstruction from the Great East Japan Earthquake*.

Saito, M. (2021) Ten years on, grief never subsides for some survivors of Japan's Tsunami. *Reuters*. Available at: https://www.reuters.com/investigates/special-report/japan-tsunami-survivors/

Smil, V. (2007) The unprecedented shift in Japan's population: Numbers, age, and prospects. *The Asia Pacific Journal* 5(4), 1–7.

Tokyo Metropolitan Government (2020) *Competition Venue*.

Ueno, H., Hida, H. and Dooley, B. (2021) Tokyo was promised glory and riches. it got an olympics in a bubble. *The New York Times*.

Vlachos, P. (2015) The 2012 olympics and small local business: A 5-year longitudinal study of south-east london. In: Poynter, G., Viehoff, V. and Li, Y. (eds) *The London Olympics and Urban Development: The Mega-Event City*. Routledge, London, pp. 176–192. DOI: 10.4324/9781315758862.

WHO (2020) *WHO Director-General's Opening Remarks at the Media Briefing on COVID-19*. World Health Organization. Available at: https://www.who.int/director-general/speeches/detail/who-director-general-s-opening-remarks-at-the-media-briefing-on-covid-19---11-march-2020

21 'Not All Fun and Games': The Case of the Victoria 2026 Commonwealth Games

Ana Delevska, Sandra Cherro Osorio and Van K. Nguyen
Business and Construction, Melbourne Polytechnic, Melbourne, Australia

Summary

In April 2022, the Australian state of Victoria was confirmed to be the host of the Commonwealth Games 2026. The proposal placed a strategic emphasis on leveraging legacy to benefit local communities, particularly through the provision of affordable and social housing and sporting facilities in regional Victoria. The Games were to take place across five regional cities to showcase Victoria globally, while generating economic growth for the regional economies, creating local jobs and boosting inclusion and participation of regional communities. However, in July 2023, the Premier of Victoria announced the cancellation of the Games, primarily justified by the increase in the initial cost of running the event. This conceptual case study provides an analysis of the factors that influenced the cancellation and identifies several significant challenges that contributed to that outcome, including lack of infrastructure, logistical complexities, and budgetary constraints, as well as the intricate dynamics among these aspects. It also discusses the potential opportunities and challenges for the state of Victoria and its brand as a host of sporting events.

The Value and Interest of the Case Study

This case study delves into the complexities of hosting a sports mega-event, especially one that aspired to be innovative but fell short in addressing crucial aspects during its conception and planning. Applying theRitchie and Crouch (2003) Conceptual Model of Destination Competitiveness, this case study demonstrates the interplay between a destination's competitiveness and sustainability in the context of mega-event planning. By examining some key factors from this framework, we pinpoint the shortcomings that led to the cancellation of the mega-event. This study prompts inquiries into the challenges and benefits of hosting such large-scale events.

Context

Hosting mega-events is a potentially strategic way to stimulate economic growth given their potential to attract significant visitation to the host regions, which in turn can generate substantial economic impacts in the host community through visitor spending and employment opportunities (Li, 2013; Allen *et al.*, 2022). However, the planning and organizing of mega-events is inherently complex, involving a myriad of logistical, infrastructural, financial, political, and socio-cultural considerations. These challenges often intersect and amplify each other. Some of the factors complicating the planning and implementation of mega-events include inadequate infrastructure, logistical complexities, budgetary constraints, limited time for implementation, political instability, public opposition, and unforeseen external events (Lauermann and Vogelpohl, 2017; Moore *et al.*, 2017; Di Vita and Wilson, 2021).

As elected host of the Commonwealth Games 2026, the Victorian government expected to deliver an innovative model in their proposal that originally included hosting the events across three of the state's largest regional cities, namely Geelong, Ballarat and Bendigo. Later, two other cities, Shepparton and Gippsland, were added (Commonwealth Games, 2023). Other key components of this multi-regional-hub plan were the use of existing sports infrastructure and venues, as well as the opportunity to improve transport and housing infrastructure in the proposed regional cities. The government's vision for the Games was to share the benefits related to the Games and their legacies across the state of Victoria.

In July 2023, the Victorian Premier announced the cancellation of the Commonwealth Games, citing the main reason as financial, with a cost increase from the initial projection of AU$2.6 billion to AU$6 billion. This raises important questions: Does the cancellation of the Commonwealth Games 2026 suggest that hosting such events is becoming more difficult to justify financially? Do the costs outweigh the benefits of hosting such a mega-event? This case study applies Ritchie and Crouch (2003) model of Destination Competitiveness and Sustainability as a comprehensive framework to guide the examination of the factors that catalysed the cancellation of the Victorian Commonwealth Games 2026. It also includes a discussion addressing the decision to breach the contract of hosting the Games in Victoria.

The Complexities of Destination Planning for Mega-Events

The cancellation of mega-events has garnered scholarly and societal attention due to the severe socio-economic consequences for the host destinations and the multifaceted challenges and factors that contribute to such outcomes (e.g. Perić and Vitezić, 2019; Flyvbjerg *et al.*, 2021; Jones *et al.*, 2022). The case of the Victorian Commonwealth Games 2026 is particularly complex, with the government acknowledging that the planned multi-hub regional model was going to require substantially more resources and funds to implement compared to traditional models of running the Games in major cities (Victorian Government, 2023). In light of this, we must consider, could the Commonwealth Games organizers have done more to assess the feasibility of the bid before committing to it?

Amidst the challenges posed by this resource-intensive approach, the decision to cancel the Victorian Commonwealth Games 2026 ultimately hinged on a myriad of economic, political and logistical considerations. One model that provides a holistic approach to assessing such considerations is the Destination Competitiveness and Sustainability framework proposed by Ritchie and Crouch (2003). Specifically, the model is used to evaluate a destination's capacity to attract visitors and generate economic benefits while maintaining a sustainable balance between environmental, socio-cultural, and economic considerations and consequences. This makes it a suitable lens to adopt in the context of this case study. While the model provides a comprehensive framework for understanding destination competitiveness and sustainability, its direct application to dissect the intricacies of a cancelled mega-event is scarce in the current literature, revealing scope for further research. The model also accounts for external forces, and competitive and comparative advantages of a destination, which will not be examined within the scope of this case study.

The destination competitiveness and sustainability framework (Fig. 21.1) has as its centre the destination's *Core Resources and Attractions* – natural, cultural and built assets that attract visitors. While the success of a destination relies on these factors, *Supporting Factors and Resources* provide a foundation for the development of a destination. These factors include accessibility, infrastructure quality, human capital, entrepreneurship, political will, and hospitality, which play pivotal roles in shaping a destination's overall success.

The lack of adequate infrastructure is a fundamental issue that affects a destination's development and sustainability when hosting

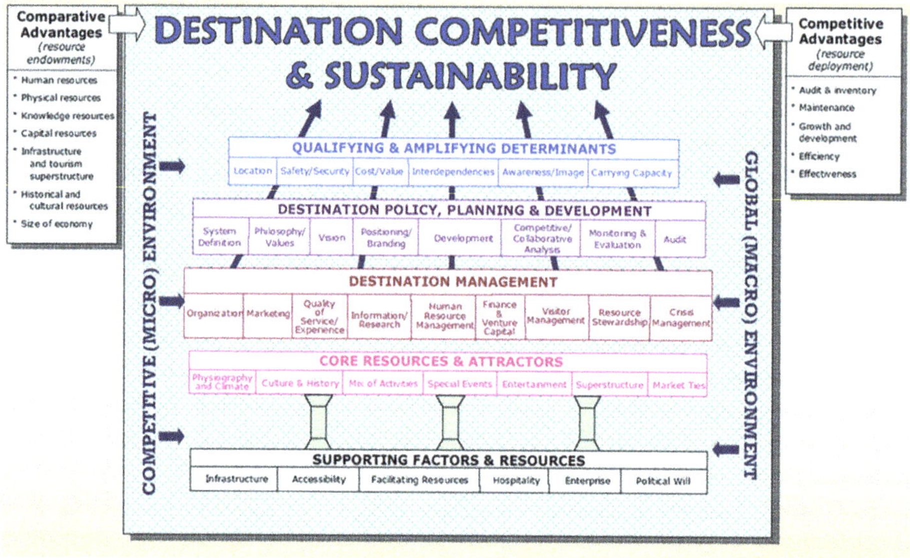

Fig. 21.1. Ritchie and Crouch conceptual model of destination competitiveness.

a mega-event, reflecting the crucial role of infrastructure in facilitating the smooth execution of a mega-event. In recent years, the organization of mega-events has moved away from infrastructure-heavy approaches to ones emphasizing reuse, adaptation and sustainability, rather than the construction of new venues (Jones, 2020; Ponzini *et al.*, 2020). In line with this, the original plan to host the Commonwealth Games 2026 aimed to control the costs by making use of existing sports venues and infrastructure (Le Grand and Hatch, 2023). Victoria's plan of hosting the Games involved the distribution of events across regional hubs, which would have created advantages in terms of efficiency, manageability of contingencies, and reduction of the level of socio-economic and spatial uncertainty while spreading the potential benefits around the state. It is suggested that such a model can serve as a strategic advantage in the planning and implementation of mega-events (e.g. Di Vita and Wilson, 2021). However, as more sports were included in the Games, the venue plan underwent significant modifications, leading to the requirement of extensive capital works and the construction of additional temporary facilities at additional expense. Therefore, the aspect that could have facilitated the successful implementation of the event had turned into one of the greatest challenges for the state and event planners. Could a more thorough assessment of potential challenges in the initial planning phase have prevented the financial strain that led to cancellation?

Another *Supporting factor* that could have played an important role in the cancellation of the games was political will, as local politics were distorted in the planning decisions related to this event (Le Grand and Hatch, 2023). The original bid documents did not estimate the costs of government services such as police requirements to ensure safety or the logistics involved in the movement of athletes and the general public between the regional hubs (Le Grand and Baum, 2023). Furthermore, while the bid to host the Games was initiated by the Victorian government, it lacked additional funding from the federal government (Yerushalmy, 2023).

Other critical considerations that may have determined the outcome of the planned Victorian Games are integrated within the strategic framework for *Destination Policy, Planning, and Development* guiding a destination's growth with specific economic, social, and societal objectives. This framework requires consensus among stakeholders on the scope of governance, influenced by the destination's unique philosophy. A clear vision of the destination's future, grounded in a thorough assessment of its attributes and competitive position, is essential. Accordingly, the *Destination Management* aspect of the model involves the daily implementation of the established policy and planning framework for destination development. It aims to enhance core attractions, improve supporting factors, and adapt to external influences. This includes various actionable activities such as destination marketing, enhancing the visitor experiences and fostering a holistic management approach, as well as logistical considerations in terms of financial and human resource needs and managing visitor impacts. The successful hosting of a mega-event relies heavily on effective logistics management to handle the intricate web of activities, movements and actors involved in the event. In this case study, the multi-hub concept was likely going to present a logistical challenge for event organizers, employees, athletes and attendees. The Victorian government estimated that 400,000 visitors would come to Victoria for the event, with authorities urging spectators not to travel between regional cities in a single day due to accommodation and transport issues (Matson, 2023). In an effort to minimize some of the logistical challenges, the government and Games officials discussed the construction of new regional railway stations to cater for the anticipated increase in visitors (Le Grand and Hatch, 2023). This shows that inadequate current *Supporting Factors and Resources* in Victoria's regions, such as event venues, accommodation, and transportation, would have hindered the destination's ability to successfully manage and accommodate the anticipated influx of visitors. The absence of suitable facilities such as accommodation could in turn have compromised the visitor experience and affected event logistics. Moreover, a lack of comprehensive planning and collaboration among event organizers, local authorities, and other relevant parties could have exacerbated logistical difficulties, making it impractical to adequately address the anticipated complexities of the event.

The group of *Qualifying and Amplifying Determinants*, often termed situational conditioners, moderate or magnify competitiveness by interacting with the other determinants. They include location advantages, interdependencies between destinations, safety and security, awareness and image, cost/value factors influenced by various economic structures, and a destination's carrying capacity. These factors can impose limits on a destination's potential, shape visitor choices, and affect the overall destination's growth and development. The *Qualifying and Amplifying Determinants* are interconnected with the three levels discussed in this section thus far. Our analysis shows that their combined impact led to a tipping point where the viability of holding the Games in Victoria became questionable. As previously discussed, several qualifying and amplifying factors had converged to create an environment that was unfeasible for the mega-event organizers and participants, leading to the cancellation of the Games. These include the relatively low carrying capacity and logistical challenges associated with the planned location of the Games, conflicting political will, limited transportation infrastructure and the cost/value perception of holding the Games in Victoria. Moreover, this mega-event required careful coordination and collaboration between multiple regional destinations, with potential issues arising from these interdependencies – such as conflicting interests, lack of efficient communication, and inability to ensure a seamless experience. These determinants could have further affected the overall success of the event. Arguably, the balance was tipped towards negative outcomes. Therefore, the cancellation of the Games could be seen as a decision to prevent potential losses and negative impacts on all stakeholders involved. This prompts us to ask, was the cancellation a failure not only of the Games but also of regional actors involved in the planning and execution? Should mega-events like the Commonwealth Games be scaled back in terms of size and scope to mitigate financial risks?

What Is Next for Victoria?

One of the key arguments of the Victorian government during the announcement of the Commonwealth Games' cancellation was opportunity cost. Since capital investment for mega-events is drawn from government funding, this money can be spent in other areas, such as improvement in the healthcare and education sectors where the host population may benefit more (Wan and Song, 2019). In effect, investment commitments to social and affordable housing and improvement of infrastructure planned to expand Victoria's carrying capacity and visitor experience during the Games are still on the agenda of the current government (Anderson, 2023).

Prior to the COVID-19 pandemic, events had been a significant driver of the Victorian visitor economy (City of Melbourne, 2020) with Victoria being recognized as a global sporting events capital for Australia (VTIC, 2021). As the state of Victoria emerges from the crisis, enhancing 'Victoria's reputation as Australia's events capital and global events state' has been prioritized among the key recovery strategies (Visit Victoria, 2021, p. 11). Nevertheless, the cancellation of the 2026 Commonwealth Games has sparked several criticisms across media channels as well as disappointment from industry bodies, such as Victoria Tourism Industry Council (VTIC), given its potential damage to Victoria's global sporting brand reputation (The Age, 2023; VTIC, 2023), which the state and industry have invested 30 years building. Looking ahead, we ponder, how can Victoria sustain its reputation and competitiveness as an events destination after the Commonwealth Games' cancellation?

Existing research has emphasized a strong reputation as one of the key success factors in bidding to host a mega or major sports event such as the Olympics and the Commonwealth Games (Westerbeek *et al.*, 2002; Geoff, 2008). Nicolas *et al.* (2012) suggest that country/destination image is an important factor in the bidding competition. Thus, the cancellation of the 2026 Commonwealth Games by the Victorian Government may affect Victoria's future competitiveness when bidding on another mega or major sports event. However, the state already hosts other sporting games, such as the Australian Football League, soccer A-league, rugby, cricket and Formula 1, that will not be affected by the cancellation. Furthermore, Melbourne has recently experienced tremendous success in co-hosting the 2023 FIFA Women's World Cup. This mega-event generated an AU$7.6 billion boost to the Australian economy. Over 1.5 million tickets were sold, with Football Australia celebrating hosting the most successful FIFA Women's World Cup ever (Football Australia, 2023). Although data on the specific economic impact of the event for each state are not yet available at the time of writing, undoubtedly, the success of the 2023 FIFA Women's World Cup in Australia as a whole and in Melbourne, in particular, has been a significant boost to the destination's reputation. Yet, whether such success can compensate for the Commonwealth Games cancellation requires empirical assessment when relevant data are available.

References

Allen, J., Harris, R., Jago, L., Tantrai, A., Jonson, P. and D'Arcy, E. (2022) *Festival and Special Event Management*. John Wiley & Sons, Australia.

Anderson, J. (2023) *Cancelling the Commonwealth Games Won't Come Cheaply – Victoria Now Faces the Legal Consequences*. The Conversation. Available at: https://theconversation.com/cancelling-the-commonwealth-games-wont-come-cheaply-victoria-now-faces-the-legal-consequences-210054 (accessed 15 August 2023).

City of Melbourne (2020) Covid-19 Reactivation and Recovery Plan. Available at: https://www.melbourne.vic.gov.au/sitecol-lectiondocuments/covid-19-reactivation-recovery-plan.pdf (accessed 19 July 2023).

Commonwealth Games Official Website (2023) Commonwealth Games 2026. Available at: https://commonwealthgames.com.au/victoria-2026/ (accessed 25 July 2023).

Di Vita, S. and Wilson, M. (2021) *Planning and Managing Smaller Events: Downsizing the Urban Spectacle*. Routledge, Abingdon.

Flyvbjerg, B., Budzier, A. and Lunn, D. (2021) Regression to the tail: Why the Olympics blow up. *Environment and Planning A: Economy and Space* 53(2), 233–260. DOI: 10.2139/ssrn.3686009.

Football Australia (2023) Football Australia Celebrates Hosting the Most Successful FIFA Women's World Cup Ever. Available at: https://www.footballaustralia.com.au/news/football-australia-celebrates-hosting-most-successful-fifa-womens-world-cup-ever (accessed 2 August 2023).

Geoff, W. (2008) *Bidding for Major Sporting Events: Key Issues and Challenges Faced by Sports Governing Bodies in the UK*. University of London UK.

Jones, Z.M. (2020) *Cultural Mega-Events: Opportunities and Risks for Heritage Cities*. Routledge, Abingdon.

Jones, Z.M., Di Vita, S. and Ponzini, D. (2022) Changing mega-events' spatial strategies and cultural policy: Scaling down, spacing out, and assembling organizations in the cases of London and Milan. *City, Territory and Architecture* 9(1), 1–11. DOI: 10.1186/s40410-022-00175-0.

Lauermann, J. and Vogelpohl, A. (2017) Fragile growth coalitions or powerful contestations? cancelled Olympic bids in Boston and Hamburg. *Environment and Planning A: Economy and Space* 49(8), 1887–1904. DOI: 10.1177/0308518x17711447.

Le Grand, C. and Baum, G. (2023) *Officials Knew about Games Cost Blowout Months Ago*. The Age. Available at: https://www.theage.com.au/national/victoria/officials-knew-about-games-cost-blowout-months-ago-20230720-p5dpx7.html (accessed 15 August 2023).

Le Grand, C. and Hatch, P. (2023) *How Victoria's Three-City Commonwealth Games Plan Ran Out of Control*. The Age. Available at: https://www.theage.com.au/national/victoria/how-victoria-s-three-city-commonwealth-games-plan-ran-out-of-control-20230727-p5drte.html (accessed 15 August 2023).

Li, S. (2013) Large sporting events and economic growth: Evidence from economic consequences of event infrastructure and venues. *Event Management* 17(4), 425–438.

Matson, K. (2023) *Commonwealth Games to Ignite Unprecedented Rural Construction Boom: Spectators Warned of Logistical Challenges at Victoria 2026 Commonwealth Games*. Herald Sun. Available at: https://www.heraldsun.com.au/news/victoria/commonwealth-games-to-ignite-unprecedented-rural-construction-boom/news-story/120fa832359a0e16d69b7f8e2df00c84 (accessed 19 August 2023).

Moore, S., Raco, M. and Clifford, B. (2017) The 2012 Olympic learning legacy agenda – The intentionalities of mobility for a new London model. *Urban Geography* 39(2), 214–235. DOI: 10.1080/02723638.2017.1300754.

Nicolas, H., Catherine, B. and Oumnia, S. (2012) Winning FIFA World cup bids: Country image analysis as a tool to understand and secure bids: A Moroccan case study. *International Journal of Sport Management, Recreation & Tourism* 10, 1–27.

Perić, M. and Vitezić, V. (2019) Socio-economic impacts of event failure: The case of a cancelled International Cycling Race. *Sustainability* 11(18), 5028. DOI: 10.3390/su11185028.

Ponzini, D., Bianchini, F., Georgi-Nerantzia Tzortzi, J. and Sanetra-Szeliga, J. (2020) *Mega-Events and Heritage: The Experience of Five European Cities*. International Cultural Centre, Kraków, Poland.

Ritchie, J.R.B. and Crouch, G.I. (2003) *The Competitive Destination: A Sustainable Tourism Perspective*. CAB International, Wallingford, UK.

The Age (2023) *Games U-Turn Damages Victoria's Reputation*. The Age. Available at: https://www.theage.com.au/national/victoria/games-u-turn-damages-victoria-s-reputation-20230718-p5dp59.html (accessed 10 August 2023).

Victoria Tourism Industry Council (2021) Formula 1 Grand Prix and Moto GP cancellations another blow to Victoria. Available at: https://www.vtic.com.au/news-item/7066/formula-1-grand-prix-and-moto-gp-cancellations-another-blow-to-victoria (accessed 20 August 2023).

Victoria Tourism Industry Council (2023) Victoria Needs Support in Wake of Cancelled Commonwealth Games. Available at: https://www.vtic.com.au/news-item/13333/victoria-needs-support-in-wake-of-cancelled-commonwealth-games (accessed 20 August 2023).

Victorian Government (2023) Joint Statement on Victoria 2026 Commonwealth Games. Available at: https://www.vic.gov.au/commonwealth-games-2026 (accessed 10 August 2023).

Visit Victoria (2021) Annual Report: 2020–21. Available at: https://assets-corporate.visitvictoria.com/documents/Visit-Victoria-Annual-Report-2020-21.pdf (accessed 20 August 2023).

Wan, S.K. and Song, H. (2019) Economic impact assessment of mega-events in the United Kingdom and Brazil. *Journal of Hospitality & Tourism Research* 43(7), 1044–1067.

Westerbeek, H.M., Turner, P. and Ingerson, L. (2002) Key success factors in bidding for hallmark sporting events. *International Marketing Review* 19(3), 303–322.

Yerushalmy, J. (2023) *Commonwealth Games 2026: Why Has Victoria Pulled Out and What Happens Now?*. The Guardian. Available at: https://www.theguardian.com/sport/2023/jul/18/commonwealth-games-2026-why-has-victoria-australia-pulled-out-what-happens-now#:~:text=The%20Australian%20state%20of%20Victoria,to%20make%20up%20the%20shortfall (accessed 25 August 2023).

22 The Bojo Algouinsan Ecotourism Initiative: Crisis Mitigation Through Community Tourism Development

Zainub Ibrahim and Paul Gemmell
School of Business and Hospitality, Algonquin College, Ottawa, Canada

Summary
Faced with many environmental threats to the ecologically sensitive Tañon Strait in the Filipino province of Cebu, the town of Aloguinsan and the village of Bojo turn to tourism to protect their community from oil drilling and cement manufacturing. Guided by Mr Joselito 'Boboi' Costas, the Bojo community has created an award-winning ecotourism destination. The aim of this case study is to highlight how a community successfully employed tourism as a tool to mitigate various risks and to provide a much-needed extra source of income to the community.

The Value and Interest of the Case Study

Globally, many communities are turning to tourism as a way to mitigate a variety of risks such as climate change and illegal primary resource extraction (Eshun and Tichaawa, 2020; Gabriel-Campos *et al.*, 2021). This is particularly relevant to destinations in developing countries that rely on extractive industries such as mining, fishing, and agriculture. Tourism can serve as part of a risk mitigation strategy as destinations aspire towards long-term sustainability (Beckent and Hughey, 2013).

The Bojo Aloguinsan Ecotourism Association has shown that through the guidance of a determined leader, a willing population, hard work, and an emphasis on education, a community can take ownership of its future. Sustainable ecotourism is a way for communities to strengthen their culture, protect their environment, and improve their financial situation. This small village, which went from a subsistence way of living to an award-winning ecotourism center, is proof of this.

Background

The Bojo River winds through the Aloguinsan province on the Filipino Island of Cebu. This narrow 1.2 km long river is found on the west side of the island and drains into the Tañon Strait. The Tañon Strait is found between the islands of Cebu to the west and Negros to the east. The strait is 160 km long and 27 km wide at the widest point but narrows to a width of just 5 km at the southern tip of Cebu Island. The river and surrounding area are a sanctuary for dolphins, fish, and over 60 species of birds. The steep rocky sides of the river near its mouth protect the animals and create stunning natural beauty.

The municipality of Aloguinsan sits on the Tañon Strait, roughly 100 km south-west of Cebu City, the largest city on the island. It is made up of 15 barangays, or villages, of which Bojo represents less than 6% of the total Aloguinsan population. The local community, called the Bojo, relies on the Bojo River for food and transportation. The river and the local people are intertwined and if anything were to happen to the river the Bojo would lose a part of their home, culture, and way of life.

In order to protect this ecologically unique and sensitive region, former president Fidel V. Ramos passed a law declaring the Tañon Strait a protected seascape. This prohibits the exploration of energy resources without approval from Congress.

Information for this case study was obtained through an interview with Mr Boboi Costas in April of 2022 and supplemented with secondary information from news articles about the events and development of the region.

Threats to the Tañon Strait

Petroleum exploration

A threat to the Tañon Strait materialized in May of 2005 when the Philippines Department of Energy (DOE) began to conduct seismic surveys without the required environmental impact assessment. Investigations revealed that the DOE had agreed to allow the Japan Petroleum Exploration Co, Ltd (JAPEX) to explore the Tañon Strait and test for the potential for drilling and extracting oil. Consequently, locals joined forces with environmental groups and sued the government. The explored region of the Strait was part of an environmentally protected area and the excavation agreement was approved only by the DOE. It had not been authorized by a general law, signed by the President, or reported to Congress, and was thus illegal.

After years of deliberations, the Philippines Supreme Court declared that the agreement between the DOE and JAPEX was unconstitutional, and 10 years later, in April 2015, the plan to allow JAPEX to drill in the strait was finally struck down.

Cement production

It was not only oil drilling in the strait that was a threat to the area, cement production in the Philippines was on the rise. The limestone rocks in the area are rich in silica and produce high-grade concrete. This growing industry had planned on building a plant near the Bojo River and using it as the primary water source. The Bojo people were able to work with the government to relocate the plant to San Fernando 30 km away from the river.

These threats to the Tañon Strait highlighted a risk to the Bojo River and consequently the Bojo community's lifestyle. Industrialization was becoming an increasing threat and the community felt the need to act. They found the help they needed in a man named Joselito 'Boboi' Costas. Together they would take this emerging crisis and turn it into an opportunity to address the environmental concerns for the river, and the social and economic ambitions of the community.

Boboi Costas

Boboi Costas, a Cebu City native, graduated from university and trained in community organization. Having grown up in the area, he knew first-hand the natural beauty that was to be found along the Tañon Strait coast and the difficult living conditions that many people in the area faced. He understood the urgency of protecting the environment and that with two major projects having been thwarted it was only a matter of time until more would come about. An analyzer of risk management, he realized the importance of a proactive approach to the preservation of the area. He thought the best approach to this problem was one where he worked with locals to create a long-term, proactive, flexible plan for the community.

In order to better understand the situation in Bojo and to gain the local's trust, in 2008 Mr Costas moved to Bojo, where he lived for 6 months. During this time, he listened to and learned from the local people. He saw and witnessed their needs. For instance, he saw that most locals were very poor and that many services and amenities needed to be improved.

Together they began to explore the idea of tourism in the region.

The idea of creating a program that would employ tourism to enrich the community upset some. They were afraid that Mr Costas was changing the status quo of the region. They tried to stop Boboi by suing him over petty trespassing charges. Efforts to deter him even escalated into death threats. Undeterred, he continued pushing his plan and worked with the local government to create the Bojo Aloguinsan Ecotourism Association (BAETAS).

The Bojo Aloguinsan Ecotourism Association (BAETAS)

Boboi began his vision to bring tourism to the region with the women of Bojo. They were more accessible than the men as they usually stayed home while the men went out and fished. Further, he realized that with the women on board, it would be easier to convince other community members of the benefits of tourism development in the community. They began small community programs that had an emphasis on environmental tours. With small community projects taking shape it was time to further push the ecotourism plan.

Mr Costas brought in professionals to educate the locals on various skills needed for tourism development, such as project management, accounting, tourism, and how to become educators themselves and share their knowledge with others. In turn, the people from Bojo taught the educators about their traditions, flora, and fauna. Teaching a community that never had any savings about finance proved challenging. Boboi recalls his realization that '[The people from Bojo] never dreamed of becoming accountants or balancing books ... you need to bring transparency and accountability especially when it comes to money. There's no end to mentoring or coaching to balancing books or auditing.'

With the financial backing and support of key tourism players such as the local government, the community, and other stakeholders, Boboi and the Bojo community created the BAETAS in 2009. BAETAS would welcome tourists from around the world and teach them about the river. With boat tours and glass bottom boats, the local population would take the tourists out and show them the unique biodiversity the region has to offer. The guides would show and teach visitors about the animals and plants of the riverway, how they play a role in protecting the river, how they supply materials for the local people, and about their history and their connection to the river. This would provide a much-needed source of income to the Bojo community.

The Boardwalk and Sweat Equity

Boboi worked with locals to build a 400-foot boardwalk paid for using a system called sweat equity. Sweat equity relies on locals to contribute to a project in any way they can. For example, it could be by investing time and labor, or by providing materials. The value of their contribution is then calculated and they receive equity in the form of a portion of tourism receipts from the admissions and tour packages.

Some locals brought supplies like bamboo or nails, others brought tools or food to feed the workers, and many put in the manual labor in building the boardwalk. Once completed, the boardwalk stretched along the Bojo River and allowed tours to intimately experience the river and learn about the animals and plants of the region. Further, having locals construct the boardwalk using natural and local materials reduced the environmental impact that using foreign materials would have caused.

Tourism Growth in Bojo

With the boardwalk, river tours in both traditional banca boats and glass bottom boats, and conservation and educational programs the region started to attract tourists. What started slowly, grew and the BAETAS quickly saw that a cap on the number of daily visitors was needed to keep the region sustainable. They decided that a cap of 60–70 tourists a day would be enforced. Most tourists were day visitors who came to enjoy the beauty of the region. Visitors who were looking to spend a few days would mostly stay at a local eco-lodge run by a nearby community. Sadly, the ecolodge was destroyed in Typhoon Odette in 2021 and at the time of writing had yet to be rebuilt.

Most of the income earned was from admission and river tour packages. A small portion of income also came from handcrafts and food souvenirs made by local people. By offering reasonable rates with an emphasis on education the BAETAS managed their guest capacity while also becoming a destination for ecotourists. This slow and steady approach yielded great results.

Not only was tourism generating revenue, but BAETAS also began winning tourism awards, including the Tourism InSPIRE Award (2015), Top 100 Sustainable Destination (2016, 2017, and 2018), and ASEAN Tourism Award – Best Community Based Tourism (2017).

In order to share their knowledge and experiences with neighboring communities, the BAETAS began hosting an annual open house and inviting neighboring leaders, children, and community members to experience their tourism offerings. They take them on riverboat rides, hold classes on the importance of mangrove and coral conservation, and answer questions. They also encourage visiting neighbors to organize their own ecotourism associations and create their own model of ecotourism in their communities.

With the Tañan Strait covering such a wide area shared by two provinces, three governors and close to 50 village heads, it can be challenging to coordinate initiatives. Thus, it is important that the small communities put the work in themselves and create a sustainable tourism concept that works at the local and regional levels.

Keeping a Good Thing Going

The BAETAS has proven to be a success. Within 10 years, the average monthly income had more than tripled and education scholarships were being distributed to the youth of the area. More money was being spent on education and conservation than ever before. However, with all this success there were some challenges that arose in making sure that tourism remained sustainable.

One challenge the organization faced as a result of increased income was the threat that many youths would leave the small village to seek a living elsewhere as expectations were raised. To address this, the BAETAS set up educational programs to teach youth about the importance of their environment to their lives and the economy. They also created internships, training, and jobs for the youth as a way to bring them more in touch with the community.

With the newfound tourism success, another risk facing the community was to their culture. It was important for the Bojo to maintain the authenticity of their culture. With this in mind, Mr Costas made a point to teach locals the importance of staying true to their beliefs and culture. The Bojo now perform a welcome ceremony for tourists that they have been performing to welcome guests for generations. They also teach tourists how to weave mats in the traditional Bojo style. These are done with the intention of economically benefitting from tourism, fostering pride in their culture, and celebrating and sharing their culture with outsiders without corrupting it.

Resilience and Risk Management

The BAETAS presents an interesting case in terms of proactive risk management. There was never a single event that caused a disaster in the region. Instead, the region and its local people faced several man-made threats. The Bojo and Mr Costas set in motion a pre-disaster planning and preparedness plan that promotes sustainability for the community and river.

Typhoon Odette in 2021

When disaster struck in the Cebu province in the form of Typhoon Odette in 2021, the tourism sector was able to recover relatively quickly. Because the boardwalk and buildings were built by locals, using locally sourced sustainable materials, they were able to be mended and rebuilt relatively easily. Further, the fact that the Bojo region's environment had not been affected by destruction from oil or issues with a nearby cement factory played a significant role in the recovery process. The local mangroves and dense Nipa fonds helped minimize coastal erosion.

COVID-19

As with other touristic regions around the globe, the COVID-19 pandemic was a test of the Bojo communities' resilience. Without tourism, locals returned to their traditional livelihoods of fishing and farming. Returning to the lifestyle that served them for years showed them that they had not lost their skills or knowledge by relying on ecotourism as a primary source of income. It also highlighted the value that sustainable tourism has in the future of the region. Working to maintain a long-term and steady form of ecotourism was a way to diversify their income sources and provide more options and opportunities.

Conclusion

With Boboi having completed his work with the BAETAS, he looks back on his time in Bojo and is proud of what they achieved, 'BAETAS is here to stay. Sustainability is already embedded in its operation and culture ... Its serious commitment to sustainability and its positive impact to other communities was the main reason they won the award[s].'

The Bojo River seems safe now, with a plan for preservation and education as a strong part of its future. The Bojo community and Mr Costas both know that they cannot stop monitoring risks and planning how best to deal with them. Risk management is a never-ending job. As the planet evolves, governments change, and tourists demand different things, the local people and the BAETAS need to stay watchful and flexible if they are to preserve their environment and lifestyle.

The implications of this case study are applicable to cases of communities in ecologically sensitive areas elsewhere in the world that face the threats to their livelihoods and lifestyles posed by extractive industries. For these communities, tourism can be used as part of a risk management strategy to provide an additional income stream to supplement traditional sources of income that may face a variety of external threats, including those caused by global warming and industrialization. Further, the additional income can be used to help the community become more resilient through the building of infrastructure and improving access to information, services, and education for locals.

Further Reading

Mancelita, R. (2020) Taiheiyo to build ₱15B new cement facility in San Fernando, Cebu. Available at: https://sugbo.ph/2020/taiheiyo-new-cement-facility/ (accessed 1 April 2023).

Miasco, M.B. (2017) Rainy weather this Saturday. The Freeman 52(177), 1–2.

Rappeler (2015) Supreme Court Stops Oil Exploration in Tañon Strait. Available at: https://www.rappler.com/environment/90661-court-stops-japex-doe-oil-exploration-tanon-strait/ (accessed 1 April 2023).

SunStar (2015) Best in Community-Based Tourism. Available at: https://www.sunstar.com.ph/article/42181/best-in-community-based-tourism (accessed 1 April 2023).

References

Beckent, S. and Hughey, K.F.D. (2013) Linking tourism into emergency management structures to enhance disaster risk reduction. *Tourism Management* 36, 77–85.

Eshun, G. and Tichaawa, T.M. (2020) Community participation, risk management and ecotourism sustainability issues in Ghana. *GeoJournal of Tourism and Geosites* 28(1), 313–331.

Gabriel-Campos, E., Werner-Masters, K., Cordova-Buiza, F. and Paucar-Caceres, A. (2021) Community eco-tourism in rural Peru: Resilience and adaptive capacities to the Covid-19 pandemic and climate change. *Journal of Hospitality and Tourism Management* 48, 416–427.

Tourism InSPIRE Award (2015). Available at: https://www.traveldailynews.asia/asia-pacific/pata-announces-winners-of-the-tourism-inspire-awards-2015/

23 Tourism Development Opportunities in Post War: A Case of Jaffna, Sri Lanka

M.T. Pathirana[1], B.C. Bigby[2], N. Scott[3] and S. Gardiner[4,5]

[1]Edith Cowan University, Sri Lanka campus, School of Business and Law, Colombo, Sri Lanka; [2]University of Waterloo, Ontario, Canada; [3]University of the Sunshine Coast, Queensland, Australia; [4]Institute for Tourism, Griffith, Australia; [5]Department of Tourism, Sport and Hotel Management, Griffith University, Gold Coast, Australia

Summary

Sri Lanka's tourism sector has undergone significant expansion, drawing global travellers with its diverse landscapes, rich cultural heritage, and welcoming locals. Despite setbacks associated with the civil war (1983–2009), the country has actively worked to rebuild its image and position itself as a secure travel destination. Infrastructure investments, including hotels, resorts, and transit networks, have played a crucial role in this development. Post-civil war, Jaffna in Sri Lanka's Northern Province has emerged as a promising tourism destination. Infrastructure upgrades, such as road renovations and the establishment of an international airport, have been pivotal. Jaffna cultural and natural treasures include the Jaffna Fort, Nallur Kandaswamy Kovil, and the Jaffna Public Library. However, further infrastructure enhancements and tourism professional training are needed for a complete guest experience. Community involvement, infrastructure investments, and effective marketing are essential for Jaffna to thrive as a tourist destination. This case study, based on stakeholder interviews, provides valuable insights for similar post-conflict areas globally, emphasizing sustainable development. It serves as a model for collaborative planning, promoting local empowerment while engaging various stakeholders. The unique experiences in Jaffna's post-civil war recovery make this case study relevant and significant for the broader South Asian and Asia Pacific region. It offers guidance for empowering local communities through tourism revitalization, focusing on inclusive and sustainable development. The study contributes to tourism research by offering practical lessons applicable to Jaffna and similar regions facing comparable challenges, fostering responsible and equitable tourism growth.

The Value and Interest of the Case Study

Due to its distinctive experiences with post-civil war rebuilding, this case study on tourism development in Jaffna, Sri Lanka, has tremendous value and should be of interest to the greater South Asian and Asia-Pacific region. It offers guidance to similar post-conflict locations around the world, helping them to empower local communities through revitalization of their tourism industries. In particular, the study emphasizes sustainable development, which fosters economic progress while protecting natural resources and cultural legacies. Additionally, it provides a useful model for group planning procedures with its inclusive approach prioritizing local empowerment while engaging numerous stakeholders. Overall, by providing suggestions and lessons gained that may be applied to Jaffna and other locations facing comparable issues, the case study advances the area of tourism research and promotes inclusive and sustainable tourism growth.

Background Analysis for Sri Lankan Tourism

The tourism industry in Sri Lanka has experienced significant growth (122%) from 2009 onwards (Dilhani and Abeynayake, 2022), attracting tourists with its diverse landscapes, rich cultural history, and hospitable people (Wickremasinghe and Ihalanayake, 2006). The government's early initiatives in the 1960s to promote tourism as a means of economic development laid the foundation for this expansion (Fernando *et al.*, 2016). However, the outbreak of the civil war in 1983 had severe repercussions, from deeply impacting local populations to damaging the country's international reputation and hindering the sector's growth (Fernando *et al.*, 2016). The end of the conflict in 2009 marked a turning point, with increased international interest and government efforts to enhance infrastructure and attract investment (Buultjens *et al.*, 2016). Investment from domestic and foreign companies, along with government support, has played a vital role in the sector's expansion (Fernando, 2016). Popular tourist destinations have emerged, such as Ella, the Cultural Triangle, Galle, Kandy, and Arugambay (Anderson *et al.*, 2007; Ranasinghe, 2015; Ranasinghe *et al.*, 2020). The tourism sector has contributed significantly to the country's economy, generating foreign exchange earnings and providing employment opportunities (Kamble and Bouchon, 2014; Fernando *et al.*, 2016). Sri Lanka has also emphasized eco-friendly tourism practices and the protection of natural and cultural resources (Arachchi *et al.*, 2015). However, challenges remain, including environmental protection and preservation, inclusive development, service quality, connectivity, infrastructure development, and maintaining competitiveness (Mustafa and Santhirasegaram, 2014; Fernando *et al.*, 2016; Fernando, 2017). Tourism has emerged as a critical accelerator for diversification and growth in the face of Sri Lanka's economic woes. Investing in the sector allows the country to produce cash, create jobs, and promote local enterprises, particularly in areas such as Jaffna. The multiplier effect of tourism can benefit communities, enhance infrastructure, and protect cultural heritage. It promotes national identity and pride in addition to economic benefits. Tourism, as a strategic solution, not only provides financial assistance but also helps Sri Lanka's towards a more resilient and culturally rich future (Sharpley, 2009). The tourism industry in Sri Lanka has become a cornerstone of economic development (Dilhani and Abeynayake, 2022).

Methodology of the Study

This study's methodology focused on inclusive and sustainable practises in order to give readers a thorough insight into Jaffna's post-civil war tourist development. The study utilized stakeholder interviews to obtain insights into the present condition of tourism in Jaffna.

Informants involved in the study included government officials, business owners in the tourism industry, members of the local community, proponents of cultural preservation, and other pertinent parties (informal sector opinion leaders representing artists, cultural leaders and societal leaders). These people were chosen in order to provide a diverse viewpoint on the difficulties Jaffna faces and its potential, based on their direct involvement in and knowledge of the tourism sector.

Sixteen interviews were conducted between December 2022 and February 2023. Every source had a different perspective on how Jaffna's tourism industry is doing right now. Due to the semi-structured nature of these interviews, a wide range of subjects pertaining to marketing tactics, community involvement, infrastructure development, and difficulties facing the tourism industry were explored. The following table depicts the participants' profiles (Table 23.1).

Thematic analysis, which entails finding patterns, themes, and commonalities among the responses, was used to analyse data. This procedure made it possible to classify the data and provide important insights into the region's potential and problems related to the growth of tourism.

Table 23.1. Participants' profiles.

Participant	Role/background	Sector/interest
1.	Government official	Tourism development
2.	Government official	Cultural preservation
3.	Business owner (Hotel)	Tourism industry
4.	Business owner (Travel Agency)	Tourism industry
5.	Business owner (Restaurant)	Tourism industry
6.	Community leader	Local community
7.	Community member	Local community
8.	Community member	Local community
9.	Proponent of cultural preservation	Cultural preservation
10.	Proponent of cultural preservation	Cultural preservation
11.	Academic/researcher	Tourism development
12.	NGO representative	Sustainable development
13.	Tour guide	Tourism industry
14.	Environmental activist	Sustainable development
15.	Tourism consultant	Tourism industry
16.	Local artisan	Cultural preservation

Source: Primary data collection.

To enhance transparency in reporting, the specific roles and backgrounds of the key informants involved in the study are acknowledged. This information helps readers assess the credibility and relevance of the insights provided. The study recognizes the importance of diverse perspectives, ensuring that stakeholders from various sectors and interests are represented in the research.

Potential of Tourism Development in the Jaffna District

Despite the positive strides taken, Fernando (2020) and Dissanayake and Samarathunga (2021) highlight several challenges that accompany the rapid growth of Sri Lanka's tourism industry. Environmental concerns, inclusive development issues, service quality, connectivity, and infrastructure development are acknowledged as areas needing attention. The aftermath of the civil war, while marking a turning point for increased international interest, raises questions about the nuanced impacts (including social displacement of local communities and cultural commodification, which can lead to a loss of authenticity and cultural erosion) on local populations and cultural heritage. Differing perspectives on the effectiveness of government initiatives and policies underscore the need for a more critical examination of their outcomes in achieving sustainable tourism development.

The northern province of Sri Lanka includes the culturally and historically significant region of Jaffna (Sivesan, 2020). Jaffna is currently experiencing tremendous development and revitalization following three decades of civil instability and conflict. Jaffna is a bustling city with a distinctive cultural history that is different from the rest of the island that is populated by the ethnic Sinhalese majority. The majority of people in Jaffna are from the Tamil community, which is well known for its distinctive food, architecture, and Hindu religious practises. Jaffna had a long and rich history as a bustling commercial port and a seat of scholarship. With the support of many development efforts and government help, Jaffna is becoming an increasingly popular tourist destination. How Jaffna differs from other Sri Lankan tourist destinations will be discussed.

Over the past three decades, Jaffna has experienced the greatest amount of civil instability and conflict in Sri Lanka, resulting in casualties, significant infrastructure damage, and economic stagnation. Both Jaffna's potential as a tourist attraction and the lives of those who lived there were seriously affected by the fighting. But after the war ended in 2009, Jaffna began a restoration and rehabilitation phase that prioritized social, economic, and infrastructure development (Sivesan, 2020).

Key tourism attractions

Distinctive tourist attractions that showcase Jaffna's rich history, culture, and scenic beauty abound in this captivating area (Dissanayake and Samarathunga, 2021). The Jaffna Fort is still standing today as a reminder of the region's colonial past. Built by the Portuguese in the 17th century and later enlarged by the Dutch, this magnificent fort offers a glimpse into Jaffna's history. Huge bastions, walls, and tunnels provide a bird's eye view of the surrounding city and exude majesty (Mathivathany and Sasitharan, 2012). Nallur Kandaswamy Kovil is a remarkable architectural work of art and one of the most significant Hindu temples in Sri Lanka. This temple honours Lord Murugan and has magnificent golden gopurams (entrance towers), vibrant murals, and intricate sculptures. Every year, the Nallur Festival draws thousands of tourists and believers to this vibrant festival of singing, dancing, and religious rituals. One of the greatest libraries in Asia, the Jaffna Public Library is a true knowledge gold mine (Aloysius *et al.*, 2020). This well-known building, with its distinctive red-tiled roof, is a paradise for book lovers, scholars, and researchers because it holds an extensive collection of books, manuscripts, and historical documents. In addition, the library serves as a focal point for cultural and educational initiatives that aid in the advancement and preservation of Tamil literature and cultural legacy.

Nestled along the pristine shoreline, Casuarina beach exudes a serene and ethereal ambience (Aloysius *et al.*, 2020). This beach's golden sand, graceful casuarina trees, and gentle Indian Ocean waves make it a perfect retreat for serenity and relaxation. Visitors can choose to sunbake, take leisurely strolls, or engage in water activities like swimming and snorkelling. Situated on the northernmost tip of the Jaffna Peninsula, the Keerimalai natural springs has spiritual and recreational value (Aloysius *et al.*, 2020). These picturesque freshwater springs are supposed to have therapeutic properties. Visitors can relax while soaking in the tranquil atmosphere by taking a plunge in the cool, crystal-clear waters. Delft Island is a little-known treasure that can be reached from Jaffna by quick ferry. With its white sand beaches and coral-rich waters, this secluded island is renowned for its natural features (Aloysius *et al.*, 2020). The wild horse population on Delft Island, which is descended from the horses that the Portuguese brought with them centuries ago, is another attraction. A tuk-tuk or bicycle tour of the island is available for visitors who want to see the ancient baobab trees and take in the tranquil atmosphere (Ranasinghe, 2018).

Along with these sights, Jaffna is also home to a large number of temples, churches, and other cultural sites that provide a window into the area's rich religious, multicultural and historical past (Ranasinghe, 2018). Visitors have the chance to fully immerse themselves in the regional customs and celebrations by attending the exciting festivals like the Jaffna Music Festival and the Hindu Vel Festival (Ranasinghe, 2018).

To accommodate the increasing number of visitors, many of Jaffna's tourism sites still need to be developed and their infrastructure improved, despite their enormous potential. Fostering Jaffna's tourism business requires enhancing accessibility, increasing lodging options, and supporting sustainable tourism practises (Sivesan, 2020).

Developing Jaffna as a tourist destination

The tourism industry in Jaffna still faces challenges, despite recent advances. The tourism attractions in the area need to be improved in order to handle the rising number of visitors (Dissanayake and Samarathunga, 2021). To improve the overall visitor experience, infrastructure upgrades, such as updating lodging, transportation systems, and tourism facilities, are crucial. To ensure sustainable tourist practises and genuine cultural experiences, there is a strong need for trained tourism experts and local community involvement (Shanmuganathan and Ramanan, 2019).

The potential for Jaffna to grow as a tourism destination is enormous. Jaffna can draw on a wide range of tourists looking for genuine and immersive experiences by protecting and marketing its cultural assets, investing in infrastructure, and offering training and support to local people (Mathivathany and Sasitharan, 2012). Jaffna stands out when compared to other tourism attractions in Sri Lanka because of its cultural and historical uniqueness, particularly seen in its Tamil heritage, architecture, and culinary customs. With Jaffna's recent growth as a tourism destination, travellers now have the opportunity to discover a less well-known region of Sri Lanka, away from the island's more travelled coastline and hill country sites (Balasubramaniam and De Zoysa, 2019).

Since the end of the civil war in Sri Lanka, Jaffna has undergone significant infrastructure improvements, including the renovation of roads, the establishment of the Jaffna International Airport, and upgrades to transportation and utility systems (Katiresu, 2004; Sivesan, 2020; Dissanayake and Samarathunga, 2021). The restoration of the A9 roadway and the availability of commercial flights at the Jaffna International Airport have enhanced connectivity and attracted more tourists (Wickramasinghe, 2020; Dissanayake and Samarathunga, 2021). The renovation of the Jaffna railway station and the reinstatement of the railway line to Colombo have improved transportation options (Katiresu, 2004; Khullar, 2022). Upgrades to electricity, water delivery networks, and telecommunications infrastructure have also benefited the region (Mathivathany and Sasitharan, 2012; Aloysius *et al.*, 2020). Additionally, public services, amenities, and historical landmarks have undergone renovation and construction (Sivesan, 2020). These infrastructure developments have contributed to economic growth, increased employment, and improved living standards (Aloysius *et al.*, 2020; Dissanayake and Samarathunga, 2021). However, further investments are needed to address remaining infrastructure gaps and ensure sustainable regional development (Sivesan, 2020). Efforts to enhance connectivity, expand public services, and promote responsible tourism will contribute to Jaffna's ongoing development as a significant tourism destination (Sivesan, 2020).

Many of Jaffna's tourism attractions are still underdeveloped and lack the necessary amenities to accommodate visitors, despite its historical and cultural significance. Among the issues that must be resolved are a shortage of skilled tourism experts, poor lodging options, and limited infrastructure (Aloysius *et al.*, 2020). However, these gaps also offer considerable chances for the area's future growth, investment, and job creation (Aloysius *et al.*, 2020). Compared to other tourism spots in Sri Lanka, Jaffna offers a unique and genuine travel experience (Mathivathany and Sasitharan, 2012). For travellers looking for an experience off the main route, its rich cultural heritage, delectable cuisine, traditional arts and crafts, and warm hospitality offer a special charm (Kangatharan, 2016). Cultural celebrations in the area, such the Thai Pongal and Nallur Festival, draw visitors from both domestic and foreign visitors (Kalimuththu *et al.*, 2018).

The public and business sectors must work together in order to transform Jaffna into a viable tourist destination (Mathivathany and Sasitharan, 2012). It is crucial to make investments in infrastructure development, which includes enhancing visitor amenities, upgrading lodging options, and strengthening transportation networks. Additionally, promoting Jaffna's attractions through efficient domestic and international marketing efforts will raise awareness and draw a greater variety of tourists.

Local communities should be empowered to take a proactive role in the development process, ensuring that tourism-related advantages are divided fairly, protecting regional history and customs and encouraging sustainable practises (Kalimuththu *et al.*, 2018). The preservation of Jaffna's cultural heritage monuments can be aided by cooperation with international organizations like UNESCO. For a number of reasons, developing Jaffna's tourism industry is essential. First, by creating job opportunities and supporting regional businesses, it helps the economy of the area thrive (Kandeepan *et al.*, 2021). Second, it fosters a sense of pride and identification among the local population by preserving and promoting Jaffna's distinctive cultural history (Kandeepan *et al.*, 2021) compared to other districts in Sri Lanka. Last, but not the least, tourism can aid in efforts to achieve peace by fostering communication, meaningful engagement and economic empowerment among various communities (Alluri *et al.*, 2014).

A thorough strategy is needed to make Jaffna a popular tourism destination. This involves spending on marketing and promotion initiatives, increasing the skills of tourism stakeholders, and implementing sustainable tourist practises (Kandeepan *et al.*, 2021). To ensure the balanced and responsible development of Jaffna's tourism industry, cooperation between the government, corporate sector, and local people is crucial (Mathivathany and Sasitharan, 2012).

Within the literature, alternative perspectives challenge the notion of tourism as a singular solution for economic diversification and growth in Sri Lanka.Calgaro *et al.* (2014) and Tsao and Ni (2016) argue for a more diversified economic strategy to mitigate the vulnerability of relying solely on tourism, pointing out potential drawbacks such as the economy's susceptibility to fluctuations in the tourism industry. Recognizing these diverse strands of thought offers a more comprehensive understanding of the complexities and nuances involved in shaping Sri Lanka's tourism landscape. It invites a balanced discussion on the multifaceted impacts of tourism, encouraging stakeholders to consider alternative pathways for sustainable development beyond the confines of the tourism sector.

Major Barriers to Tourism Development

In the Jaffna region, there has been a lack of confidence in the development of tourism. A spirit of caution and scepticism has developed among the local population as a result of the conflict's memories and experiences, as well as the loss and displacement they experienced. For tourism to develop sustainably in the area, trust must be established and issues relating to how tourism affects cultural heritage, social dynamics, and economic rewards must be meaningfully addressed. Data analysis indicates a range of barriers that impact and in some cases hinder or delay tourism development in Jaffna. The main challenges to tourism development in the Jaffna district include cultural concerns, societal challenges in a post-conflict context, migration, a lack of investment, political representation issues, problems with government departments, a lack of awareness, and active security measures.

> I'd like to highlight the issues faced by insufficient investments in Jaffna's tourism development. The protracted battle and its aftermath generated an atmosphere of uncertainty and instability, discouraging private investors from investing in the region's tourism infrastructure. As a result, vital services like housing, transportation, and other necessities have been curtailed, limiting the district's ability to adequately welcome and serve to tourists. This lack of investment has hampered Jaffna's overall development and appeal as a booming tourism destination. (Interview 1)

> I'd like to emphasize the challenges that the Jaffna district faces, particularly its lack of political representation in parliament. The region's fight for proper political power and representation within the national government has created barriers to obtaining critical financing and support for the expansion of the local tourism industry. This lack of representation has sadly led in a lack of focus on critical tourism-related infrastructure and policies, impeding the region's overall growth and potential. (Interview 4)

The case study on Jaffna's post-civil war recovery holds significant relevance for the broader South Asian and Asia-Pacific region. It provides guidance for empowering local communities through tourism revitalization, focusing on inclusive and sustainable development. The study contributes to tourism research by offering practical lessons applicable to Jaffna and similar regions facing comparable challenges, fostering responsible and equitable tourism growth.

The following are the main barriers identified from the data collection that may challenge the development of tourism in Jaffna.

1. **Cultural Concerns:** Diverse cultural perspectives, particularly those of the predominantly Hindu population, significantly influence tourism development in Jaffna. Strong religious beliefs and cultural customs

create challenges in introducing and adjusting tourism-related activities. Striking a balance between respecting cultural traditions, as seen in events like the Nallur Kandaswamy Temple festival, and meeting the desires of non-Hindu tourists require careful cultural understanding, defined viewing locations, and explicit scheduling.

2. **Societal Challenges in a Post-Conflict Context:** The prolonged closure of communities due to the civil war has hindered the preparedness of the local populace and infrastructure for tourism development. Limited access and opportunities have created challenges in welcoming tourism, impacting the region's ability to capitalize on the industry's growth.
3. **Migration:** The significant population displacement during the conflict has led to a decline in skilled workers and potential tourists. The resulting shortages have adversely affected the expansion and development of Jaffna's tourism industry.
4. **Lack of Investment:** The aftermath of the civil war created an atmosphere of unpredictability and instability, dissuading private investors from contributing to the tourism infrastructure. This lack of investment has restricted the district's ability to adequately cater to tourists, limiting its overall appeal as a tourism destination.
5. **Political Representation Issues:** The Jaffna district's minimal political influence and representation within the national government hinders the necessary funding and assistance required for tourism growth. This lack of representation has contributed to neglect and slower development of tourism-related infrastructure and policy efforts.
6. **Bureaucratic Roadblocks and Delays:** Political factors, including bureaucratic roadblocks and delays, have impeded the necessary permissions and approvals for tourism-related projects. Inefficient government procedures and a lack of coordination among various offices have discouraged potential investors and slowed the expansion of the tourism sector.
7. **Lack of Awareness:** The local community's lack of awareness of the advantages of tourism growth poses a significant obstacle. Overcoming this challenge requires inclusive, community-driven approaches to tourism development, along with initiatives to raise awareness about the positive effects of tourism.
8. **Aggressive Security Measures:** Stringent security measures, a response to the area's turbulent past, pose a challenge by potentially intimidating visitors. Striking a balance between security precautions and fostering a peaceful, welcoming environment is crucial for the sustainable growth of tourism.

Recommendations and suggestions

This section presents recommendations to be implemented. The following proposals and ideas are put forth to improve Jaffna tourism over the course of the next 5 years based on the challenges and development opportunities for the tourism industry in Jaffna.

This case study suggests to actively involve the local community in tourism decision-making through workshops, seminars, and cultural exchange programs. By engaging residents in decision-making processes, a sense of ownership and pride can be fostered, aligning with the goal of empowering communities to take charge of their tourism industry.

> I would emphasize the need of including the local population in tourist decisions and raising knowledge of the benefits of tourism. We may foster a sense of ownership and pride through workshops, seminars, and cultural exchanges, encouraging active engagement in tourism projects. (Interview findings 9)

Empowering local people with the necessary knowledge and skills not only enhances the quality of services but also contributes to long-term livelihoods in tourism, thus supporting the goal of sustainable development.

> I would emphasize the importance of training and development for tourism personnel in order to improve their abilities and provide quality service. This helps to preserve livelihoods, provide excellent guest experiences, and promote Jaffna's tourism potential. (Interview findings 7)

This case study recommends strengthening the collaboration between local authorities, community members, and government organizations to ensure a peaceful and stable environment. This directly addresses political difficulties and maintaining security for the growth of tourism. By fostering a conducive atmosphere for tourism, the region can attract more visitors and foster economic progress while protecting cultural legacies.

Investing in building tourist-friendly amenities, improving lodging options, and upgrading transportation infrastructure will improve Jaffna's tourism appeal through infrastructural upgrades. Enhancing infrastructure not only enhances visitor experiences but also lays the groundwork for sustainable tourism development, supporting economic progress while preserving natural resources.

Implementing extensive awareness programs targeting both domestic and foreign audiences through social media, travel journals, and tourism fairs facilitates responsible and equitable tourism growth. By showcasing Jaffna's unique attractions and cultural heritage, these efforts can attract a diverse range of tourists while promoting local empowerment and engagement.

> I'd emphasize the importance of strong awareness campaigns to highlight Jaffna's unique attractions and culture. This means reaching out to both domestic and foreign audiences via social media, travel literature, and tourism events. (Interview findings 10)

This case study recommends that government institutions encourage private investment through incentives, tax cuts, and streamlined procedures. By facilitating the expansion of tourism-related businesses, the region can create additional opportunities for economic growth and job creation while diversifying tourism offerings.

It is crucial to enhance air and sea connectivity by establishing agreements with smaller airlines for direct flights and improving ferry services. Improved accessibility not only increases Jaffna's appeal to tourists but also fosters economic growth and cultural exchange by facilitating easier travel to and from the region.

> the value of building agreements with smaller airlines for direct flights to Jaffna. Improved air connectivity and ferry services would bring in more tourists, increasing Jaffna's accessibility and appeal. (Interview findings 11)

Informing local communities about changes and advancements in Jaffna's tourism industry will instil trust and confidence in local tourists and encourage more local visitors. By promoting awareness and pride in local attractions, these efforts can boost the local economy and foster a sense of belonging among residents.

Finally, this case study recommends establishing and enforcing quality standards for lodging, dining, guides, and transportation. Consistency in quality not only enhances the overall visitor experience but also elevates Jaffna's tourism profile, attracting more tourists and supporting sustainable growth.

> I'd like to emphasize the significance of establishing and enforcing quality standards for lodging, food, guides, and transportation. Consistency guarantees better guest experiences and raises Jaffna's tourism profile. (Interview findings 6)

By implementing these recommendations, Jaffna can begin to address the challenges identified in the case study and work towards achieving its objectives for sustainable and inclusive tourism development.

Conclusions

With its varied landscapes, rich cultural legacy, and welcoming people, Sri Lanka's tourism business has seen substantial expansion throughout the years. However, the devasting impacts of civil war had a negative effect on the travel industry, harming the nation's image and infrastructure and impacting its people. Sri Lanka concentrated on reconstruction and marketing itself as a secure and alluring tourism destination after the conflict, which sparked more interest from visitors from abroad.

The growth of Sri Lanka's tourism industry has been greatly aided by government backing as well as investment from both domestic and foreign businesses. Travellers now have more accommodation alternatives because of infrastructure development, which has included the building of hotels, resorts, and transportation networks. The tourism industry has made a substantial economic contribution to Sri Lanka by bringing in foreign cash and creating employment opportunities across a range of sectors.

The nation has also placed a strong emphasis on sustainable tourism practises and the preservation of natural and cultural resources.

The northern province of Sri Lanka's Jaffna district has potential to develop its tourism offerings, but faces challenges. After three decades of conflict, Jaffna is enjoying revitalization thanks to its distinctive cultural heritage. The Jaffna International Airport and other infrastructure upgrades have improved connectivity and accessibility. The Jaffna Fort, Nallur Kandaswamy Kovil, Jaffna Public Library, Casuarina beach, and Keerimalai natural springs are among the district's most popular tourist destinations.

Additional infrastructure and amenity investments are required for Jaffna to grow as a tourist destination. It is vital to improve hotel options and accessibility and encourage sustainable tourism strategies. Due to its distinct Tamil culture and unique history, Jaffna stands out from other tourist attractions in Sri Lanka and offers visitors a distinctive experience. Most importantly, the development of the tourism business in Jaffna must be balanced and responsible, and this requires cooperation between the public and private sectors as well as meaningful leadership by and inclusion of the local community.

References

Alluri, R.M., Leicher, M., Palme, K. and Joras, U. (2014) Understanding economic effects of violent conflicts on tourism: Empirical reflections from Croatia, Rwanda and Sri Lanka. In: Wohlmuther, C. and Wintersteiner, W. (eds) *International Handbook on Tourism and Peace*. Centre for Peace research and Peace Education of the Klagenfurt University, Austria, pp. 101–119.

Aloysius, N., Yousaf, A. and Saira, M.S. (2020) Challenges and opportunities for ecotourism in district Jaffna, Sri Lanka. *Journal of Wildlife and Ecology 4*, 122–129.

Anderson, R., Green, J. and Souter, C. (2007) Galle harbour maritime archaeological impact assessment report for Sri Lankan department of archaeology. *Western Australian Museum Department of Maritime Archaeology* 235, 1–33.

Arachchi, R.S.S.W., Yajid, M.S.A. and Khatibi, A. (2015) Ecotourism practices in Sri Lankan eco resorts: A supplier perspective analysis. *Journal of Tourism and Hospitality Management* 3(9–10), 169–180.

Balasubramaniam, K. and De Zoysa, M. (2019) Sustainable ecotourism development: A case study. In: *Mandaitivu Island*. ISA, Jaffna, p. 31.

Buultjens, J.W., Ratnayake, I. and Gnanapala, W.A.C. (2016) Post-Conflict tourism development in Sri Lanka: implications for building resilience. *Current Issues in Tourism* 19(4), 355–372.

Calgaro, E., Lloyd, K. and Dominey-Howes, D. (2014) From vulnerability to transformation: A framework for assessing the vulnerability and resilience of tourism destinations. *Journal of Sustainable Tourism* 22(3), 341–360.

Dilhani, E.V.D. and Abeynayake, N.R. (2022) How strong Is the linkage between tourism and economic growth in Sri Lanka; evidence from 1971 to 2020. *Journal of the University Of Ruhuna* 10(1), 17–24.

Dissanayake, D.M.M.I. and Samarathunga, W.H.M.S. (2021) Post-war city to a tourism city: the perspectives of local stakeholders on post-war city tourism development in Jaffna, Sri Lanka. *International Journal of Tourism Cities* 7(3), 602–621.

Fernando, S. (2016) Managing the post-war tourism development in Sri Lanka. *International Journal of Business and Social Science* 7(11), 90–100.

Fernando, S. (2017) The tourism-led development strategy in Sri Lanka. *Journal of Business and Technology* 1(1), 38–52.

Fernando, I.N. (2020) Tourism competitiveness by shift-share analysis to way-forward Destination Management: A case study for Sri Lanka. *Journal of Tourism and Services* 11(21), 88–102.

Fernando, S., Bandara, J.S. and Smith, C. (2016) Tourism in Sri Lanka. In: *The Routledge Handbook of Tourism in Asia*. pp. 251–264.

Kalimuththu, S., Abeysekara, N. and Gamini, L.P.S. (2018) Tourists satisfaction towards heritage tourism in Jaffna Peninsula: Sri Lanka. In: *Proceeding of the 15th open University Research Sessions*. The Open University of Sri Lanka, Nugegoda, Sri Lanka, pp. 397–400.

Kamble, Z. and Bouchon, F. (2014) Tourism planning and a nation's vision: A review of the tourism policy of Sri Lanka. *Procedia-Social and Behavioral Sciences* 144, 229–236.

Kandeepan, V., Seevaratnam, T. and Somasuntharam, R. (2021) Impact of the tourism development on the quality of life of the people: A case in Jaffna municipal area. *Shanlax International Journal of Management* 9(2), 73–83.

Kangatharan, G. (2016) The Study of Pilgrim Tourists' Perception of Jaffna as a Tourist Destination (Doctoral dissertation, Uva Wellassa University).

Katiresu, S. (2004) A Hand Book to the Jaffna Penninsula and a Souvenir of the Opening of the Railway to the North. Asian Educational Services.

Khullar, S. (2022) Journeys with Yal Devi: War, peace, and contemporary art in Sri Lanka. *Feminist Studies* 48(2), 502–546.

Mathivathany, V. and Sasitharan, P. (2012) Potentials for regional development of tourism industry-Post war at Jaffna district in Sri Lanka. Conference: 2nd International conference on Sustainable Built Environment (ICSBE 2012), University of Peradeniya, Sri Lanka.

Mustafa, A.M.M. and Santhirasegaram, S. (2014) Empirical investigation of the relationship between tourism receipts and sustainable economic growth in Sri Lanka. *Journal of Emerging Trends in Economics and Management Sciences* 5(7), 131–137.

Ranasinghe, R. (2015) Strategic myopia of tourism development in Sri Lanka. *A Critique International Journal of Multidisciplinary Research and Development* 2(2), 604–609.

Ranasinghe, R. (2018) Cultural and heritage tourism development in postwar regions: concerns for sustainability from northern Sri Lankan capital. *Jaffna Journal of Tourism and Recreation* 4(1), 1–18.

Ranasinghe, J.P.R.C., Danthanarayana, C.P., Ranaweera, R.A.A.K. and Idroos, A.A. (2020) Role of destination smartness in shaping tourist satisfaction: A SEM based on technological attributes in Sri Lanka. *IOP Conference Series: Earth and Environmental Science* 511(1), 012001.

Shanmuganathan, K. and Ramanan, M. (2019) Tourists' satisfaction and word of mouth communication-with special reference to Delft Island. In: *Challenges, Trends and Opportunities of Technical and Vocational*. Jaffna, Ratmalana, Sri Lanka, p. 184.

Sharpley, R. (2009) *Tourism Development and the Environment: Beyond Sustainability?* Earthscan, London.

Sivesan, S. (2020) Sustainable tourism development in Jaffna District. *Journal of Tourism and Hospitality* 9(3), 1–5.

Tsao, C.Y. and Ni, C.C. (2016) Vulnerability, resilience, and the adaptive cycle in a crisis-prone tourism community. *Tourism Geographies* 18(1), 80–105.

Wickramasinghe, A. (2020) Civil aviation industry in Sri Lanka: challenges and solutions. In: *Aviation Law and Policy in Asia*. Brill Nijhoff, Netherlands, pp. 326–341.

Wickremasinghe, G.B. and Ihalanayake, R. (2006) The Causal Relationship Between Tourism and Economic Growth in Sri Lanka: Some Empirical Evidence. WP2006, 10.

Section 5: Critical Studies in Tourism

This section explores power imbalances in tourism through cases from Bangladesh, Malaysia, Nepal, the Philippines and India. It covers issues like indigenous rights, unequal opportunities, wartime legacies, animal exploitation, and post-colonial impacts, emphasizing the need for more ethical and inclusive tourism practices.

24 The Growth of Tourism and 'Development' in an Indigenous Setting: Exploring a Case of the Chittagong Hill Tracts, Bangladesh

Sabiha Yeasmin Rosy
Department of Women and Gender Studies, University of Dhaka, Dhaka, Bangladesh

Summary

This case study examines the process of tourism development in the Chittagong Hill Tracts of Bangladesh, where multidimensional concerns related to Indigenous communities are coming to the fore. This area contains valuable natural resources that provide huge potential for the growth of tourism. However, accounts of discrimination and violations by the state towards Indigenous peoples indicate the potential for inequality and exploitation while developing tourism. Eleven Indigenous groups live in the Chittagong Hill Tracts, a historically conflict-ridden area where there is tension between the dominant ethnic Bangalee and the Indigenous groups due to the latter's limited control over their way of life and the environment. Structurally, the state endorses mass tourism, where powerful actors – government, military, politically influential individuals, and companies – are engaged in determining the 'development' for the Indigenous people and modernizing the place as a part of capitalist ventures to gain profits. Given that state actors or private investors wield power and control over tourism, exploitation ranges from Indigenous peoples' lack of control over the planning to limited participation in tourism. Therefore, this case study explores the implications of tourism development for Indigenous peoples, analysing the state's economic intervention in favour of the Bangalee people, who are an ethnically dominant group in Bangladesh. Moreover, this case study raises questions about how the 'development' is taking place and identifies the practice of 'development' in an Indigenous setting and how the Indigenous peoples are situated in this capitalistic venture in a post-conflict setting. The case study is based on empirical findings collected through an ethnographic study as part of a PhD. The study followed ethical guidelines approved by the Macquarie University Ethics Committee in 2017.

Background

Tourism is a growing business all over the world, generating multiple opportunities and benefits for the state and the different people involved in this process. Tourism is substantially connected to the physical environment, and economic aspects, where the environment, society, culture, and economy experience change due to the infrastructural development in the host community and capitalist modernized growth (Saarinen and Nepal, 2016). Bangladesh, being an emerging country in the tourism sector, holds the potential to enhance the development of tourist destinations with a focus on economic development due to its picturesque landscapes and diversified culture. Generally, domestic tourism still prevails in Bangladesh, with many visiting Chittagong Hill Tracts. Chittagong Hill Tracts is a region with mountains and green terrain where indigenous livelihood, culture, and nature create a complex nexus, mainly due to the history of conflict between Bangalee and Indigenous peoples (Rasul and Thapa, 2007). With the recent growth of tourism in a post-conflict Indigenous setting, it becomes important to understand the way tourism is taking place, who is mainly involved, and the impacts involved on the Indigenous peoples. This is the focus of this case study.

The Chittagong Hill Tracts region, resting in the south-east of Bangladesh, comprises three hill districts named Rangamati, Khagrachari, and Bandarban (Gain, 2000). This region shares the border with India and Myanmar, covering about 10% of the country's total land surface, where 11 Indigenous groups live. They are known as 'tribal' or 'small ethnic groups' people instead of Indigenous people in official documents of Bangladesh, which has contributed to the ignition of identity politics and the violation of the human rights of Indigenous people. In contrast, Indigenous people prefer to be identified as '*adivasi*' (Indigenous), '*pahari*' (hill people), and '*jumma*' (slash and burn farmer). This case study uses the term 'Indigenous people', as this is how the people generally prefer to be recognized. However, the state has not accepted the way the Indigenous people of the Chittagong Hill Tracts want to be known, preventing them from having an equal identity as the dominant ethnic Bangalee people, who represent 99% of the total population of Bangladesh (Gerharz, 2014; Chakma and Chakma, 2023). To protest the denial of identity and threat of possible annihilation, Manabendra Narayan Larma, an Indigenous political leader, formed a political platform named the Parbattya Chattagram Jana Samhati Samiti (PCJSS) in 1972. Aside from PCJSS, one armed wing was also developed, the *Shanti Bahini* (peace forces). From the 1980s on, there was a violent conflict between the military and peace forces, until the signing of the peace accord in 1997. However, the denial of fulfilling the promises of the peace accord still holds the Indigenous people in a situation where they cannot exercise their autonomy over their land or the development projects taking place in the Chittagong Hill Tracts.

Though a post-conflict zone, Chittagong Hill Tracts has become a popular tourist destination over the years due to its potential to showcase natural beauty, Indigenous peoples, and culture. The way tourism is developing in the Chittagong Hill Tracts reflects modernized development ideas, which mainly focus on economic profits through the commercialization of nature. The development of tourism follows mass capitalist expansions, which shape the natural environment as a gateway for tourists from different cities through the commercial development of hotels and resorts. The intention behind modernized tourism is to 'develop' the region economically and introduce the Indigenous peoples to new ways of life and livelihoods. However, in this process of making tourism, Indigenous peoples are under the control of different powerful groups and remain marginalized in the decision making process, although they are mainly the host groups for tourists. Thus, this case study highlights how tourism is growing, who is mainly involved, and what the impacts of tourism are on Indigenous peoples and their environments.

This case study is built on the work of my PhD that employed an ethnographic research methodology to conduct fieldwork in Bandarban in 2017, with a short visit to Rangamati. The study followed the ethical guidelines of the Macquarie University Ethics Committee. As a Bangalee, the researcher entered the Indigenous community through Aunt Naly, who served as a gatekeeper. She is the founder of a local non-governmental organization in Bandarban, and she assisted the researcher in locating a native research assistant and arranging lodging with a Marma Indigenous family. Over time, the researcher's frequent visits to the study locations and interactions with local Indigenous people helped her become familiar with the traditions and practices of many Indigenous populations. She took part in and observed the participants' daily lives. Individuals from different Indigenous communities helped her reach participants by using their networks.

'Development' Efforts and Growth of Tourism

Development involves planning and implementation following modernization to progress economic growth, livelihoods, and the well-being of people. The modernization theory of development is debated, yet it is a persistent form of development implicated mainly in promoting capitalistic ideas of growth. Due to the economic focus, it has never been denied. In the case of Chittagong Hill Tracts, the history of development and the economic focus dates back to the pre-independence era of Bangladesh, which started in the 1950s (Rosy, 2020).

From the 1950s on, Chittagong Hill Tracts experienced growth-based development projects and interventions for economic development by the then government. In 1951, Karnaphuli Paper Mills (KPM) was established in Chandaghona, Chittagong – an administrative area of Kaptai within Rangamati Hill District. This was a step towards development during the Pakistan era, before the independence of Bangladesh. Later, the construction of the Kaptai dam (1957–1962) was another development initiative to produce electricity for the country. Both projects had different impacts on people and the environment. The wastage from the paper mills polluted the river Karnaphuli, resulting in a reduction in fish production as well as the land fertility of surrounding areas. Similarly, 100,000 indigenous people were evicted from their ancestral land as the dam inundated about 40% of the total cultivable land of the Rangamati district (Parveen and Faisal, 2002). These people had to move to different places without receiving proper relocation or compensation for the loss of land and other assets. After the independence of Bangladesh in 1971, the then President General Zia formed the Chittagong Hill Tracts Development Board (CHTDB) in 1976 to initiate development programs on a large scale in terms of road construction, telecommunication, electrification, and moving hill people into 'cluster' villages. Through this process, the government took control of the land and 'developed' it through their preferred ways of modernization (Gain, 2013).

Similarly, the development of tourism started to come to the forefront after the signing of the peace accord in 1997 in the Chittagong Hill Tracts. All three districts under the Chittagong Hill Tracts have mountainous landscapes, reserve forests, rivers, waterfalls, and Indigenous peoples, rightly 'fitted' to grow tourism for commercial purposes. Hence, the Chittagong Hill Tracts' growing tourism industry serves as an example of mass tourism, which is driven by technological advancements and corporate interests. Everything related to tourism has the potential to be commercialized, which has an adverse effect on people, culture, and natural areas. The major focus of tourism development is on infrastructural development through the construction of parks, hotels, resorts, and restaurants (Rosy, 2022). To take a break from mundane city life, tourists come to the Chittagong Hill Tracts to enjoy nature and visit some common tourist destinations such as *Chimbuk Hill, Golden Temple, Meghla, Nilgiri,* and *Nilachol.* While entrance to these locations was free a few years ago, visitors now have to pay to enter. These places went through renovations and have a new look now to accommodate more tourists with modern seating arrangements (Fig. 24.1), refreshments, parking facilities, and resorts to stay at night with prior permission from the authorities. Tourists also buy cultural commodities – such as shawls, mufflers, bags, blankets, and bed sheets – as souvenirs (Fig. 24.2). Indigenous women make these products using hand looms, which attract many Bangalee tourists as these products represent Indigenous culture and are unavailable in city areas. Overall, the expansions with modern amenities and culture have been helping to generate the attention of city-based tourists in Bangladesh for the last decade, making people more interested in exploring new places. As a result, the government or military expanded the areas gradually, allocating more lands to these destinations that used to be under Indigenous people's control.

Private investors have been building luxurious resorts following ethnic styles of housing with modern amenities. This allows the tourists to experience serene nature in a quiet place, alongside the hustle of crowded common tourist destinations, with a glimpse of Indigenous people, their crafts, and their lives. Due to the fact that roads pass through the communities where Indigenous people reside, tourists have the opportunity to witness Indigenous people doing their daily chores or in the *jum* field. The construction of roads and bridges contributed to the easy access to these destinations. Over the years, military and government involvement shaped the tourism industry in different ways,

Fig. 24.1. *Nilgiri* in Bandarban with a modern sitting arrangement.
Source: Photograph taken by the author during the fieldwork.

Fig. 24.2. Ethnic crafts at *Shoilopropat* (waterfall) in Bandarban.
Source: Photograph taken by the author during the fieldwork.

from giving the places a modern turn to creating complexities for the Indigenous people and environment.

The Actors Involved

There are two ways to understand if Indigenous people can participate in the development of tourism – (1) participation in decision making; and (2) direct participation in tourism. Power is a defining factor in understanding how the relationships between different actors provide access to both of these options. In the case of Chittagong Hill Tracts, the resources – physical, natural, and cultural – attract outsider actors to concentrate on making tourism. The main tourist destinations are developed by various actors – civil administration, military, and individuals. Moreover, government administrative offices organize the places and build small cottages or resorts applying their control over these things.

According to the participants of this research, among many actors, Indigenous people remain the least prioritized actors in this whole process due to their limited access to power, money, and other facilities. The participation of different actors – Indigenous people, government institutions, the military, politically powerful individuals, or business groups – in tourism development planning highlights their uneven roles. The government and military develop tourism plans without consulting with local Indigenous people. This was criticized by the people interviewed in Bandarban, particularly development and human rights workers, for not consulting Indigenous institutions or people before implementing development plans. While answering why the government institutions do not consult Indigenous people, one government employee mentioned that Indigenous people 'are incapable of making decisions for their own development', which is why they take care of the development needs of the people and their surroundings. However, this assumption has proven to be wrong. Back in 2016, when the Bangladesh government decided to acquire 699.98 acres of land in Khagrachhari district for developing tourism, the Indigenous communities started protesting against it (Barua and Dewan, 2016). They feared losing *jum* (slash and burn cultivation) land and horticultural land, destroying their only source of livelihood. With the protests coming from community members, Indigenous rights groups, and activists, the government had no choice but to withdraw the plan. Thinking that Indigenous people lack knowledge about their needs or cannot fathom the potential implications of tourism within their environment reflect unequal power dynamics existing between multiple actors.

The involvement of Indigenous people in tourism is quite limited, as they rarely own hotels or resorts. At best, they work as hotel staff and sell fruits and ethnic crafts at tourist places. Their participation in tourism is in the lower tier, with a lack of decision making rights, where profit hardly reaches. The powerful actors invest money and share the profits by making a chain of 'development'. The participants of this research discussed the liaison between government administration, militaries, and private investors in order to grab land, exclude Indigenous people from broader development, and take control over their nature. One Bawm Indigenous participant from Bandarban informed –

> The conflict regarding our land rights is still unsolved. Through tourism, our land can be taken away, and we can become evicted from our place. It happened in Nilgiri, where militaries grabbed the land, developed tourism, and restricted Mro Indigenous people's mobility. Moreover, the central government controls the tourism department, although they are supposed to hand over the responsibilities to us. If we get the responsibilities, then we might get some profits out of tourism. As we are not involved, you can understand who is benefiting.

The quote indicates that among all the actors involved in tourism, the Indigenous people who are the hosts are the most deprived ones. When a place is developed as a tourist destination, the issues related to Indigenous people should be taken care of. In contrast, in the name of 'development', these aspects are overlooked.

The Impacts

Despite the government's justification, development initiatives have varying effects on Indigenous people and nature due to exploitation by powerful actors. Indigenous people live near tourist areas, and their livelihood depends on *jum* or shifting cultivation. Shifting cultivation is the prime livelihood strategy in the Chittagong Hill Tracts, which is also a part of Indigenous culture, and is at risk of alteration and change. With the creation of eco-parks, sanctuaries, and resorts, the land they use for agricultural production is turning into a recreational area for others, which has an impact on livelihood patterns. These people have to look for other sources of income when the state, military or other powerful actors take their lands. Thus, many of them switched from shifting cultivation to horticultural gardening.

Furthermore, as capitalist modernist development grows as a means of attracting large numbers of tourists, land grabbing is discovered to be a significant issue that affects people's connection to places and dependency on the environment. The loss of resources and limited access to the environment (Rasul, 2007) are the results of development projects underpinning the violent power of powerful actors. Indigenous peoples have been displaced from their ancestral lands due to the expansion of development projects over the last few decades. In 2015, the Daily Star published an article about the threats faced by 100 Indigenous families in Bandarban by the Gazi Group, a well-known consortium of private companies that had obtained 700 acres of land illegally for rubber plantations with the help of the Union Parishad, a local government unit, and the Upazila sub-district administration. According to the Indigenous people, the Gazi Group had illegally taken over more than 2000 acres of land. In addition, in Bandarban, the last 6 of 42 Marma Indigenous families were recently forced to leave their lands as they were claimed by Jasim Uddin Mantu, chairman of Sylvan Wye Resort and Spa Ltd (Barua, 2018). The amount of land claimed was approximately 100 acres of *Jum* land, of which a portion was also shared with the Bandarban Superintendent of Police, who was a government employee. Due to strong administrative support and political backing for the planned five-star resort, Indigenous people were unable to resist eviction in this area. These evictions and displacements from their own places have impacts on people's lives and livelihoods. Shedding light on this issue, one Tripura development worker from the Rural Livelihoods and Climate Change Adaptation in the Himalayas (HIMALICA) project said, 'when *Jum* cultivation is restricted by grabbing acres of Indigenous peoples' lands, these people have to shift elsewhere to maintain their livelihoods'. Thus, these evictions, when done for the commercial or development purposes of tourism, create great challenges for Indigenous people by destroying their traditional livelihoods.

The government justifies development not only for the purpose of promoting economic growth but also in order to assert territorial control over the land, primarily in order to reap financial benefits. According to Roy (2002), in order to maintain control over or supervision of Indigenous people's land, powerful actors have constructed infrastructure to reach remote locations. As the military expanded their camps all over the region, more infrastructural development occurred so that they could visit any place. However, constructing roads and bridges facilitates outsider's access to remote places. The various actors can reach untouched forests, causing the destruction of natural resources. Thus, deforestation is an outcome of infrastructural development, which can cause landslides, water crises, and an unbalanced environment. The effects are complex, necessitating planning in order to grow tourism or other infrastructure in a way that protects Indigenous people and the environment.

Concluding Remarks

This case study focused on a post-conflict setting where not only militarization still prevails, but also powerful actors have control over land, forests, and other resources, as well as Indigenous people. These actors focus on modernized development for commercial purposes in this post-conflict setting. The setting does not restrict tourists' entry into the region, except for foreigners who need approval from the government in advance. Unfortunately, the ongoing subtle conflict limits the participation of Indigenous people in the development of tourism. The government is building roads, dams, military camps, and tourist destinations by grabbing lands, destroying forest resources, and evicting them forcefully from their own places. The experience of Chittagong Hill Tracts reflects, as documented elsewhere, how the exploitation of resources for tourism involves dispossession, as documented in Guatemala (Devine, 2017), and in West Kalimantan, Indonesia (Peluso, 2008). Whatever is done in the name of development has further business interest and is a booster for the growth of tourism. However, only the powerful actors negotiate over development projects for self-interest, marginalizing Indigenous people in the decision making process. Power reflects not only on the decision making process but also on the idea that Indigenous people are less capable of making decisions on their own. This way of thinking minimizes their autonomy and involvement in the process of development, with the exception of low-paying jobs in hotels and the sale of fruits and crafts made by ethnic people. In contrast, Bangalee people control the main business chain. In addition, impacts on the environment, loss of livelihood, and eviction from ancestral lands are taking place due to infrastructural development. Thus, under the current arrangement, there are greater risks associated with tourism for Indigenous people than benefits. As examined in other settings (Devine, 2013; Colucci and Mullett, 2016), this case study invokes a look at the complexities and internal politics that determine power dynamics in controlling tourism and initiating violence against Indigenous people. Therefore, in order to ensure that Indigenous people have equal participation and share in the benefits, this case study suggests understanding the history of conflict with a focus on valuing everyone's concerns, including those of Indigenous people, and the environment.

References

Barua, S.K. (2018) Last six Marma families evicted: Land grabber plans to build 5-star hotel, resort in Bandarban Marma Para. *The Daily Star*. Available at: https://www.thedailystar.net/news/country/last-six-marma-families-evicted-1617844 (accessed 10 January 2019).

Barua, S.K. and Dewan, S. (2016) What Comes First: Tourism or livelihood?. *The Daily Star*. Available at: http://www.thedailystar.net/backpage/what-comes-first-tourism-or-livelihood-1293169 (accessed 10 January 2019).

Chakma, P. and Chakma, B. (2023) The Indigenous World 2023: Bangladesh. Available at: https://www.iwgia.org/en/bangladesh/5110-iw-2023-bangladesh.html (accessed 8 December 2023).

Colucci, A. and Mullett, A. (2016) Maya as commodity fetish: Accumulation by dispossession and eco-tourism in the Yuca-tan Peninsula. In: Nepal, S. and Saarinen, J. (eds) *Political Ecology and Tourism*. Routledge, New York, pp. 149–162.

Devine, J.A. (2013) *Tourism and Territory in the Mayan World*. (Doctor of Philosophy). University of California, Berkeley, CA. Available at: https://escholarship.org/uc/item/5727x51w.

Devine, J.A. (2017) Colonizing space and commodifying place: tourism's violent geographies. *Journal of Sustainable Tourism* 25(5), 634–650. DOI: 10.1080/09669582.2016.1226849.

Gain, P. (2000) Life and nature at risk. In: Gain, P. (ed.) *The Chittagong Hill Tracts Life and Nature at Risk*. Society for Environment and Human Development (SEHD), Dhaka, Bangladesh, pp. 2–41.

Gain, P. (2013) The Chittagong hill tracts: An ecological disaster. In: Gain, P. (ed.) *The Chittagong Hill Tracts Man-Nature Nexus Torn* (Vol. 17–63). Society for Environment and Human Development (SEHD), Dhaka, Bangladesh.

Gerharz, E. (2014) Recognising indigenous people, the Bangladeshi way: The United Nations declaration, transnational activism and the constitutional amendment affair of 2011. *Indigenous Policy Journal* 24(4).

Parveen, S. and Faisal, I. (2002) People versus power: the geopolitics of Kaptai Dam in Bangladesh. *International Journal of Water Resources Development* 18(1), 197–208.

Peluso, N.L. (2008) A political ecology of violence and territory in West Kalimantan. *Asia Pacific Viewpoint* 49(1), 48–67. DOI: 10.1111/j.1467-8373.2008.00360.x.

Rasul, G. (2007) Political ecology of the degradation of forest commons in the Chittagong Hill Tracts of Bangladesh. *Environment Conservation* 34(2), 153–163. DOI: 10.1017/S0376892907003888.

Rasul, G. and Thapa, G. (2007) The impact of policy and institutional environment on costs and benefits of sustainable agricultural land uses: The case of the Chittagong hill tracts Bangladesh. *Environmental Management* 40(2), 272–283. DOI: 10.1007/s00267-005-0083-8.

Rosy, S.Y. (2020) Development and culture in the context of Chittagong Hill Tracts (CHT), Bangladesh. *Social Science Review* 37(2), 87–103.

Rosy, S.Y. (2022) *An Ethical Political Ecology of Tourism in the Chittagong Hill Tracts of Bangladesh: Engaging with Indigenous and Gender Concerns* (Doctor of Philosophy). Macquarie University, Australia. Available at: https://figshare.mq.edu.au/articles/thesis/An_ethical_political_ecology_of_tourism_in_the_Chittagong_Hill_Tracts_of_Bangladesh_engaging_with_Indigenous_and_gender_concerns/20501520.

Roy, R.D. (2002) *Land and Forest Rights in the Chittagong Hill Tracts*. International Centre for Integrated Mountain Development (ICIMOD), Kathmandu, Nepal.

Saarinen, J. and Nepal, S. (2016) Conclusions: Towards a political ecology of tourism -Key issues and research prospects. In: Nepal, S. and Saarinen, J. (eds) *Political Ecology and Tourism*. Routledge, Oxon & New York.

25 Equal Opportunities in Rural Tourism Entrepreneurship: Challenges in Malaysia

Joo-Ee Gan
School of Hospitality and Service Management, Sunway University, Bandar Sunway, Malaysia

Summary
Malaysia's rural demography is multi-racial, with a blend of indigenous sub-groups. To advance the socio-economic status of rural communities, the government introduced various incentives and protectionist policies, including targeted measures in tourism development. As rural Malaysia is predominantly home to the Malays, the indigenous Orang Asli and the natives of Sabah and Sarawak, these policies are consistent with the government's constitutional responsibility to safeguard these groups' welfare. Notwithstanding the equal opportunity premise of the tourism policies, the government's initiatives to advance rural tourism entrepreneurship arguably benefit some groups more than others. In West Malaysia, this is due to the indigenous Orang Asli's lack of entrepreneurial readiness, which means that they are less able to capitalize on these opportunities. Consequently, the Malay entrepreneurs reap the benefits of protectionist policies and government aid, while the Orang Asli lag behind in tourism participation. This is arguably the outcome of a broad-brush approach in capacity building, with limited differentiation between the rural communities.

The Value and Interest of the Case Study

This case study examines rural tourism entrepreneurship in Malaysia. It highlights the pitfalls of equal access tourism policies that overlook unequal entrepreneurial readiness between different rural communities.

Background

Entrepreneurship can invigorate the tourism sector. Post-COVID-19 pandemic, entrepreneurship helps to drive intelligent and innovative solutions that accelerate tourism recovery (UNWTO, 2020). In Malaysia, where 97.2% of business entities are small or medium-sized businesses (SMEs) (OECD, 2022), entrepreneurism is vital to the country's growth. Incentivizing rural tourism entrepreneurship creates jobs, provides additional income sources and reduces rural-urban migration (Chen *et al.*, 2022). Hence, the Malaysian government encourages rural tourism entrepreneurship through favourable policies. For instance, the Ministry of Tourism and Culture (MOTAC) offers training on entrepreneurship and facilitates apprenticeships on tourism-related businesses as a form of capacity-building (Upskill Malaysia, 2021).

By virtue of Malaysia's rural demography, entrepreneurial capacity-building and assistance are usually channelled to three racial/ indigenous groups – the Malays and the indigenous Orang Asli in West Malaysia, and the natives in East Malaysia (the term 'natives' is used in Malaysia's Federal Constitution to denote the groups that are indigenous to the States of Sabah and Sarawak). It should be noted that the Orang Asli constitute only 0.7% of Malaysia's total population of 33 million. By contrast, the natives represent the majority of people in East Malaysia's Sabah and Sarawak states (IWGIA, n.d.). Collectively, the Malays and the natives of Sabah and Sarawak are known as the 'Bumiputera' – whose welfare the government is constitutionally required to safeguard. The Federal Constitution also enjoins the government to safeguard the welfare of the indigenous Orang Asli (Whah and Guan, 2017).

The extant literature shows that the Orang Asli seriously underperforms economically. The explanation lies partly in the political landscape of the country (Aiken and Leigh, 2015; Subramaniam, 2015; Whah and Guan, 2017; Hassan and Nordin, 2020). The Malays have the advantage of political dominance that comes with being the largest racial group. In comparison, the high concentration of the natives in Sabah and Sarawak enables them to influence the political agenda of these states. By contrast, the small population size of the Orang Asli means that they are less able to exert political influence (Hassan and Nordin, 2020). The imbalance of power between Malaysia's indigenous groups is also observable from the treatment of land rights. In Sabah and Sarawak, the customary land rights of the natives are recognized in state ordinances (albeit imperfectly enforced). But in West Malaysia, the Orang Asli's customary tenure is not statutorily recognized but subject to common law (Aiken and Leigh, 2011; IWGIA, n.d.).

Rural Tourism Entrepreneurial Support

Since the 1980s, Malaysia has introduced various measures to support the tourism sector, notably under the country's successive Malaysia Plans (Marzuki, 2010). However, the potential of rural tourism was only recognized in 2001, with the inception of the Rural Tourism Master Plan (RTMP) – formulated pursuant to the 9th Malaysia Plan (2006–2010) (Razzaq *et al.*, 2011). Through the RTMP, the government sought to support innovative tourism projects and allocate resources to human capital development in the rural tourism sector. More recently, under the National Tourism Policy 2020–2030, the government expressly reiterated its commitment to create a pro-investment environment through targeted business facilitation, including capacity-building for tourism SMEs (MOTAC, 2020b). To stimulate the recovery of tourism SMEs post-COVID-19, the government provided tourism loan guarantees, matching grants for the repair or refurbishment of budget hotels and matching grants for cultural events and the creative sector (Loong and Wan Usamah, 2022).

How and to whom entrepreneurial support is channelled partly depends on the government's conception of 'rural' tourism. In this regard, the RTMP defines rural tourism as 'tourism that provides opportunities to visitors to visit rural areas and rural attractions, and to experience the culture and heritage of Malaysia, thereby providing socio-economic benefits for local communities ... the proximity of many of these rural areas to the hinterland and rainforest also offers visitors an opportunity to extend their holiday and enjoy those unique natural resources' (as cited in Hamzah, 2004). While the definition appears reasonably broad, the RTMP's main strategy was the promotion of homestays (Nair *et al.*, 2014). Established by MOTAC in 1995, the Homestay Programme was enhanced through the RTMP and is generally considered successful (The New Straits Times, 2022). In 2012, Malaysia's Homestay Programme was awarded the UNWTO Ulysses Award for Innovation in Public Policy and Governance (Tourism Malaysia, 2023).

However, the Homestay Programme is arguably of limited relevance to the Orang Asli. This is because MOTAC's list of 450 homestay kampongs and 3328 homestay properties (as of April 2023) primarily showcase Malay culture, with the exception of homestays in Sabah and Sarawak (MOTAC, 2020a). This strong portrayal of idyllic kampongs and traditional Malay lifestyle in the West Malaysia homestay narrative was highlighted by Tavakoli *et al.* (2017). The current model of homestay business is that a local host should be closely involved in the operation of the homestay, to provide tourists with authentic lived-in experiences. Once a local host is registered with MOTAC as a homestay entrepreneur, they may be eligible for training, technical expertise and grants for the refurbishment of homestay properties. Logically, capacity-building and financial assistance necessarily benefit the communities where rural homestays are offered. Since the RTMP homestay development takes place in Malay communities – entrepreneurial support has been invariably channelled to Malay entrepreneurs.

Interestingly, the RTMP definition of rural tourism considers nature-based tourism as the progressive extension of a visitor's travel itinerary, as opposed to a facet of rural tourism (Hamzah, 2004). Consequently, ecotourism entrepreneurship (a segment that the Orang Asli can develop by virtue of their presence in the hinterland) did not come under the RTMP. Thankfully, ecotourism entrepreneurship support is available through other channels. The National Ecotourism Plan 2016–2025 highlights the need for local investments (Ministry of Tourism and Culture Malaysia (MOTAC), 2016), and there are initiatives to address funding challenges. For instance, The Habitat Foundation Sustainable Tourism Grant supports capacity-building initiatives to strengthen ecotourism and conservation in Malaysia (The Habitat Foundation, 2023).

Politically, entrepreneurial development is consistent with the Malaysian government's strategies to increase Bumiputera's participation in the economy. Where tourism is concerned, a target of 30% Bumiputera participation was set by the Third Malaysia Plan (1976–1980) (Din, 1982), with similar iterations in other policy documents. From this perspective, the RTMP is certainly laudable. Unfortunately, the narrow rural tourism narrative in the RTMP effectively excludes the Orang Asli from the Homestay Programme. Aside from the unequal access to entrepreneurial support between the Malays and the Orang Asli, the exclusionary narrative lends greater legitimacy to Malay culture.

Belum-Temengor Forest Reserve

The Belum-Temengor Forest Reserve (Belum-Temengor), located in the northern region of West Malaysia, is a 320,000 ha touristic protected area that is home to the Orang Asli. In 2007, 117,500 ha of Belum-Temengor was accorded additional protection through the establishment of the Royal Belum State Park (Schwabe *et al.*, 2015). At the fringe of Belum-Temengor are Malay villages that come under the Gerik Municipality, Perak. Touristic sites are only accessible through the jetty at Banding Island, which is linked to the Gerik Municipality by a bridge. In Belum-Temengor, the Orang Asli villages are important destinations in the ecotourism itinerary. Fig. 25.1 shows a picture of Lake Temengor. Fig. 25.2 shows an Orang Asli tourism village in the forest reserve. Fishing and the extraction of forest produce for sale are the main sources of livelihood. Tourism merely supplements the income of some touristic villages.

Methodology

Three qualitative studies were conducted at Belum-Temengor between 2015 and 2019, involving eight fieldtrips where semi-structured interviews were held with government officials, NGO members, nature guides, tourism operators and the Orang Asli. Between 2015 and 2016, 29 respondents were interviewed through a combination of purposive sampling and snowball sampling. The focus at the time was to examine the environmental governance structure of Belum-Temengor. In 2017, a second study delved deeper into the socio-economic benefits of tourism to the Orang Asli. To gain more insights into incidental data obtained during the first study, iterative sampling was applied, whereby six respondents were interviewed for the second time. Ten more respondents were interviewed through purposive sampling. During this period, tourism development funds were channelled for the construction of a new jetty and tourism amenities at Banding Island. Upon the completion of the tourism development project in early 2019, a third study was conducted to analyse the perception of Belum-Temengor's tourism stakeholders. In connection with this, a total of 22 respondents were interviewed, one of whom was interviewed for the third time.

This case study analyses Belum-Temengor's tourism governance, the dynamics between Malay tourism entrepreneurs and the Orang Asli, and the latter's participation in tourism during these three periods. The three studies

Fig. 25.1. Lake Temengor.

Fig. 25.2. Orang Asli tourism village in Belum-Temengor.

were not designed with a unified longitudinal approach (Vogl, 2023). Consequently, the respondents varied between the studies, although iterative sampling was applied in some instances. However, the cross-sectional data obtained has facilitated a longitudinal observation of the Orang Asli's tourism involvement in Belum-Temengor over a 5-year period. Strauss and Corbin's (1998) thematic analysis technique of coding and memoing was applied to analyse the three data sets afresh, cognizant of the potential influence of new data on the classification of themes and sub-themes and the perspectives that they present.

Equal Opportunity in Tourism Entrepreneurship

The main thrust of the Gerik Municipal Council's (GMC) tourism governance is to provide opportunities to locals so that foreign operators and non-locals do not corner the tourism sector in Belum-Temengor (Gan, 2022). To encourage tourism entrepreneurship in Belum-Temengor, the GMC introduced various incentives and protectionist policies. For instance, only residents of Gerik Municipality are eligible to apply for boathouse and/or nature guide licences. Similarly, capacity-building activities and entrepreneurship funding are for residents only.

Although these protectionist policies apply equally to the Malays and the Orang Asli, the former benefitted disproportionately. At the policy level, there was no overt discrimination by GMC. The licence criteria were transparent and applicable to all residents of the Gerik Municipality. However, the Orang Asli could not avail themselves of the opportunities due to poor entrepreneurial readiness. For instance, according to Perak State Parks Corporation's (PSPC) list of nature guides for 2014/2015, the 108 registrants were predominantly Malays. Similarly, all houseboat operators were Malays.

The Orang Asli's lack of formal education and tourism skills are obstacles that prevent them from applying for tourism licences and funding. Where tourism training was provided, unfamiliarity with bureaucratic processes prevented the Orang Asli from capitalizing on their qualifications. In 2017, two Orang Asli respondents stated that their nature guide certificates were not beneficial as they did not lead to any income. The interviews identified that the respondents were unsure of the administrative actions required to turn their eligibility into business opportunities (Gan, 2022). Low participation in tourism employment also means that Orang Asli youths lack documented experience – a precondition for entrepreneurial funding.

The enormity of the Orang Asli's lost opportunities in tourism can be glimpsed from the growth of tourist arrivals, as recorded by the Royal Belum State Park. Within a decade, there was a fourfold increase in tourist arrivals – from 6736(2008) to 27,805 (2018) (PSPC data as reported in Gan, 2023). The economic benefits from this growth were primarily enjoyed by Malay entrepreneurs. The 2019 data set revealed that some Malay entrepreneurs were frustrated at the documentary hurdles of compiling a funding/licence application. Nevertheless, the respondents concerned seemed knowledgeable in terms of GMC bureaucratic channels and the broader tourism governance in Perak state. This suggests that these entrepreneurs could, with additional capacity building, overcome such challenges. However, interviews with the Orang Asli respondents indicate that such obstacles were deemed insurmountable. Overall, the lack of contenders among the Orang Asli enables Malay entrepreneurs to monopolize tourism businesses in Belum-Temengor (Gan, 2022).

While an in-depth exploration of the macro-environment factors that influence Orang Asli's tourism entrepreneurship is beyond the purview of this case study, two inter-related forces (political and demographic) should be highlighted. The demographic imbalance between the Orang Asli and other races in Malaysia arguably contributes to the Orang Asli's political disadvantage. According to JAKOA (2022), only 2.4% (or 61,225 people) of Perak's population consists of the Orang Asli Department of Statistics Malaysia, 2023. By contrast, in Sabah, the native groups constitute 58.6% (or 2,233,100 people) of the state's population of 3,813,200. Whereas in Sarawak, the native groups make up 70.5% (or 1,932,600 people) of the state's population of 2,707,600 people (IWGIA, n.d.). Being the majority in Sabah and Sarawak, the natives are politically and economically stronger compared to the Orang Asli (Hassan and Nordin,

2020). The likelihood of the former benefitting from state support for community-based tourism in Sabah and Sarawak is considerably higher. For instance, Sabah Tourism Board is seeking to develop off-the-beaten-path tourist destinations with environmental attractions (BIMP-EAGA, 2022). The Sabah natives are almost certain to be the beneficiary group, by virtue of their population size and demography. In comparison, community-based tourism or rural tourism initiatives may not necessarily benefit the Orang Asli, as the Homestay Programme entrepreneurship shows.

Conclusion

In the Malaysian rural tourism context, this case study highlights the disjuncture between policies and outcomes. More specifically, tourism policies premised on equal opportunity do not necessarily provide equal access to every racial/indigenous group. This equality challenge must be addressed to provide sustainable livelihoods to the indigenous people, which in turn preserves heritage and incentivizes conservation (Quang *et al.*, 2023).

How should this gap between the policy discourse of equality and equitable outcomes be addressed? A targeted empowerment programme is necessary to promote Orang Asli's tourism participation. Additionally, entrepreneurship assistance must recognize the socio-economic gulf between the Malays and the Orang Asli. This is because systemic obstacles such as the lack of formal education, skills and knowledge are almost certain to impact the Orang Asli's progress.

More broadly, the concept of Malaysian rural tourism should be inclusive. Although the RTMP was formulated almost two decades ago, the narrative legacy of Malay kampong homestays remains strong (Mura, 2015). A Google search with the keywords 'rural tourism' and 'Malaysia' invariably returns websites of must-see kampongs. It is time to reimagine an encompassing rurality that embraces the cultural heritage of the Orang Asli.

References

Aiken, S.R. and Leigh, C.H. (2011) Seeking redress in the courts: Indigenous land rights and judicial decisions in Malaysia. *Modern Asian Studies* 45(4), 825–875.

Aiken, S.R. and Leigh, C.H. (2015) Dams and indigenous peoples in Malaysia: Development, displacement and resettlement. *Geografiska Annaler: Series B, Human Geography* 97(1), 69–93.

Brunei Darussalam-Indonesia-Malaysia-Philippines East ASEAN Growth Area (BIMP-EAGA) (2022) Sabah Nurtures Rural Tourism with the Help of Local Communities. Available at: https://bimp-eaga.asia/article/sabah-nurtures-rural-tourism-help-local-communities (accessed 20 June 2023).

Chen, P., Clarke, N. and Hracs, B.J. (2022) Urban-rural mobilities: The case of China's rural tourism makers. *Journal of Rural Studies* 95, 402–411.

Department of Statistics Malaysia (2023) Quarterly Demographic Statistics. Available at: https://www.dosm.gov.my/portal-main/release-content/8c705dac-ef05-11ed-96d5-1866daa77ef9 (accessed 30 June 2023).

Din, K.H. (1982) Tourism in Malaysia: Competing needs in a plural society. *Annals of Tourism Research* 9, 453–480.

Gan, J.E. (2022) Ecotourism and Malaysia's Orang Asli: The social construction of indigeneity in community boundary-drawing. *Journal of Ecotourism* 21(1), 54–70. DOI: 10.1080/14724049.2021.1918131.

Gan, J.E. (2023) Convert 'good' to 'excellent' tourism governance: Improve indigenous engagement. In: Dixit, S.K. and Piramanayagam, S. (eds) *Teaching Cases in Tourism, Hospitality and Events*. CAB International, Wallingford, UK, pp. 69–80. DOI: 10.1079/9781800621022.0007.

Hamzah, A. (2004) Policy and planning of the tourism industry in Malaysia. In: *Proceedings of the 6th ADRF General Meeting on Policy and Planning of Tourism Product Development in Asian Countries, Bangkok, Thailand*.

Hassan, M.S. and Nordin, R. (2020) Political participation of aboriginal people (Orang Asli) in Peninsular Malaysia: Examining the compatibility of the Aboriginal Peoples Act 1954 with

UNDRIP standards. *International Journal on Minority and Group Rights* 28(3), 1–21. DOI: 10.1163/15718115-BJA10010.

International Work Group for Indigenous Affairs (IWGIA) (n.d.) Indigenous People in Malaysia. Available at: https://www.iwgia.org/en/malaysia.html#:~:text=The%20Indigenous%20World%202021%3A%20Malaysia&text=million%20national%20population.-,They%20are%20collectively%20known%20as%20Orang%20Asal.,of%20Peninsular%20Malaysia%20(31%2C950%2C000) (accessed 30 June 2023).

Jabatan Kemajuan Orang Asli (Department for the Advancement of Orang Asli) (JAKOA) (2022) Bilangan penduduk mengikut jantina dan negeri (Population according to gender and state). Available at: https://www.jakoa.gov.my/orang-asli/bilangan-penduduk-mengikut-jantina-dan-negeri/ (accessed 30 June 2023).

Loong, Y.S. and Wan Usamah, W.A. (2022) The Malaysian economy and Covid-19: Policies and responses from January 2020–April 2021. United Nations Conference on Trade and Development 2022, UNCTAD/BRI PROJECT/RP 30. Available at: https://unctad.org/system/files/official-document/BRI-Project_RP30_en.pdf (accessed 8 May 2023).

Marzuki, A. (2010) Tourism development in Malaysia. A review on federal government policies. *Theoretical and Empirical Researches in Urban Management* 8(17), 85–97.

Ministry of Tourism and Culture Malaysia (MOTAC) (2016) National Ecotourism Plan 2016–2025. Available at: https://www.motac.gov.my/en/download/category/86-pelan-eko-pelancongan-kebangsaan-2016-2025 (accessed 28 March 2023).

Ministry of Tourism and Culture Malaysia (MOTAC) (2020a) Laporan statistik program pengalaman Homestay Malaysia (April 2023) (translation: Statistics Report on Homestay Malaysia Programme). Available at: https://www.motac.gov.my/program/pelancongan/homestay (accessed 30 June 2023).

Ministry of Tourism and Culture Malaysia (MOTAC) (2020b) National Tourism Policy 2020–2030. Available at: https://www.tourism.gov.my/files/uploads/Executive_Summary.pdf (accessed 28 March 2023).

Mura, P. (2015) Perceptions of authenticity in a Malaysian homestay – A narrative analysis. *Tourism Management* 51, 225–233.

Nair, V., Munikrishnan, U.M., Rajaratnam, S.D. and King, N. (2014) Redefining rural tourism in Malaysia: A conceptual perspective. *Asia Pacific Journal of Tourism Research* 20(3), 314–337.

OECD (2022) Financing SMEs and Entrepreneurs 2022: An OECD Scoreboard – Malaysia. Available at: https://www.oecd-ilibrary.org/industry-and-services/financing-smes-and-entrepreneurs-2022_3bc2915c-en#:~:text=SMEs%20are%20the%20backbone%20of,employment%20for%207.3%20million%20people (accessed 28 March 2023).

Quang, T.D., Dang, V.Q., Alang, T. and Nguyen, H.V. (2023) Can tourism enhance inclusivity for indigenous peoples? Cham perspectives on tourism benefit sharing at living heritage sites in Vietnam. *Equality, Diversity and Inclusion: An International Journal*. DOI: 10.1108/EDI-08-2022-0243.

Razzaq, A.R.A., Hadi, M.Y., Mustafa, M.Z., Hamzah, A., Khalifah, Z. and Mohamad, N.H. (2011) Local community participation in homestay program: Development in Malaysia. *Journal of Modern Accounting and Auditing* 7(12), 1418–1429.

Schwabe, K.A., Carson, R.T., DeShazo, J.R., Potts, M.D., Reese, A.N. and Vincent, J.R. (2015) Creation of Malaysia's royal belum state park: A case study of conservation in a developing country. *Journal of Environment & Development* 24(1), 54–81.

Strauss, A. and Corbin, J. (1998) *Basics of Qualitative Research: Grounded Theory Procedures and Techniques*. Sage, Thousand Oaks, CA.

Subramaniam, Y. (2015) Ethnicity, indigeneity and indigenous rights: The 'Orang Asli' experience. *QUT Law Review* 15(1), 71–91. DOI: 10.5204/.

Tavakoli, R., Mura, P. and Rajaratnam, S.D. (2017) Social capital in Malaysian homestays: exploring hosts' social relations. *Current Issues in Tourism* 20(10), 1028–1043.

The Habitat Foundation (2023) The Habitat Foundation Sustainable Tourism Grants. Available at: https://habitatfoundation.org.my/the-habitat-foundation-sustainable-tourism-grants/ (accessed 28 March 2023).

The New Straits Times (2022) Homestay Programme Should Be Revitalised (26 February 2022). Available at: https://www.nst.com.my/opinion/letters/2022/02/774993/homestay-programme-should-be-revitalised (accessed 30 June 2023).

Tourism Malaysia (2023) Malaysia Homestay Programme Lauded as Prime Example of Collaborative Tourism Initiative. Available at: https://www.tourism.gov.my/media/view/malaysia-homestay-programme-lauded-as-prime-example-of-collaborative-tourism-initiative (accessed 30 June 2023).

United Nations World Tourism Organization (UNWTO) (2020) Calling on Innovators and Entrepreneurs to Accelerate Tourism Recovery. Available at: https://www.unwto.org/calling-on-innovators-and-entrepreneurs-to-accelerate-tourism-recovery (accessed 30 June 2023).

Upskill Malaysia (2021) Ministry of Tourism and Culture (MOTAC). Available at: https://upskillmalaysia.gov.my/agencyMotac (accessed 30 June 2023).

Vogl, S. (2023) Mixed methods longitudinal research. Forum: Qualitative Social Research 24(1), Art 21. DOI: 10.17169/fqs-24.1.4012.

Whah, C.Y. and Guan, B.T.C. (2017) Malaysia's protracted affirmative action policy and the evolution of the Bumiputera commercial and industrial community. *Sojourn* 32(3), 336–373.

26 Elephant-Based Tourism in Nepal: The Key to Reducing Human Poverty or a Tool to Commodify Endangered Species?

Michelle Szydlowski
University of Exeter, St. Luke's Campus, Exeter, UK

Origin: Adapted from: Szydlowski, M. (2024) Elephant Tourism in Nepal: Historical Perspectives, Current Health and Welfare Challenges, and Future Directions. CABI.

Summary
Tourism in Nepal has been politically and academically linked to poverty reduction since the 1960s, but the actual benefits of tourism practice have failed to trickle down to local families, especially those belonging to marginalized communities. The main producers of tourism income are national park entry tickets, followed closely by income from privately run, elephant-backed safaris. The use of captive animals as transport for the viewing of wild animals has been a common practice in Nepal for the last 50 years. However, changing human perspectives on the use of animals in tourism, especially among tourists arriving from nations in the Global North (an alternative term for those previously deemed 'Westernized' nations), have led to widespread debates surrounding the use of endangered elephants to promote tourism practices. The perception of tourism practitioners is that engagement with captive animal-based activities by tourists from so-called western countries has greatly declined over the last decade, and in response, so has the total number of privately held tourism elephants. However, the fall in participation by visitors from the Global North has been offset by increasing tourist arrivals from Asian countries. Asian tourists are now the largest consumers of elephant-backed safari in Nepal and, according to research, may hold a different view of such activities. This changing face of tourism may increase the commodification of endangered species while also failing to provide needed income for marginalized communities.

The Value and Interest of the Case Study

This case study discusses how the benefits of tourism practice have yet to materialize for local families, and how the use of animals in tourism may instead be increasing socio-economic gaps. Likewise, the use of captive animals as vehicles for the viewing of wildlife may increase the commodification of elephants and inadvertently damage wildlife due to disturbance or transmissible disease. The desire to link poverty reduction to conservation practice may instead result in a loss of native wildlife, further commodification of marginalized communities of endangered individuals (see Abdullah *et al.*, 2023), and increased poverty and human-wildlife conflict in Nepal. As tourism within Asia continues to grow, how will differing perspectives on wildlife and wild spaces affect the numbers of wild and captive animals?

Background to the Case

Nepal sandwiched between the Chinese Autonomous Region of Tibet and India, is home to over 30 million humans. Nepal has made strides in reducing poverty since the 1960s, but faces ongoing challenges due to increasing population growth, loss of income from COVID-19 and related lockdowns, and the continued lack of access to services or housing for marginalized communities. Thus, Nepal remains among the least developed countries (LDCs) in the world (United Nations, 2023).

Nepal is a small federalist republic, covering only 141,000 km^2, but is rich in biodiversity. Due to early protections enacted by the former monarchy, nearly 24% of the nation's land has been set aside within protected areas (PAs) (Government of Nepal (GoN), 2017). Beginning in the 1960s, the monarchy sought to connect landscape and species preservation with poverty-alleviation efforts. These efforts resulted in an increasing nature-based tourism market which coincided with the birth and expansion of elephant-backed tourism safari within protected areas (Mishra, 2008). Today, the entire tourism market employs over a million Nepalese and provides >7% of Nepal's GDP (World Bank, 2018). In Chitwan, Nepal's first – and busiest – national park, elephant-backed tourism represents the second-highest tourism revenue producer.

Earlier research in Nepal suggested that conservation efforts and marginalized communities might benefit from increases in tourism (Bookbinder *et al.*, 1998; Nyaupane and Thapa, 2004). However, ongoing studies have indicated that rather than a panacea for poverty reduction, tourism has resulted in the siphoning of income from local peoples, widening gaps between socio-economic groups, and increased mistreatment of marginalized communities (Kellert *et al.*, 2000; Roka, 2012; Kandel *et al.*, 2020). In addition, some studies have shown tourism practices have increased damage to PAs (for example, in the form of greater water usage, increased trash, increased traffic in sensitive areas, and damage to flora and fauna), offered limited or low-wage positions for community members (especially those from marginalized groups), failed to provide equal economic benefits, and created the sense that parks are more important than local peoples (Carrier and Macleod, 2005; Subedi, 2010; Fletcher, 2015). Likewise, an increase in tourists' desire to experience wild spaces is increasing the commodification of captive safari pachyderms, and negatively impacting their health and welfare (Szydlowski, 2022a, Szydlowski, 2024).

This case study is based upon ongoing research investigating the interplay between tourism and marginalized communities of humans and elephants, between species, and between captive and wild elephants (*Elephas maximus*), which began in 2014. This case study serves as the foundation for a critical examination of elephant tourism and its perceived benefits in Nepal.

Elephant-Backed Tourism Safaris

Tourist elephant safaris involve one to one and a half-hour excursions through the protected areas (community forests) surrounding the national parks of Nepal. The majority of these safaris are undertaken near Chitwan National Park (see previous section). During these safaris, up to six guests at a time ride in enclosed platforms (see Fig. 26.1) placed upon elephants, accompanied by an elephant driver (in Nepal, they are referred to generally as *mahout*). Because elephants are indigenous to the forests of Nepal, they are able, according to locals, to move through the forest quietly while masking the smell of accompanying humans. The use of elephants is thought to prevent disturbance to other wildlife while allowing tourists to better view native wildlife *in situ*.

These safaris began when Nepal had few tourists (<6000 per year), mainly visiting royalty on hunting excursions (Mishra, 2008). Government forestry elephants provided transport for early outings, but as tourism grew, the government of Nepal pivoted herd members exclusively towards forest management activities. By the 1970s, more than 70,000 tourists

Fig. 26.1. A tourist safari elephant wearing a *howdah* (a special saddle to hold tourists).

visited Nepal, and local businessmen quickly purchased elephants from India to fulfil tourists' requests for safari (removing elephants from the wild in Nepal was illegal). Before COVID-19 and its subsequent travel lockdowns struck in 2019, Nepal was on track to see 2 million annual tourists, with more than 60% of international tourists visiting at least one protected area (Government of Nepal (GoN), 2019).

The government of Nepal, with the help of numerous foreign governments and private entities, has continued to focus its efforts on endangered species conservation and to market nature-based tourism as the key financial support for these efforts (Government of Nepal (GoN), 2015). However, the keeping of captive elephants, themselves members of an endangered species, within Nepal has become a point of contention between non-governmental organizations and governing bodies. Citing concern for elephant health and safety, local guide groups, community business owners, researchers, and advocacy organizations have become involved in attempts to evaluate and improve captive endangered species housing, nutrition, protection, and use in both tourism and conservation practices.

Local Activism

Over the last decade, grassroots organizations in communities surrounding Nepal's national parks have begun to question the use and treatment of non-human animals in tourism practice. For the last decade, in the village of Sauraha (the primary entry point for visitors to Chitwan National Park) tourism operators, nature guides, hoteliers, local businesspeople, and students have rallied against elephant-based tourism practices (Szydlowski, 2024). Citing the need to protect endangered individuals, many Nepalese tourism offices refused to sell tickets to, or advertise for, elephant safari rides. Elephant owners and mahouts faced a backlash against events which promoted the use of elephants in tourism practice, such as elephant-backed polo, elephant beauty pageants, elephant soccer, and other types of marketing aimed at increasing tourist-elephant engagement. Grassroots organizations participated in protests at these events, marches, and community education campaigns. Like global efforts which resulted in the cessation of ticket sales to animal-based venues by Trip Advisor (among others), local activism led to the end of (public) elephant polo matches and a perceived drop in safari ride consumption by guests from the United States, Europe, and Australia (see ABTA, 2020; Tripadvisor, 2020). However, some academics and practitioners blame the refusal of guests from the Global North to support elephant-human engagement at tourism venues on 'paternal' attitudes, claiming that a refusal to support so-called traditional practices was simply neo-colonialism (Suter, n.d.). However, tourist safari rides cannot be considered traditional practice within Nepal, having only begun mid-century (Mishra, 2008), and involving only a small number of community members. Because elephant purchases (up to US$100,000) and care costs are well outside what most Nepalese can afford, most elephant owners represent the most socio-economically advantaged groups (Szydlowski, 2024).

While external forces may be pressing for an end to human-elephant interaction at venues in other Southeast Asian countries, in Nepal the drive seems to have arisen locally. Nepalese organizations have since allied with international groups like People for the Ethical Treatment of Animals (USA), World Animal Protection (UK), and Wildlife SOS (India) to improve animal health and decrease commodification. Local animal advocacy organizations have since received financial and social support from Global North tourists, and several venues now list 'ethical elephant activities' on their ticket boards. These activities include differing concepts of 'appropriate' activities such as: touching or petting elephants, touch-free observations at stables and within protected areas, elephant-led jungle walks, hand feeding, and scrubbing or rubbing elephants during their bathing time. While marketed as more ethical options for captive elephants, there is no academic evidence to determine whether these activities lessen stress or improve welfare within Nepal. Likewise, nature guides reported hesitance in recommending these activities because they are unsure whether they will further commodify elephants. Mahout and owners reported that these activities provide opportunities for mahout tips and are comfortable supporting them. Because mahouts traditionally belong to marginalized communities and receive very low wages while also working extended hours (see following sections), tips are a key financial incentive for continued participation (see Szydlowski, 2024; for further discussion).

The Changing Face of Tourism

Tourists from other parts of Asia represented a large market share of tourism to Nepal prior to 2019 and its associated COVID-related travel restrictions. Statistics suggested that post-COVID tourism to Nepal rebounded more quickly from so-called western countries such as Australia, the USA, and Europe (especially France, the UK, and Germany) (GoN, 2023). However, by July 2023 guests from Asian nations again represented more than half the visitors to Nepal. Tourism from India continues to show the highest numbers, followed by China and the USA. Tourist arrivals grew exponentially after Nepal was added to China's 'allowed travel' list in mid-March 2023, and by April 2023 showed a fourfold increase over the same period in 2022.

Asian tourism has experienced rapid growth over the last 25 years, thanks in part to the deregulation of certain industries and growing regional affluence. Continued increases in Asian nature-based travel may be related to increased diversification in tourism markets. Recently, Asian consumers have shifted their focus from less expensive or local visits to more expensive venues that require longer flights and overnight stays. Cultural changes surrounding Asian tourists' desire to actively engage with nature may also be responsible for increased visitation to biodiverse areas like Nepal (Cochrane, 2007; Wen and Ximing, 2008).

Changing cultural perspectives may also be impacting the practice of elephant-based tourism in Nepal. Visitors from the Global North were generally perceived by elephant owners, mahouts, and advocacy organizations as supportive of ending elephant-backed safari, while tourists from within Asia were perceived as supportive of its continuation. The reasons for differences between Asian and non-Asian tourist participation could be due to a variety of factors including religious considerations of non-human animals, differing levels and/or types of environmental education, the promotion of elephant interaction as positive in media, and differing concepts of wilderness or definitions of ecotourism. Differing perceptions have been documented in both research and popular media, with 'western' tourists' concept of nature requiring protection from humans, and Asian tourists' desire for interactivity with nature dominating (Wen and Ximing, 2008; Long, 2019).

Increasing numbers of visitors from the regions supportive of captive animal interactive tourism may increase existing challenges within Nepal. For example, the population of captive safari elephants in Nepal is now stable, due to increasing enforcement of legislation which forbids the sale or trade of endangered species for commercial use across country borders. In 2021, the government of Nepal instituted mandatory registration of captively held elephants, which further limited the ability of owners to offload non-working individuals (a practice which traditional entailed returning them to India). What this means, however, is that if demand for elephant safari returns to pre-COVID levels, the 37 working safari elephants in Sauraha will be responsible for providing ever-increasing numbers of rides for guests. Pre-COVID, some working elephants provided nine rides per day during the high tourist season (up to 54 guests total per elephant per day). The health of Nepal's working elephants is already precarious. Most of them suffer from long-term diseases, ongoing injuries, joint disorders, foot infections, and a lack of proper nutrition, among other challenges (Shrestha and Gairhe, 2006; Varma and Ganguly, 2011; Mandal and Khadka, 2013; Szydlowski, 2024). If tourist demand grows, these elephants could face further decreases in their health and/or greater negative welfare impacts. Likewise, the financial gains from elephant safaris (see above) might encourage interested owners to find ways to obtain more elephants illegally.

Other Challenges

Nepal has been extremely successful in reducing poaching, increasing numbers of endangered species, and meeting their goals for landscape protection. For example, Nepal has tripled their number of tigers (*Panthera tigris tigris*) and increased rhino (*Rhinoceros unicornis*) populations sixfold. However, successes in species conservation and decreases in poaching have led to several unintended consequences. Due to a lack of space or connected habitat for self-relocation, record numbers of rhinos are dying from territorial disputes, human-wildlife conflict, and falls into wells or septic tanks (Thapa *et al.*, 2016; see also Szydlowski, 2022b).

Disease transmission is also on the rise. Because captive elephants are housed with humans and domesticated livestock, they are exposed to diseases such as anthrax, tuberculosis, and foot and mouth disease (Shrestha and Gairhe, 2006; Mikota *et al.*, 2015). Captive elephants share landscapes with wild individuals, creating opportunities for disease transmission. These diseases, along with elephant endotheliotropic herpesviruses (EEHV), are working their way into wild populations (Szydlowski, 2022a). In addition, academics are concerned that EEHV may be zoonotic, but this is yet unproven (Murray, n.d.).

Rhinos are dying at unprecedented rates within the protected areas of Nepal, more than 40 annually (Mandal, 2019). Both rhinos and wild elephants have been found dead from diseases likely transported into shared spaces by captive elephants. Transmission along the captive-wild interface is also increasing through the presence of wild elephants in human-dominated areas. Wild bulls are drawn in by highly palatable crops, food stores, and captive mates. The habituation of wild males to human-dominated areas may have arisen from the practice of encouraging wild male elephants to approach captive females in government stables. This is the traditional, and only, breeding method used within governmental facilities (Gairhe, 2012). Such encouragement is now leading wild bulls to approach private tourism stables which are located within villages, increasing the potential for disease transmission and human-elephant conflict (Szydlowski, 2023).

Park-People-Wildlife Conflict

Marginalized or landless human communities are often located alongside national park boundaries for a variety of reasons, with many families having been forcibly removed from the park during past biodiversity preservation initiatives. Because of their location further from the tourist gates, these communities do not benefit equitably from tourism-based income. In addition, they have fewer opportunities for alternative income production and are less likely to be involved in land-use or decision planning, and may have no land or only enough for a small subsistence farm. Most mahouts in Nepal come from these marginalized communities, and like others in the area face daily social, health, and financial challenges.

Economically or socially vulnerable groups have historically faced greater impacts from human-wildlife conflict, including crop loss, livestock depredation, human casualties and fatalities. In addition, it is these communities which most heavily rely upon community forest products for food and fuel, increasing their chances for negative interactions with wildlife during daily provisioning activities. Communities have faced increases in human-wildlife conflict *outside* the protected areas as populations of animals increased and left protected areas in search of unoccupied territory.

These communities are also more likely to be reliant upon illegal provisioning from *within* protected areas as well and are therefore more likely to face human-wildlife conflict within the park. Any human-wildlife conflict which occurs within the national park is not subject to reimbursement from the government, making these necessary incursions even more of a risk.

Conclusions

The government of Nepal has long believed that tourism is a key to poverty reduction. However, tourism has yet to provide benefits to marginalized communities of humans, and may, in fact, be increasing socio-economic gaps. Likewise, nature-based tourism has been marketed as a way to provide income for conservation projects within Nepal. Such tourism, in this case, reliant upon the use of captive endangered species as transport, may also result in the loss of members of the same, and other, endangered species. This is especially concerning as wild charismatic species, like elephants, are the very animals who serve to draw nature-based visitors to protected areas.

Improving health and increasing the positive welfare of marginalized groups of captive endangered species is perhaps the first step. Limiting work hours, ensuring rest breaks, locating stables outside human-dominated areas, and developing further legislation would go far to improve living conditions in the short term. Maintaining healthier captive elephants should decrease the risk of disease transmission to wild individuals. Improving conditions for captive elephants is also likely to improve the health and welfare of mahouts.

There is a need to balance single species and landscape conservation with the needs of local communities, including communities of non-human animals. Rather than assessing how tourism impacts specific species populations (including humans) in isolation, perhaps the focus should rest on the interactions between individuals within multispecies communities. An interdisciplinary approach that considers the health, welfare, and societal needs of all stakeholders is needed. Such an approach could provide opportunities to prevent further commodification of endangered species while reframing the marketing of nature-based tourism in ways that could benefit sustainable development efforts. This marketing may need to address the desires of growing numbers of tourists interested in interactions with other species and reframe how these interactions are portrayed in media. Otherwise, the continued loss of biodiversity married with the lack of support for marginalized groups of humans could result in social and environmental disaster for Nepal.

Further Reading

Chaudhary, S., McGregor, A., Houston, D. and Chettri, N. (2018) Environmental justice and ecosystem services: A disaggregated analysis of community access to forest benefits in Nepal. *Ecosystem Services* 29, 99–115. DOI: 10.1016/j.ecoser.2017.10.020.

Ellis, S. and Talukdar, B. (2019) Rhinoceros Unicornis. *The IUCN Red List of Threatened Species 2019* e.T19496A18494149. DOI: 10.2305/IUCN.UK.2019-3.RLTS.T19496A18494149.en.

Hayward, G.S. (2020) Highly Distorted Picture of Elephant Herpesvirus Disease. EEHV Advisory Group Website. Available at: https://eehvinfo.org/eehvinformation/highly-distorted-picture-of-elephant-herpesvirus-disease/ (accessed 20 April 2021).

Lamichhane, B.R., Persoon, G.A., Leirs, H., Poudel, S., Subedi, N. *et al.* (2018) Spatio-temporal patterns of attacks on human and economic losses from wildlife in Chitwan National Park Nepal. *PLoS ONE* 13(4), e0195373. DOI: 10.1371/journal.pone.0195373.

McLean, J. (1999) Conservation and the impact of relocation on the Tharus of Chitwan, Nepal. *Himalaya, the Journal of the Association for Nepal and Himalayan Studies* 19(2), 38–44.

Puri, J. (2019) Socio-economic impacts of wildlife tourism in Kasara, Chitwan. *Journal of APF Command and Staff College* 2(1), 70–79.

Sedhain, J. and Adhikary, A. (2016) Human-Rhino conflict: Local people's adaptation to impacts of Rhino. *Journal of Forest and Livelihood* 14(1), 53–63. DOI: 10.3126/jfl.v14i1.23162.

Shrestha, S. (2019) 41 Rhinos Dead in Nine Months at Chitwan National Park. My Republica/My City. Available at: https://myrepublica.nagariknetwork.com/news/41-rhinos-dead-in-nine-months-at-chitwan-national-park (accessed 1 August 2023).

Shrestha, S. (2023) Nepal Receives More than 300,000 Foreign Visitors in Last Four Months. Available at: https://myrepublica.nagariknetwork.com/news/nepal-receives-more-than-300-000-foreign-visitors-in-last-four-months/ (accessed 15 August 2023).

Shrestha, U., Shrestha, B. and Shrestha, S. (2010) Biodiversity conservation in community forests of Nepal: Rhetoric and reality. *International Journal of Biodiversity and Conservation* 2, 98–104.

Stronza, A., Hunt, C. and Fitzgerald, L. (2019) Ecotourism for conservation? *Annual Review of Environment and Resources* 44, 229–253. DOI: 10.1146/annurev-environ-101718-033046.

Tripadvisor (n.d.) Tripadvisor: Improving Animal Welfare in Tourism. Available at: https://www.tripadvisor.com/blog/animal-welfare-education-portal/ (accessed September 2023).

UNWTO (2006) Tourism Market Trends. World Tourism Organization, Madrid. Available at: www.world-tourism.org

Wondirad, A. (2019) Does ecotourism contribute to sustainable destination development, or is it just a marketing hoax? Analyzing twenty-five years contested journey of ecotourism through a meta-analysis of tourism journal publications. *Asia Pacific Journal of Tourism Research* 24(11), 1047–1065. DOI: 10.1080/10941665.2019.1665557.

References

Abdullah, T., Lee, C. and Carr, N. (2023) Conceptualising human and non-human marginalisation in tourism. *Tourism and Hospitality Research* 23(2), 254–268.

Association of British Travel Agents (ABTA) (2020) Animals in Tourism. Available at: https://www.abta.com/industry-zone/raising-standards-in-the-industry/animals-in-tourism (accessed 1 September 2023).

Bookbinder, M., Dinerstein, E., Rijal, A., Cauley, H. and Rajouria, A. (1998) Ecotourism's support of biodiversity conservation. *Conservation Biology* 12(6), 1399–1404. DOI: 10.1111/j.1523-1739.1998.97229.x.

Carrier, J. and Macleod, D. (2005) Bursting the bubble: The socio-cultural context of ecotourism. *The Journal of the Royal Anthropological Institute* 11(2), 315–334. DOI: 10.1111/j.1467-9655.2005.00238.x.

Cochrane, J. (2007) Ecotourism and biodiversity conservation in Asia: Institutional challenges and opportunities. In: Higham, J.E. (ed.) *Critical Issues in Ecotourism: Understanding a Complex Tourism Phenomenon*. Routledge, London, pp. 287–307.

Fletcher, R. (2015) Nature is a nice place to save but I wouldn't want to live there: Environmental education and the ecotourist gaze. *Environmental Education Research* 21(3), 338–350. DOI: 10.1080/13504622.2014.993930.

Gairhe, K.P. (2012) Veterinary care and breeding of elephants in Nepal. *Gajah* 37, 27–30.

Government of Nepal (GoN) (2015) *Chitwan National Park and Its Buffer Zone Management Plan*. Kathmandu. Ministry of Soil and Forest Conservation. Department of National Parks and Wildlife Conservation. Available at: https://chitwannationalpark.gov.np/index.php/document-repository/publications-chitwan-national-park/28-management-plan-printed/file (accessed 14 September 2023).

Government of Nepal (GoN) (2017) *Wildlife Sanctuaries*. Kathmandu. Available at: http://nepal.gov.np:8080/NationalPortal/view-page?id=82 (accessed 10 June 2019).

Government of Nepal (GoN) (2019) *Visit Nepal 2020 Campaign Information*. Ministry of Culture, Tourism and Civil Aviation, Department of Tourism, Kathmandu. Available at: https://lohanivishnu.wordpress.com/goals-and-objectives/ and http://www.tourismdepartment.gov.np (accessed May 2019).

Government of Nepal (GoN) (2023) Arrival/Departure Data 2022 (Annual). Department of Immigration. Available at: https://www.immigration.gov.np/post/arrival-departure-data-2022-annual (accessed 1 August 2023).

Kandel, S., Harada, K., Adhikari, S. and Dahal, N.K. (2020) Ecotourism's impact on ethnic groups and households near Chitwan National Park, Nepal, *Journal of Sustainable Development* 13(3), 113–127.

Kellert, S.R., Mehta, J.N., Ebbin, S.M. and Lichtenfeld, L.L. (2000) Community natural resource management: Promise, rhetoric, and reality. *Society & Natural Resources* 13(8), 705–715. DOI: 10.1080/089419200750035575.

Long, M.J. (2019) Ecotourism reconsidered: Chinese and western participation in the Thai Elephant Industry. Honours Thesis. Bowdoin College.

Mandal, C.K. (2019) Rising rhino deaths in Nepal put conservationists on the horns of a dilemma. *The Kathmandu Post*. Available at: https://kathmandupost.com/national/2019/01/20/rising-rhino-deaths-put-conservationists-onthe-horns-of-a-dilemma (accessed 26 December 2023).

Mandal, R. and Khadka, K.K. (2013) Health status of captive Asian elephants in Chitwan National Park, Nepal. *Gajah* 39, 37–39.

Mikota, S.K., Gairhe, K., Giri, K., Hamilton, K., Miller, M. *et al*. (2015) Tuberculosis surveillance of elephants (Elephas maximus) in Nepal at the captive-wild interface. *European Journal of Wildlife Research* 61(2), 221–222.

Mishra, H. (2008) *The Soul of the Rhino: A Nepali Adventure with Kings and Elephant Drivers, Billionaires and Bureaucrats, Shamans and Scientists and the Indian Rhinoceros*. The Lyons Press, Connecticut.

Murray, S. (n.d.) Novel Platelet-Derived Treatment for EEHV. Smithsonian's National Zoo & Conservation Biology Institute. Available at: https://nationalzoo.si.edu/global-health-program/novel-platelet-derived-treatment-for-elephant-endotheliotropic-herpesvirus (accessed 26 May 2020).

Nyaupane, G.P. and Thapa, B. (2004) Evaluation of ecotourism: A comparative assessment in the Annapurna conservation area project, Nepal. *Journal of Ecotourism* 3(1), 20–45. DOI: 10.1080/14724040408668148.

Roka, K. (2012) Are NGOs making a difference? A community approach to measuring non-governmental organizations (NGOs) effectiveness in Nepal. PhD dissertation, Pennsylvania State University, Pennsylvania.

Shrestha, S. and Gairhe, K. (2006) Nepal. In: Fowler, M.E. and Mikota, S.K. (eds) *Biology, Medicine, and Surgery of Elephants*. Blackwell Publishing, Ames, Iowa, pp. 465–467.

Subedi, K.D. (2010) Tourism Industry in Nepal and Destination Chitwan: Current Issues and Challenges (March 2, 2010). No issue number. DOI: 10.2139/ssrn.1562686.

Suter, I. (n.d.) Activism or Paternalism? The case of Elephant-based Tourism in Southeast Asia. ACES. Available at: https://www.elephantstandards.com/post/activism-or-paternalism-the-case-of-elephant-based-tourism-in-southeast-asia (accessed 1 August 2023).

Szydlowski, M. (2022a) Elephants in Nepal: Correlating disease, tourism, and welfare. *Journal of Applied Animal Welfare Science* 25(2), 1–13. DOI: 10.1080/10888705.2022.2028628.

Szydlowski, M. (2022b) Renegotiating citizenship: Stories of young rhinos in Nepal. *Journal of Ecotourism* 21, 1–18. DOI: 10.1080/14724049.2022.2139833.

Szydlowski, M. (2023) One Stable's novel approach to mitigating human-elephant conflict Near Chitwan National Park, Nepal. *Gajah* 56, 23–29.

Szydlowski, M. (2024) Elephant Tourism in Nepal: Historical Perspectives, Current Health and Welfare Challenges, and Future Directions. CAB International, Wallingford, UK (in press).

Thapa, J., Paudel, S., Sadaula, A., Shah, Y., Maharjan, B., Kaufman, G.E. and Nakajima, C. (2016) Mycobacterium orygis–associated tuberculosis in free-ranging rhinoceros, Nepal. *Emerging Infectious Diseases* 22(3), 570.

Tripadvisor (2020) Tripadvisor: Improving Animal Welfare in Tourism. Available at: https: //www.tripadvisor.com/blog/animal-welfare-education-portal (accessed 1 June 2020).

United Nations (2023) Least Developed Country Category: Nepal Profile. Available at: https://www.un.org/development/desa/dpad/least-developed-country-category-nepal.html (accessed 15 July 2023).

Varma, S. and Ganguly, S. (2011) *Captive Elephants in Bardia National Park, Nepal, Investigations into Population, Management, Welfare and a Review of Elephant Training by Working Elephant Programme of Asia (WEPA) and WWF Finland at Bardia Hattisar.* Elephants in Captivity: CUPA/ANCF – Occasional Report no.18. Compassion Unlimited Plus Action (CUPA) and Asian Nature Conservation Foundation (ANCF), Bangalore, India.

Wen, Y. and Ximing, X. (2008) The differences in ecotourism between China and the West. *Current Issues in Tourism* 11(6), 567–586. DOI: 10.1080/13683500802475927.

World Bank (2018) Project Information Document/Integrated Safeguards Data Sheet. Nepal Sustainable Tourism Value Creation. *World Bank Website*.

27 Haunting Memory of Forgotten History: A Case Study of World War II 'Comfort Stations' in the Philippines as a Tourism Resource Featured in Selected Filipino Films

Aileen C. Ani
College of Asia Pacific Studies, Ritsumeikan Asia Pacific University, Beppu, Oita, Japan

Summary
The focus of this case study is the World War II 'comfort stations' in the Philippines where the war victims during the Japanese occupation were held. For this case study, the scope is centralized on two main sites in the Philippines. Characters, settings, and the narratives of 'comfort women' and 'comfort gays' were analyzed from the two full-length Filipino films entitled *Bahay na Pula* and *Markova: Comfort Gay*. Within this context, the study assesses the potential of the featured 'comfort stations' and other war-related sites in the films as a tourism resource that will educate tourists about war.

The concepts of *remembering* and *forgetting* are essential when understanding critical areas of history. Analysing what specific events, persons, or memories people choose to retain or erase, reveals different factors and motivations that push them to do so. This case study argues that it is necessary to remember important historical sites in the Philippines, particularly the information signs that have been commonly taken for granted and are slowly deteriorating. It uncovers forgotten sites, restores the memory, and transforms the historical significance of the 'comfort stations' featured in Filipino films by analysing the story being told, the kind of place being depicted, and the relevance of preserving them.

The Value and Interest of the Case Study

The unfortunate physical and cultural devaluation of some of the historical sites in the Philippines is quite alarming as it may reflect how local people value history and memory. Aside from underrepresentation in textbooks or history books of the tragic story of war victims, discussion on war sites like the 'comfort stations' during the Japanese occupation in the Philippines between 1942 and 1945 is lacking. The knowledge gap of the disconnection between people and their history is the main target of the contribution of this case study. To better preserve and memorialize the learning from the war, it is necessary to identify, document and rehabilitate these historical landmarks. The study utilizes the media as a tool to explore interdisciplinary approaches to studying war memories and tourism.

Context: *Forgotten History*

World War II 'comfort women' during the Japanese occupation is still a pressing issue between formerly colonized nations and the Japanese government. The memories of war continue to haunt the current generation, with many longing to restore the forgotten history. Presently, the Korean 'comfort women' are most active in engaging the public in discourses about who they are and sharing their history of struggles and misfortunes, which consequently made most people think that 'comfort women' were only Korean (Cabusao-Silva, S. 2023, online communication). The opposing views about the existence of 'comfort women' lead to the deliberate forgetting of war memories, and that includes the sites related to colonization. There were approximately, 50,000–100,000 'comfort women' from a range of countries, including Korea, China, the Philippines, Indonesia, Malaysia and Japan (Asian Women's Fund, n.d.). Although some arguments have been made that these women were paid prostitutes, this is a contentious claim and certainly does not apply to Filipinas from 9 years old up to teenage years who were victimized (Henson, 1999). Moreover, the former 'comfort stations' in the Philippines where the events took place are not recognized or documented in history or textbooks. The grim condition of these important historical landmarks manifests the disconnection between local people's memory and their history. For this case study, the link between history (*memory*), media (*representation*), and tourism (*sites*) serves as the passageway to explore the darker yet relevant tourism resources of the war in the Philippines.

By performing content analysis, this case study focused on the film elements of *Markova: Comfort Gay* (2000) and *Bahay na Pula* (2020). This includes identifying the major themes of the films, particularly the central topic discussed (Viva Films, 2022). Characters and the characterization itself are briefly discussed. The analysis summarized the symbolic representations of each character, reflecting the realities during the war. Settings composed of the depicted time or period and the place where the scenes are located were also analysed. The film production's crucial selection of these aspects was also examined as it connects with important historical events. Finally, since films are considered part of the creative line of expression, a series of metaphors and icons have been interpreted that signify key messages whether an obscured or explicit illustration of their essence. Observations are guided by short insights from interviews conducted in September 2023 with a tourist and an independent tour guide operator-historian.

Discussion: *Haunting Memory*

In the modern era, *film-induced tourism*, sometimes called *cinematic tourism* or *media pilgrimage*, a niche where a site visitation was inspired by the featured destination from a TV show, film, or other form of media, is becoming a more prevalent way of explicating the nexus between media and tourism. Consequently, through the destinations being featured in two films: *Fort Santiago* in Manila and *The Bahay na Pula-Ilusorio Ancestral House* in Bulacan, their popularity was enhanced after the release of the said movies.

For example, a frequent visitor-tourist of war-related sites emphasizes the added attraction of tourist sites (or in her case, 're-visit' sites) after being featured on the big screen. Whether it is a war film or not, it leaves an impact on moviegoers:

> It became more interesting to see it again, having a new perspective of the site, you will feel the difference now that we acquired some, deeper knowledge about it. Aside from accessibility, historical sites became more appealing to tourists if it is free. In addition, former war battlefields are quite unique for me, personally, that's why I usually visit these kinds of places. (SR, 34, local tourist)

In this regard, the tourist acknowledged the importance of media, both mainstream and social networking sites:

> Depending on the time of visit, not all historical sites have an available tourist guide that can help us learn more things about the place. Therefore, accurate information about these landmarks given by the media is helpful. It feeds us input that is not commonly heard in school. For movies, referencing is common, especially for the red house or grandmother's house, which almost became equivalent to a 'haunted house'. Visiting various 'haunted houses' and the real 'Bahay na Pula' are also trendy nowadays, especially for vloggers until it became redundant, actually. (SR, 34, local tourist)

Considering this scenario, the responsibility of the viewer-tourist was also highlighted as it became necessary to possess critical and analytical skills to identify which contents are factual and which are just artistic representations of historical narratives. On the other hand, taking the perspective of a tourist guide:

> Making films about our history greatly helps in terms of its popularization. Films have a wide reach as part of popular culture and media. Multiple times, I have documented the nature of films and historical documentaries because I wrote the official study guide of historical films like *Heneral Luna* (2015), *Ang Hapis at Himagsik ni Hermano Puli* (2016), *Goyo, Ang Batang Heneral* (2018), and *The Kingmaker* (2022). In the case of tourism, if the site is not yet featured in films, images and maps are important visual aids. Based on my experience, if the oral narratives are well-researched and interesting, accompanied by visual representation, they become more effective. Tourists can have a better grasp and deeper understanding of the places they visit. (AC, 36, tourist-guide/historian)

He further explained:

> Before I offered the [Intramuros group] tour, its content and practice were already included in my teaching as a history professor. It took a lot of time for me to research on this and in fact, up to this time, the oral narratives are still evolving and changing. I designed it in such a way that it encapsulates Philippine history. For example, the Fort Santiago, I summarize the history of Manila and the Philippines. It covers the discussion about prehistory, ancient civilization, the Spanish colonial period, Americans, World War II, Japanese Occupation, and the Contemporary Period. I also summarize the history of structures and buildings in that way. I emphasize and connect the essence of each site, church, building, house, headquarters, and other local, national, regional, and world history.
>
> In terms of my personal approach, it depends on my client, most of the time, it's a customized tour. But if there are tour preferences that need to be met, I discuss this with clients via e-mail. If there are special topics that want to be included, or specific spots that would like to visit, it can be communicated ahead of time. As with historical discussion, the tour guide needs to choose which to tackle or not. My advantage compared to another typical tour guide is I am a historian and a teacher of history. So, I have a better judgment in terms of the breadth and depth of the discussion. Discussions are structured and can be deepened depending on the client's interest. (AC, 36, tourist-guide/historian)

Aside from the significant contribution of media, the role of the tourist guide in historical sites is undeniably valuable. Flexibility, openness, and reflexivity are necessary skills for tour guides, particularly in sites that are rich in history, and sometimes controversies.

Case No. 1: The bitter memories of the 'Red House'

The horror film *Bahay na Pula* (Red House) was released in 2022. It was directed by the premiere Cannes Film Festival-awardee Director Brillante Mendoza. The film focuses on a newlywed couple who are both in the field of house construction. Raffy (husband) is an architect, and Jane (wife), is an interior decorator. They aim to sell the ancestral house once previously owned by Jane's grandparents. A few eerie things happened during their stay along with Aling Ising, the supposed house caretaker. Dealing with their own personal issues, the transaction of selling the old house was halted due to a newly passed resolution protecting the local ancestral houses. A sinister feeling was experienced by pregnant Jane inside the creepy house. The story inside the mysterious house ends tragically when it unravels the unfortunate experience of Filipino women during the Japanese occupation of the Philippines (Figs 27.1 and 27.2).

The movie was released a few years after the removal of the 'comfort women'

Fig. 27.1. Actual 'Bahay na Pula' located in San Ildefonso, Bulacan.

monument. It was removed from its original location on Roxas Boulevard in the capital city of Manila and subsequently reportedly stolen. The plot revolves around the red house which refers to Raymundo's ancestral house situated in Kabilang Ibayo, Batuhan, Pola, Oriental Mindoro. It is one of a series of vintage houses in Oriental Mindoro Province. Hence, as previously mentioned, if based on the storyline and movie title, the inspiration for this is possibly the real *'Bahay na Pula'* in San Ildefonso, Bulacan Province which became a Japanese garrison and 'comfort station' during World War II (see Fig. 27.1).

Various themes were touched on in the *Bahay na Pula* movie (see Fig. 27.3). *Restoring important historical landmarks* is a serious concern in the Philippines since many of the

Fig. 27.2. Remains of the 'Bahay na Pula'.

sites are forgotten and no longer receive proper attention from different stakeholders (see Fig. 27.2 and Videos 1-3: https://www.cabidigitallibrary.org/doi/book/10.1079/9781800628687.0000). Interestingly, the backstory chosen in the movie was a *love story between members of the Japanese Imperial Army and Filipinos*. This sensationalized narrative is based on a real-life love story of two Japanese soldiers, Sakae Narioka and Fujiro Takahashi, who married Filipina

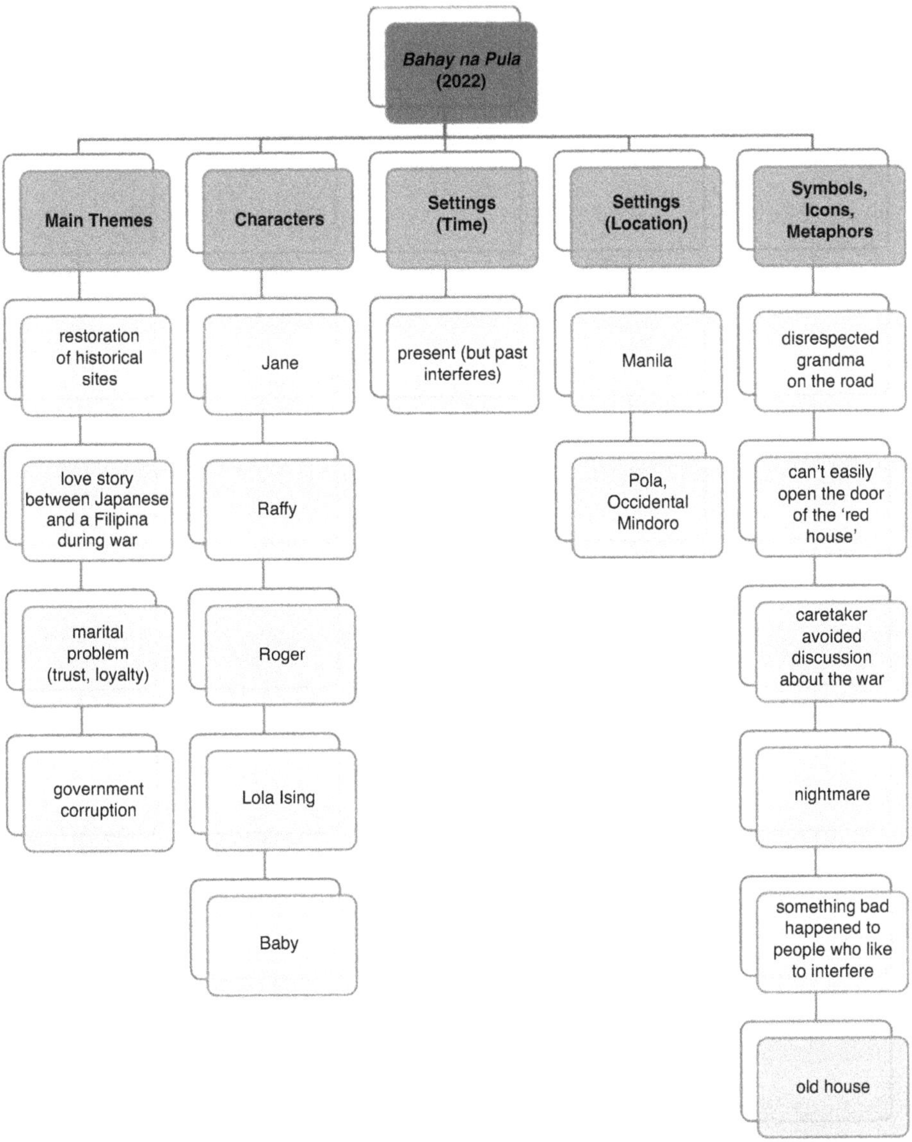

Fig. 27.3. Film elements of the *Bahay na Pula* movie.

locals named Belen Castillo and Adela Tolentino, respectively. This bond between the Japanese and Filipina was recognized to be the main reason why Vigan City, a well-restored UNESCO World Heritage Site, was saved from possible destruction by American bombs during World War II (Dayrit, 2009). Aside from the historical aspect, themes like *marital problems* that involve issues of trust and loyalty between couples were tackled. Finally, the existence of *corruption in the government* was also featured. By depicting how under-the-table transactions are made to fast-track permit applications, the movie successfully portrays the major problem in the country.

Distinct persona were represented by the characters of the movie. *Jane* represents the war victims during the Japanese occupation. *Raffy* represents the government or Filipino people who do not listen to or recognize the problem. *Roger*, represents the oppressor, while *Lola Ising* is the one who is expected to protect the house and the welfare of the family but does the opposite. The *baby* represents the continuous haunting of war traumas in the next generation. As long as the issues of war reparations are not addressed, the film reminds its audience that the war still continues.

Though the *present time* captured the majority of the narrative sequence, the interference of the past was highly noticeable and directly affected the current generation. The central capital of *Manila* and a distant province of *Occidental Mindoro* were used as the location sites. Both these places were important sites during World War II. Indeed, there was the Battle of Manila and Battle of Mindoro which killed not only thousands of people but also destroyed residential and important structures.

Finally, using effective and highly symbolic icons, the film creatively rendered subtle messages about Filipino experiences during the Japanese occupation. For example, a scene where *Raffy almost hits an elderly woman across the street* signifies how Filipino 'comfort women' were treated by society. Raffy shouted and showed aggression to the grandma who just crossed the street. Second, the film metaphorically used the scene where Jane and Raffy were having a hard time *opening the 'red house'* and they had to break the door lock just to enter. It shows how difficult it was to revisit the issue of war victims during the war. In fact, even just to initiate a dialogue or enter an academic discussion about 'comfort women' remains taboo (Ramos, 2023). Controversies and complex issues related to war reparations and responsibilities were avoided. In the film, it was depicted by the *subtle avoidance of Aling Ising*, the old house caretaker, when she was asked by Jane about what happened in the house during the occupation. Similar to the narratives, if someone wants to interfere in selling the old red house something bad happens to them. The recurring symbol used in the movie was the *nightmare* of Jane when she experienced being physically abused by a Japanese soldier. It worsened when she got killed after the soldier forcibly opened her womb to get the baby, symbolizing the political power structure even in narrating and passing historical narratives that shape the next generation.

Case No. 2: The discomfort of 'comfort gays' in the Philippines

Markova: Comfort Gay, a biographical film of the life of Walter Dempster Jr, was a box-office success in 2000. It narrates the life of a Filipino gay from her childhood dealing with family problems, survival during the war, to her post-war struggles. Markova got some inspiration when she saw Lola Rosa (Maria Rosa Henson, the first Filipino 'comfort woman' to testify) on national TV speaking about her experience during World War II (Asian Women's Fund, n.d.).

Various important sites in the movie *Markova: Comfort Gay* are not just relevant to Markova's life but to Filipino people in general, especially during the war. Care homes for elderly gay people, Catholic churches, different former pubs and clubs across the Manila area, an unnamed internment camp, barracks, and evacuation hospital were presented (ABS-CBN Star Cinema, 2023). Fort Santiago, where different fights took place against the Japanese, Spanish and Americans, also became a witness of war (see Figs 27.4–27.12). This encompasses not only Filipino-American soldiers but also Filipino women and children during the Japanese occupation. As a matter of fact, this popular historical site, which is regularly visited as a former battleground, was also a place of physical violence against Filipinas, something that is not commonly known. It became the headquarters, holding area, detention center, and 'comfort station' of the Japanese army. 'Comfort women' survivors recall that even though there were heavy bombings and airstrikes on the Fort, Japanese soldiers continued to abuse them (Dolgopol and Paranjape, 1994).

The movie *Markova: Comfort Gay* mainly focuses on one major theme, the *life and struggle of LGBTQIA+* whether it is during the war or not (see Fig. 27.13). The social realities of how a Filipino homosexual is experienced and treated by individuals and institutions were featured. Characters, particularly *Markova's*

Fig. 27.4. Main gate of Fort Santiago in ManilaWomen's Active Museum on War and Peace (n.d.).

circle of friends – Carmen, Sophie, Anita, and Minerva, represent the sexually discriminated against sector in Philippine society. From the smallest unit like their families, denial had been the default treatment. Loren Legarda, portraying herself as a highly cynical yet curious journalist who was interested in finding the truth, represents a skeptical citizen to his/her fellow Filipinos, especially regarding the shared experience of 'comfort gays'Watanabe (2020).

The movie was set during the present time but is constantly interrupted by the traumatic past. Several sites were highlighted that had a

Fig. 27.5. Entrance of the dungeon inside Fort Santiago.

significant role during World War II. The *Home for the Golden Gays* was depicted but not the actual residence for the elderly gays of Manila. The original house was built by Justo Justo, a former councilor and journalist. The El Hogar Filipino Building, and churches like Quiapo Church, Sta. Cruz Church, and Binondo Church were part of the daily lives of Markova, who was a faithful Catholic. These churches witnessed the war but were fortunately spared. Dewey Boulevard, performing arts and night clubs like Tsubaki Club, Dragon, Clover, Gala, Greenlight Theatre, Burlesque Theatre, and Avenue became a site of entertainment and

Fig. 27.6. Information sign of the dungeon.

silent protest for Filipino audiences during the war. These sites became a creative hub for Filipino performers seeking an avenue for self-expression during the difficult time. It is also necessary to pay attention to the unidentified barracks, garrisons, and other makeshift Japanese headquarters as these sites were underrepresented and commonly taken for granted in history books (e.g. classic references like *History of the Filipino People* by Teodoro Agoncillo and *The Philippines: The Continuing Past* by Renato and Letizia Constantino) (Constantino and Constantino, 1982; Agoncillo, 1990). The historical significance of

Fig. 27.7. Interior of the dungeon with a human figure presented as a soldier.

Fig. 27.8. White Cross inside Fort Santiago.

these landmarks provides an opportunity to let the war victims and their perspectives be heard. The well-preserved Fort Santiago, a frequent battleground, could explore other types of war that Filipinos experienced, including gender-based violence.

Finally, different metaphors and references were used in the movie. To contextualize the conflict, *real video clips from the war archives* were embedded in the narration. *Camote*, a root crop, was an icon during the war signifies the extreme poverty the Filipino people experienced. *Markova's dream or nightmare* continues to haunt her and describes the level of anxiety she had which triggered her willingness to publicize war experiences. Hence, the irony of *Markova's perception of how she reconciles herself* as a war victim from the past and as a Japayuki instructor (in Filipino culture, it commonly refers to Filipina entertainers trained to seek jobs in Japan) in the present. This tug-of-war represents how difficult it is to tackle controversial issues like 'comfort women' if it involves a formerly colonized-developing country that depends on the socio-economic support of a former colonizer-developed country.

Fig. 27.9. White Cross information sign.

Conclusions: *What Is Next?*

This case study attempts to fill in the gap in the existing literature about the Filipina 'comfort women' by offering a new approach that combines aspects of media and history fields together with tourism. This transdisciplinary approach, to media representations of war memories, explores the interactions of different social representation forms, ranging across movies, stories, people, and sites. This is the major contribution of this case study – linking personal stories (plot/historical narratives) to creative remembrance (films), and physical remembrance (tourist sites) Runquist (2020).

Framing the two films as fitting sample case studies with how the interconnection of media, tourism, and history is viable illustrates the

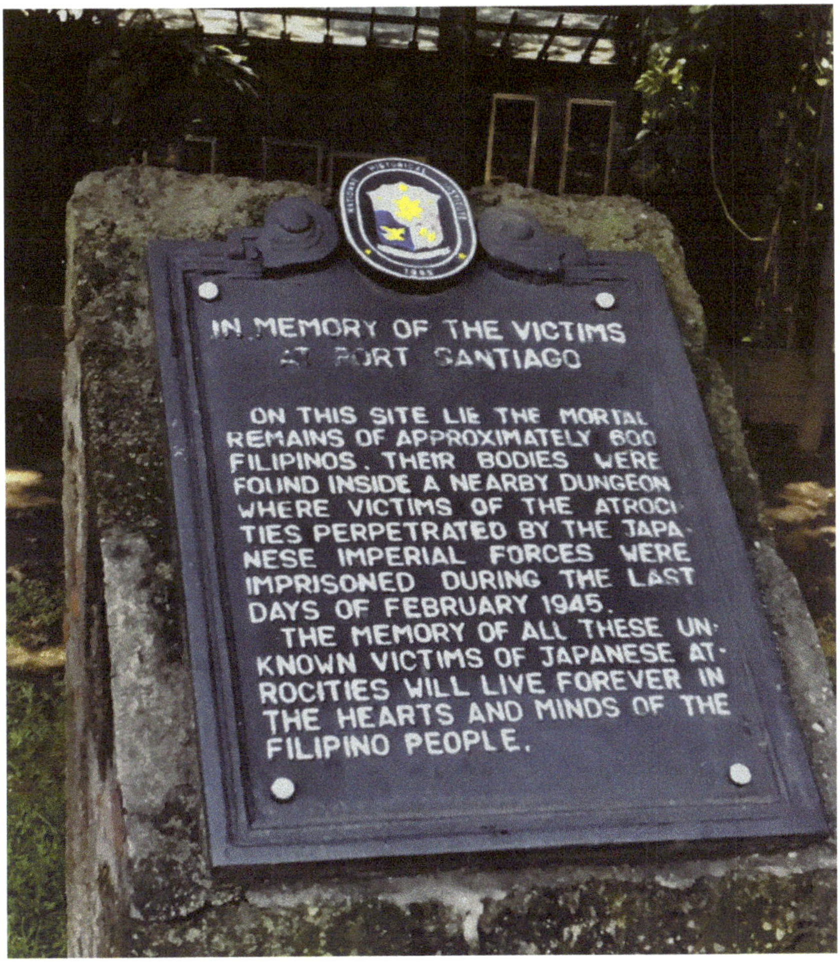

Fig. 27.10. Official marker of the National Historical Institute installed in the White Cross.

Fig. 27.11. Rajah Sulayman Theatre, a former military barracks.

Fig. 27.12. Rajah Sulayman Information sign.

political power structure in Philippine society. The remembered and forgotten aspects of war memories also demonstrate the dominant power in historical writing. Narrating the story of Markova ('comfort gay') and Lola Rosa ('comfort woman') remains an effective tool to socially represent the violence while the sites related to their life and war experiences immortalize their life stories and the Filipino stories. This is what dark tourism wants to tell us. By converting the memories and related sites, these tourism resources can relive the haunting past and may help the current and future generations learn from it. The proposed tour (see Figs 27.14 and 27.15) serves as an initiative to commemorate the life and struggle of Filipina 'comfort women and gays' during World War II.

Two recurring metaphors used in both films were identified. The continuous interruption of the past to the present in the storyline and the constant use of nightmares of the main characters: Jane (representing 'comfort woman') and Markova (representing 'comfort gay') AWHRC (1992)manifests the war trauma and profoundly hints that the war is not yet over.

Further readings were listed to help the readers access relevant information about the full context of the issue.

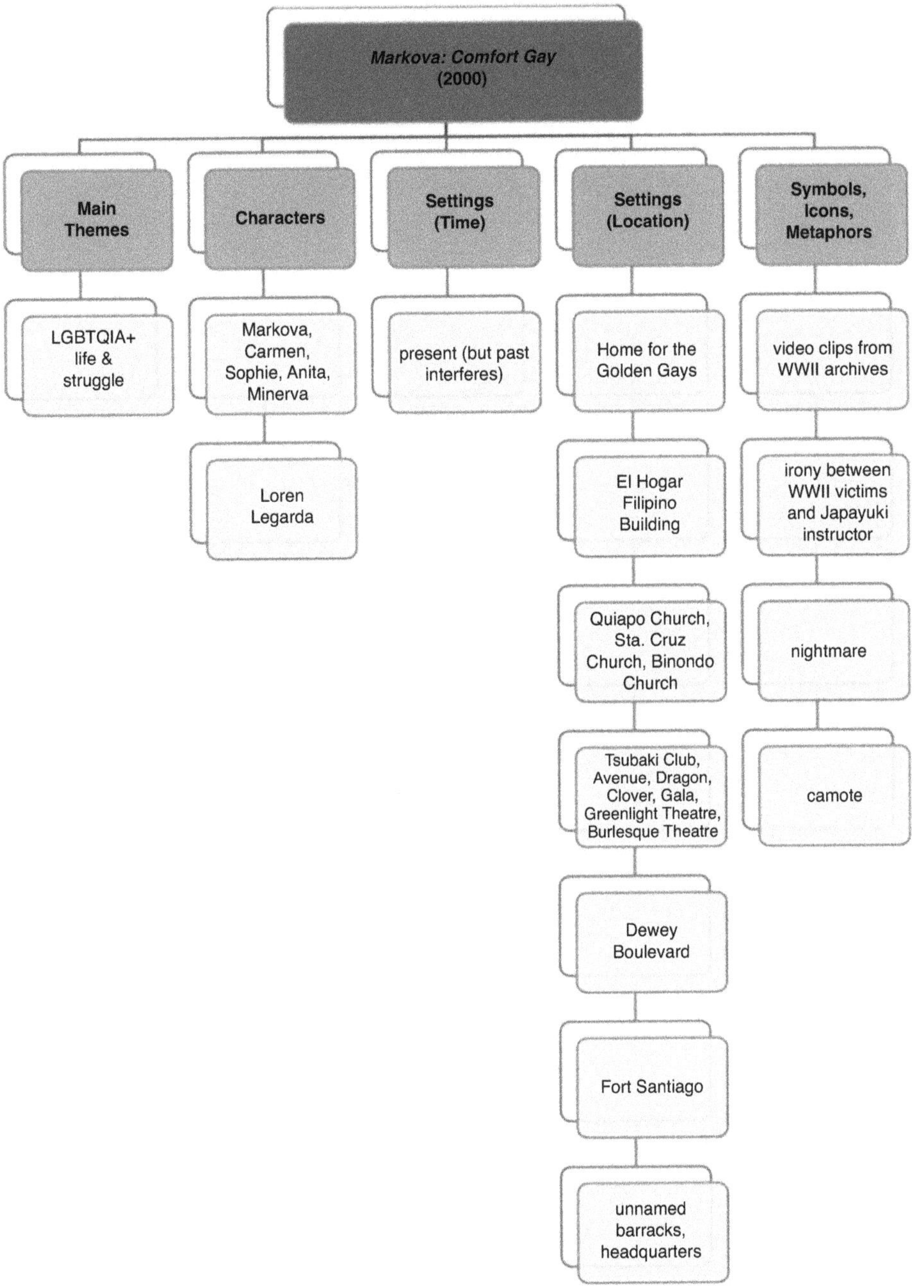

Fig. 27.13. Film elements of *Markova: Comfort Gay* movie.

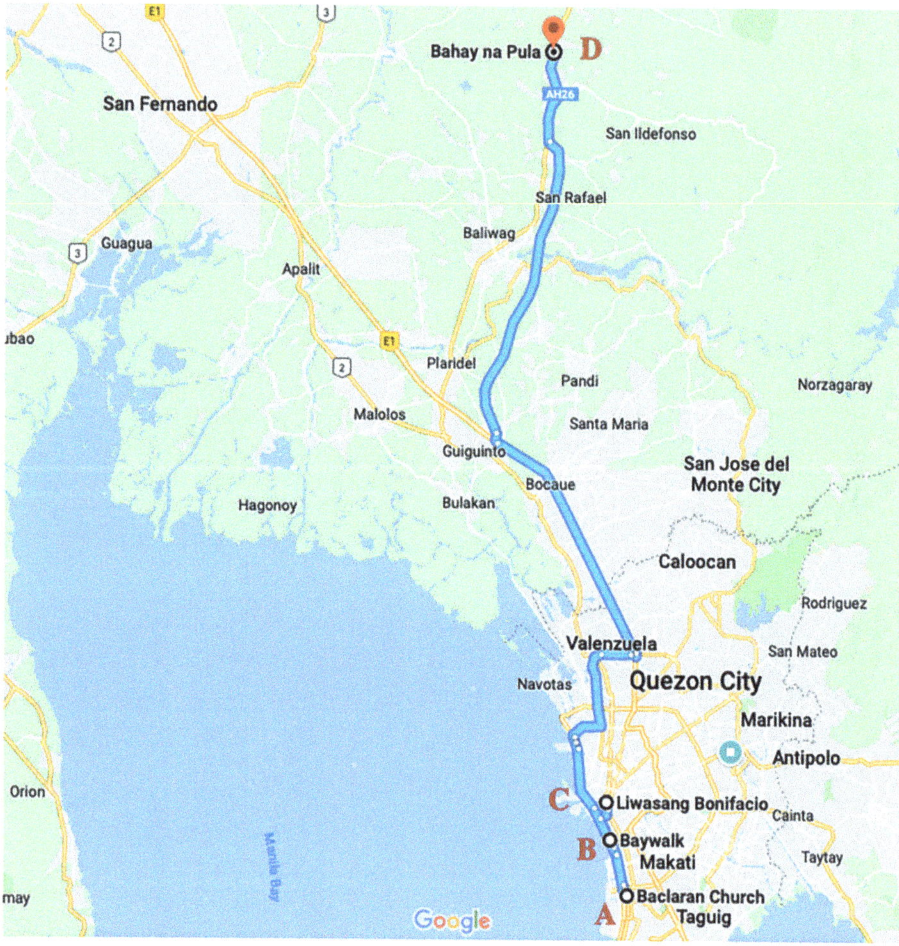

Fig. 27.14. Proposed 'Lola's Tour' to commemorate Filipino 'comfort women' during World War II [also Second World War]. Map of the Philippines tracing [A] Baclaran Church with 'comfort women' sign, [B] Baywalk in Roxas Boulevard, the former site of the Filipino 'comfort women' monument, [C] Liwasang Bonifacio with 'comfort women' sign, and [D] the real 'Bahay na Pula' in San Ildefonso, Bulacan.

Fig. 27.15. Proposed 'Markova's Tour' to commemorate Filipino 'comfort gays' during World War II [also Second World War]. Map of Manila tracing the Home for the [A] Golden Gays in Pasay, [B] Dewey Boulevard in Roxas Boulevard, [C] Quiapo Church, [D] El Hogar Building in Escolta, [E] Sta. Cruz Church, [F] Binondo Church, and [G] Fort Santiago.

Acknowledgement

My deepest gratitude to my editor and reviewers for helping me to improve my work. I would also like to thank my mentors Prof Bui and Prof Progler for sharing their time and wisdom, APU Research Office, Mark John, Ana-Liza, Yodo san, and Nathan for their patience, unending support, and inspiration, and Ms. Cabusao-Silva, SR, and AC for their interview responses. Nevertheless, I take all the responsibility for any correction in this case study.

Further Reading

AWHRC (1992) Primer on Filipino 'comfort women'. Questions and answers. Available at: http://freedomarchives.org/Documents/Finder/DOC46_scans/46.FilipinoComfortWomen.pdf (accessed 1 January 2023).
Beeton, S. (2006) Understanding film-induced tourism. *Tourism Analysis* 11, 181–188. DOI: 10.3727/108354206778689808.
Cardoso, L, *et al*. (2017) Film induced tourism: A systematic literature review. *Tourism & Management Studies* 13(3), 23–30. DOI: 10.18089/tms.2017.13303.
Gjorgievski, M. and Trpkova, S. (2011) Movie-induced tourism: A new tourism phenomenon. *UTMS Journal of Economics* 3(1), 97–104.
Naoko, K. (2016) *The Comfort Women: Historical, Political, Legal and Moral Perspectives*. International House of Japan, Tokyo.
Runquist, T. (2020) Comfort Women: Korean, Chinese, and Filipina survivors and their portrayals in the media. *Western Illinois Historical Review* 11. Available at: http://www.wiu.edu/cas/history/wihr/pdfs/WIHR%20spring%202020%20Runquist%20final%20version.pdf (accessed 11 January 2023).
Tanaka, Y. (2001) *Japan's Comfort Women: Sexual Slavery and Prostitution during World War II and the US Occupation (Asia's Transformations*). Routledge Taylor & Francis Group.
Tzanelli, R. (2007) *The Cinematic Tourist: Explorations in Globalization, Culture and Resistance*. Routledge Taylor & Francis Group.
Watanabe, M. (2020) Recalling the strong voices of 'comfort women' survivors. *Positions Politics*. Available at: https://positionspolitics.org/nothing-about-us-without-us/ (accessed 1 January 2023).
Women's Active Museum on War and Peace (n.d.) The Japanese military 'comfort station' Map. Available at: https://wam-peace.org/ianjo/ (accessed 1 January 2023).
Yoshiaki, Y. (2002) *Comfort Women: Sexual Slavery in the Japanese Military during World War II*. Columbia University Press.

References

ABS-CBN Star Cinema (2023) Markova: Comfort Gay Full Movie. Available at: https://www.youtube.com/watch?v=ixDfrlY-S6U (accessed 7 May 2023).
Agoncillo, T. (1990) *The History of the Filipino People*. C & E Publishing, Inc, Metro Manila.
Asian Women's Fund (n.d.) How did the Comfort Women Issue Come to Light? Available at: https://www.awf.or.jp/e3/oralhistory-00.html (accessed 8 May 2023).
Constantino, R. and Constantino, L. (1982) The Philippines: The Continuing Past. The Foundation for Nationalist Studies, Quezon City.
Dayrit, C. (2009) We've only just Vigan. *Philstar Global*. Available at: https://www.philstar.com/lifestyle/travel-and-tourism/2009/03/01/444013/weve-only-just-vigan/amp/ (accessed 8 May 2023).
Dolgopol, U. and Paranjape, S. (1994) *Comfort Women: An Unfinished Ordeal*. International Commission of Jurists, Geneva.
Henson, M. (1999) *Comfort Woman: A Filipina's Story of Prostitution and Slavery under the Japanese Military*. Rowman & Littlefield Publishers, Inc, Washington, DC.
Ramos, M. (2023) For Filipino wartime rape survivors, a last hope for reparations. *The Japan Times*. Available at: https://www.japantimes.co.jp/news/2023/09/25/asia-pacific/social-issues/filipino-survivors-reparations-hope/ (accessed 28 September 2023).
Viva Films (2022) Bahay na Pula trailer. Available at: https://www.youtube.com/watch?v=ayzClMejJ-E (accessed 7 May 2023).

28 'We Invite You for Lunch': Post/ Colonial Subjectivities and Performances of Tourism in Puducherry, India

Aindrila Chakraborty
Department of Anthropology, College of Arts and Sciences, University at Albany, SUNY, New York, USA

Summary

Tourism is contested, as Edward Bruner suggests. These contestations may not always take the form of overt conflicts between actors and/or institutions. They may operate in silent pauses and gaps, performances of the actors involved, their identifications and agencies and constructed meanings. In my research, I focus on the town of Puducherry, a French colony merged with India in 1962, to engage with the discourse of authenticity in which the hosts take part. By positioning myself as both tourist and ethnographer, I delve into the complexities of post/colonial subjectivities and performances of identity in Puducherry, with a focus on the Franco-Pondicherrian community who, through their performances reproduce the authentic and reclaim their collective identity. In the process, the members of the community, who are also the hosts in this context, become both the gatekeepers of authenticity and the toured. Using the idea of staged authenticity in a social constructionist frame, I engage with how hosts stage cultural performances in the intimate space of their home which works to make the tourist aware of their cultural heritage and abets their identity practices.

The Value and Interest of the Case Study

The town of Puducherry in Southern India is a lucrative tourist destination, for both domestic and international tourism. Being a former French colony, the city is promoted through a mix of Franco-Indian culture, heritage, and architecture (MacCannell, 1973; Bruner, 2005).

This case engages with the complex cultural identity production at tourist sites, especially where colonial histories and postcolonial subjectivities are produced and negotiated through tourism. It demonstrates how the hosts use tourism as a means and mode to perform collective identity.

Background

Nestled in corner of South India, the town of Puducherry (the official name since 2006, but still largely known by its colonial name Pondichéry/Pondicherry) borders the Bay of Bengal Sea and is one of the top tourist destinations in India (see Fig. 28.1). While characterized by multiple beaches, these are not the sole attraction of the town. In postcolonial India, tourism is highly controlled by the state. The state plays a huge role in what can be consumed for touristic purposes and how. The town is reproduced through its lingering coloniality, a product which is manufactured and marketed by the Indian state like it has done with no other town, which has become the

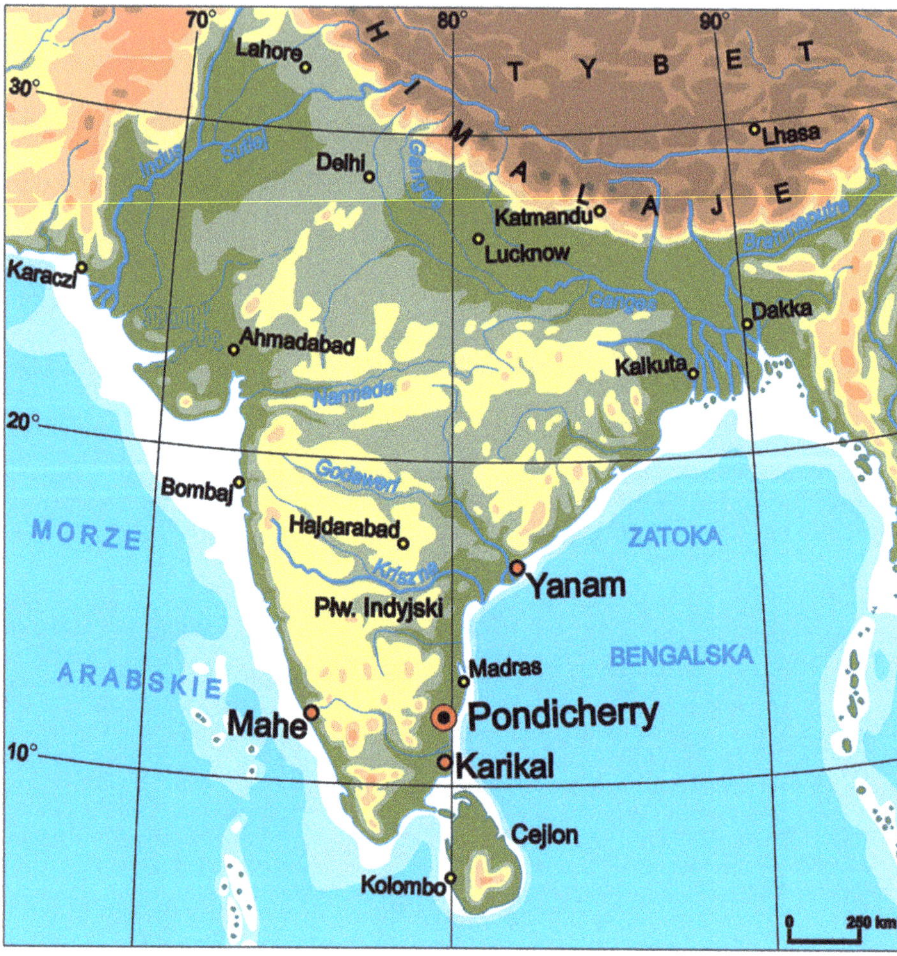

Fig. 28.1. Puducherry, India.

major force behind the construction of the tourist imagination of Puducherry. The production of Puducherry, as seen through the government website and reflected through tourists' travel blogs, is as an 'escape from time,' as 'a place which takes one back to the pages of history.' It is also seen through the reproduction of colonial nostalgia like Puducherry as the 'French Riviera of the East.' Puducherry is produced as the last vestige of colonial charm, a marginal pocket which was and continues to remain tranquil and pristine, untainted even through years of colonial occupation. The fantasy produced here is not of a precolonial purity but of colonial originality. The union territory of Puducherry, with the districts of Puducherry (the capital), Mahe, Yanam and Karaikal, was a French colony since the 1700s and was merged with India (de jure) in 1962. Among the four districts, Puducherry is a major tourist attraction. During its postcolonial existence, Puducherry has been a host to primarily international tourists (drawing tourists from Western Europe, primarily France and UK, and the USA). It is increasingly becoming a destination preferred by Indian tourists. The colonial racial segregation of the city into *Ville Blanche* (white town) and *Ville Noire* (Black/native town) is still largely maintained, not on racial lines but to sustain the tourism infrastructure. The white town marks the French-built colonial heritage, the numerous French institutions and consulate and the beachfront, with French/colonial-themed heritage hotels and restaurants. Heritage actors and the government are currently in the process of getting the city listed as a UNESCO World Heritage site and the colonial spatiality contributes to its uniqueness. Along with the colonial presence as a major force of tourist attraction, Puducherry draws in a host of secular pilgrims to Aurobindo Ashram, founded by Aurobindo Ghosh, an Indian nationalist who escaped to Pondicherry from British India and withdrew from politics and resorted to Hindu mysticism and its offshoot establishment Auroville, located on the outskirts of the town. The latter is an international township complex founded by the Frenchwoman and Ghosh's spiritual companion Mirra Alfassa (invested with the title 'The Mother') which is also a UNESCO World Heritage Site and stands on histories of indigenous dispossession of land (Namakkal, 2021).

French colonization of India is much neglected in both South Asian and French historiography, and the postcolonial Indian state constructs itself and its subjects against an ostensibly British colonial past. The community that I am engaging with bear testimony to the distinct post/colonial subjectivities of British and French Indian subjects and how this percolates into a postcolonial Indian imaginary. In the late 1800s, French Indian subjects were granted French citizenship based on renunciation of their communal identity, whereby they would fall under the French civil code. A marginal proportion of the native community opted for renunciation (thereby known as renouncers), some of whom also migrated to other French colonies through serving in the colonial army and administration. In 1962, when the treaty of cession for the merger of Pondichéry with India was ratified by France, it provided an option to retain/opt for French nationality to the subjects, thereby creating a distinct and ambiguous group who lay in the hyphenated space between two postcolonial nation-states. Today, there remains a minor proportion of Franco-Pondicherrians in Puducherry while a substantial proportion is settled in France, forming the Franco-Pondicherrian diaspora. Mostly of Tamil origin, they represent a syncretic mix of years of colonial contact and ethnic roots.

This research is a part of my PhD research on heritage and tourism in Puducherry. Ethnographic data was collected during the field trip conducted between May and July 2022. Data collection was undertaken with the consent of the participants and the analysis is my own.

Ethnographic Case: *'Chez Pushpa'*

One of the most insightful openings into Puducherry's Franco-Pondicherrian community was through the de Canaga family, whom I met on my field trip to Puducherry in June 2022, and established a relationship beyond our short but profound encounter. Although there is a recent interest in some Indian tourists to search for creole pockets in India, which can be observed from travel blogs published in the recent past, for an average Indian tourist traveling to Puducherry, families like that of de Canaga do not always feature in their tourist imagination of the Pondicherrian 'Other.' I chanced upon the de Canaga family due to my own research on the Franco-Pondicherrian community. Had I been in Puducherry for solely touristic purposes, I may not have known about them, let alone booking a meal with them at their house. One must contact them in advance, showing your interest in having a meal at their house and reserving one's spot. One meal (unlimited in quantity and consisting of ten dishes) costs INR2000, which is around US$24 at the current exchange rate. My interest in having lunch at their house was redirected to Anita de Canaga, daughter of Mrs Pushpa de Canaga and Mr Noel de Canaga, who oversees tourist interactions. I had a long conversation with Anita where we spoke about my research interests regarding her community, and she explained to me, in detail, the concept of chez Pushpa. The concept of chez Pushpa is La Table d'Hôte, wherein guests are invited to have a Pondicherrian meal with the family. Anita explained to me what goes behind planning the menu and what I should expect. 'We are inviting you to have lunch with us in our Franco-Pondicherrian household. Mumma [Mrs de Canaga] goes to the bazaar [market] in the morning and decides what she is going to get for us to eat, so you will not know the menu beforehand. This also means that all the food you are going to eat will be fresh and healthy!'. On asking further about the story behind this venture, Anita delved into why it is not only important to her mother and her family but also to the entire community. 'Most people visiting Pondichéry today know nothing about the history and culture of the region. It is becoming a studio for Instagram content and people go to fancy and pricy places in the white town in the name of French cuisine. But Pondichèry cuisine is different. It is a mix of French, Tamil and in our case, also Indo-China cuisines because of our history.' Implicit in Anita's conversation was her notion of authenticity. Indeed, the term 'authentic' came up in our conversation a few times. Understanding that I was interested in not only having a Pondicherrian meal at their house but also in her community's story, she made her presentation of her sense of identity, which in this case is reproduced through this food venture, visible.

The de Canagas' are one of those families who have migratory histories linked to former French Indochina and, at present, are settled in Puducherry, moving back and forth between France and Puducherry. For most tourists traveling to Puducherry, the history of colonial contact and how it affected and continues to shape group identities does not occupy a space in their imagination. Puducherry, as a physical space (conceived space), has a visible semblance with France, at least for the domestic tourists. Indian tourists mostly comprise middle-class Indians who more often only long for travel beyond their country but do not have the cultural-economic capital to do so. However, little interest is taken by the tourists in the perceived space that the natives and locals occupy. The conceived space is formed through the palatable aspects of the history underscored by the Indian state through its Indianizing mission of former pockets of European control. Drawing from Lefebvre's (2012) production of space, I observed that the dialectical tension between the conceived and perceived spaces is reproduced through touristic encounters in Puducherry, which produces the touristic borderzone. This is, according to anthropologist Bruner (2005), a point of encounter between tourists and hosts in structured ways, a creative space of collaboration and liberation, wherein cultures are invented. The Franco-Pondicherrian community consists of those who have (both historically and in the present society) occupied a marginal space in the region's cultural milieu. The present marginality arises not from a lack of cultural capital but because their collective, hyphenated identity stigmatizes them as a cultural minority.

Shifting the scene to their house, I now underscore my experience of La Table d'Hôte (see Fig. 28.2). My younger brother (who accompanied me on the trip), and I were welcomed by the family into their home in a manner that suggested I had been, in essence, invited. This, I think, is what MacCannell (1973) would call 'staged authenticity.' Mr Noel contacted me as I got close to the house so he could give directions to the exact location and stood outside to welcome us. Mrs Pushpa peeked from the semi-enclosed space of her kitchen; the place where the magic happens! Anita mentioned that 'Mumma is more comfortable in the kitchen space while we have guests because she can only speak Tamil and she is very shy.' I could witness this as most of the conversation was driven by Mr Noel, while Mrs Pushpa moved back and forth between the dining/living rooms. The intimate space of her kitchen could pass as the backstage, which was also open to the tourist gaze but over which she had complete authority.

The dining room, more specifically the dining table was, in Bruner's (2005) term, our touristic borderzone. The dining table was where we sat throughout, relishing not only the food but our presentations of our 'Selves.' Shortly after we arrived, the menu for the day was handed out to us, hand doodled by Mr Noel himself, which added a personal touch to our acts. The menu consisted of ten items in total, including drinks and desserts. Mrs Pushpa, who was reluctant to converse at the beginning, gradually opened up to us as she sat with us while we ate. I asked if they could also eat with us, to which they replied they had already had their lunch. However, they emphasized the fact that the food was cooked for all of us, including them, and that they served the guests what they themselves would eat. As Anita remarked 'For you it (the food) might look different and fancy, but this is our regular meal. This is what we eat on a regular basis, only some items are for special occasions.' The menu had dishes unique to the Franco-Pondicherrian community, as underscored by the family multiple times. Dishes

Fig. 28.2. Lunch at the de Canaga house (courtesy of the author).

such as Mutton Sambar and Pork Vindali were emphasized as they mark the difference from the larger Tamil (Indian) community. Sambar is a stew made with lentils and vegetables, other spices, and a hint of tamarind (also used largely in the Tamil cuisine). Mrs Pushpa emphasized the fact that 'we make it differently because we add meat to our sambar. Only Pondicherrian cuisine has meat in sambar.' Anita further added that the 'Pork Vindali [curry] you find only in some Franco-Pondicherrian households. The Hindu/Muslim Franco-Pondicherrian households abstain from having pork.' This was particularly interesting given the condition which was attached to the formation of this community and varied ways culture gets negotiated and reproduced within single communities. The other specific which marked their boundaries through food was the serving of Spring Rolls as a starter. They emphasized that this is a staple that is found only in those Franco-Pondicherrian households that have roots in Indochina. Furthermore, the food contains the secret ingredient: Vadouvan, an intense mix of French and Tamil spices (12 spices in total form the ingredient) which marks the uniqueness of Franco-Pondicherrian cuisine.

Mr Noel de Canaga was born to a Franco-Pondicherrian family in Vietnam. Mrs Pushpa is not a Pondicherrian herself as she was born in the neighboring state of Karnataka. She was married into the Franco-Tamil family of de Canaga and acquired her culinary skills from her mother-in-law, who, as she remarked 'was great at cooking!' While our conversations centered around food, it ranged from an exchange of our different cultural norms to their community's history. Mr Noel provided interesting insights on the popular sentiment of Pondicherrians during the time of merger, the disappointment about a referendum to join India or remain a French oversees territory was never held, and how Mr Noel's father was involved in making Pondicherrians aware that they could retain their French nationality. Intertwined with food, as often happens in tourism, was the reclaiming of a collective identity (Giampiccoli and Kalis, 2012; Sammells, 2016; Montero, 2020) and inserting it within the tourist imagination about Puducherry and Pondicherrians.

Tourists from all over the world (if they are willing to do some background research on Pondicherrian society) can end up having a meal with the de Canagas. What is particularly interesting in the context of the de Canagas' cultural performance is that it is intertwined with their concerns of authenticity. Helle Jørgensen (personal communication), based on her ethnographic study of Puducherry, notes that the complex decolonization of the region and the post/colonial mobile circuits between France and Puducherry, coupled with growing tourism infrastructure propagated by the state leads to complex insider-outsider questions in the region. There are the old Pondicherrians: The Franco-Tamils (the de Canaga family is an example), the creoles and the optants who opted for French citizenship in 1962. There are also migrant laborers who locally reside in the region. There is also a floating French population (from metropolitan France) who work at the numerous French institutes in the region. In the conversation I had with the de Canaga family, the insider-outsider positions and their relationship to authenticity was, at times, explicit. Mr Noel nostalgically noted the changing social scape of the town, due to commercialization through tourism. Anita, who is involved in the tourism industry, remarked how people often mistake authentic Pondicherrian food with the 'French named cafes in the white town which are French only in name but are driven by outsiders.' She noted, 'We came up with this idea (chez Pushpa) because we wanted people to know what authentic Pondicherrian food is, from Pondicherrians themselves. When people think of Pondichèry, they only think of the beaches, the Ashram, Auroville and yellow buildings in front of which they take photos, without even knowing the names of most buildings. That is not authentic Pondichèry.'

From these conversations, it can be observed that the construction of authenticity through the concept of chez Pushpa is intricately and intimately tied to their collective identity. Intertwined with and working behind the concern for authenticity are the multiple located processes of re-membering and re-presenting. The act of re-membering, as Fortier (2020) in the context of Italian migrants in Britian notes, is constitutive of not only remembering what was, but is also about manufacturing bodies that member the space of remembrance, to occupy such spaces and perform remembrance to give it definition and meaning. Touristic encounters at the de Cannga household produce a space re-membered by the Franco-Pondicherrians collectively and the tourist is called upon to witness the re-membering through their re-presentation.

Conclusions

Cultural performances in the intimate space of the home in touristic encounters and concerns with authenticity are not unique to the de Canaga family or Pondicherrians. Walter Little (2000) discussed how the home becomes a space for the exhibition of culture in Maya communities in Guatemala. He noted how the town of San Antonio has been incorporated into global flows through tourism development. Kaqchikel Mayas have incorporated these global flows into their households wherein homes have been opened to tourists for exhibition and performance of culture and laying collective claims to it, such that the domestic sphere is opened not only to the public sphere but also becomes part of the global. The performance of household activities such as weaving not only abets collective identity claims but also makes tourists aware of Mayan identity in a theatrically mediated way. Similarly, Pierre-Walter (2017) engaged with how a cooking school in Thailand has become a site of staging authenticity for the tourist experience and of living history. While the stage is not someone's home in this context, the staged performance in the spatial-temporal setting, from grocery shopping, cooking to dining, tourists are made aware of the history and culture of Thai food through embodied experience of authenticity. Cooking classes offered by Pushpa to tourists, wherein they are engaged in a similar process, can be seen to echo a similar idea. In a distinct yet interrelated line, Gabriela Spears-Rico (2015) engaged with how Mestiza/o tourists tour indigenous communities in Mexico, leading to commodification and appropriation of the living and dead indigenous peoples. However, the indigenous communities work within the tourism infrastructure, engaging in performances of their identity to preserve their community's intimacy, controlling the ways they are perceived, consumed, and toured. In the process, they forge their resistance as not just being passive hosts who are toured but active partners in the production of tourism. The concept meal at the de Canaga household is somewhat similar, wherein the tourist and the host perform their respective roles to co-construct and experience the authentic. In the case of the hosts, it can be observed that the construction of authenticity through the concept of *chez Pushpa* is intricately and intimately tied to their collective, subjective experience of being a cultural minority, which is increasingly diasporizing (Chakraborty, 2022) in their own home. The need to construct the 'authentic' is driven by the marginal space the Franco-Pondicherrian community occupies in the tourist imagination of Puducherry and its culture. Ann Reed (2014), in her work on diasporic Africans visiting the Elmina Castle in Ghana, extends Appadurai's concept of 'scape' to include 'memory-scape.' It is what diasporic groups reside in, in order to forge and reproduce their ancestral ties to their ethnic homelands. The staging of their home as a space where the tourists are welcomed and presentation of their culture to understand what the authentic Puducherry is, is the memory-scape they reside in and sustain in the face of remaking of identities and subjectivities and ideas of home in the postcolonial and the fading Franco-Tamil and creole communities (due to emigration) from the socio-cultural scape of Puducherry. The Franco-Pondicherrian community in Puducherry, through their cultural performances, tends to reproduce this memory-scape and tourist encounters to provide a fertile ground for such performances. I base my observations not only on my encounter with the de Canaga family but also the other Franco-Pondicherrian households that I have visited and the members of the community with whom I have conversed. From the hosts' side, in this case, the discourse of authenticity becomes important as it is through this they present their culture and heritage for tourist consumption while simultaneously laying their collective claims to it. In the process, they negotiate the power dynamics that determine what is authentic, and by working within the dynamic, they bring forward what is omitted in the production of the region to their tourists. In the process, they willingly become the toured, while simultaneously situating themselves as the 'authentic Other.' At this point in the tourist encounter, 'The Host Moment' (to draw from Cary's (2004) Tourist Moment) is visible if one looks closely.

The de Canaga house is a stage set for tourist consumption. From the point of entry till the moment one leaves, the house itself plays multifarious roles; from a space of intimacy where people reside to a human repository where Franco-Pondicherrian cultures and identities are stored to a stage where it is reproduced. In this house, the people (the guests and the hosts) play their particular parts, in such a manner that the border is not breached. However, in the house, new stages are produced, with the more open dining room and the intimate kitchen space peeking from behind the stage, only partially open to the tourist gaze. As part of this, Mrs Pushpa moves back and forth between the kitchen and the dining room, breaching the divisions between front and back, while Mr Noel reiterates the division by emphasizing that they have had their lunch before we (the guests) arrived. In these new, constructed stages, the roles are in process, coproduced through interactions, while the border shifts but remains. Here, the existential authenticity which might get lost in public roles (Reisinger and Steiner, 2006; drawing from Berger, 1973) is restaged and performed in public to reclaim it. The public/private dichotomy is agentially blurred to reproduce the authentic.

Walter Little (2019 in Bruner *et al.*, 2019), in his work on Maya vendors' encounters with tourists in Guatemala, noted that authenticity is subjective to the host and guest and that the two authenticities can differ. It may not always be a concern for the host to tweak the guests' ideas. In this case, I observe that presenting authenticity becomes important for the host. This concern can be attributed to the increasingly marginal space they occupy in the socio-cultural milieu of the region. This can also be attributed to the fact that they hold the cultural capital to be concerned with authenticity and willingly be toured. Here, authenticity can be driven by what lies beyond an objective authenticity according to Bruner (2005); performance of a collective identity. Chez Pushpa may not yet be known to a large number of tourists traveling to Puducherry, but her reach is far and wide among some renowned Indian chefs and food enthusiasts.

Engaging with such performances of identity within the space of tourism reveals the postcolonial complexities of the region, taking shape through the lingering coloniality and the colonial nostalgia through which not only the metropole but even the former peripheries experience the 'Self' (Jørgensen, 2022). All of which is further complicated through the production of the region for tourism (Jørgensen, 2017).

Through this research, I have underscored how hosts can be active agents in coproducing the authentic through cultural performances. In most cases, such performances are driven by the need to preserve and claim collective identity and control the ways culture gets mediated within the tourism infrastructure. Such cultural performances reveal the social structure in which communities are embedded and the power relations within. This case also reveals how communities caught between post/colonial affective cartographies actively engage in constructing a postcolonial identity and how tourism can become a fertile ground for such constructions. Students and researchers engaged in the anthropology of tourism, cultural studies, working on identity and performance in tourism and those studying post/colonialism can gain insights from this research.

References

Berger, P.L. (1973) 'Sincerity' and 'authenticity' in modern society. *The Public Interest* 31, 81.

Bruner, E.M. (2005) *Culture on Tour: Ethnographies of Travel*. University of Chicago Press.

Bruner, E.M., Swain, M.B., Di Giovine, M.A., Graburn, N., Harrison, J. *et al.* (2019) *The Ethnography of Tourism: Edward Bruner and Beyond*. Rowman & Littlefield.

Cary, S.H. (2004) The tourist moment. *Annals of Tourism Research* 31(1), 61–77.

Chakraborty, A. (2022) Citizenship, belonging and the sense of ambiguity among French Tamils of India. *Migration Letters* 19(4), 547–556.

Fortier, A.M. (2020) *Migrant Belongings: Memory, Space, Identity*. Routledge.

Giampiccoli, A. and Kalis, J.H. (2012) Tourism, food, and culture: Community-based tourism, local food, and community development in Mpondoland. *Culture, Agriculture, Food and Environment* 34(2), 101–123.

Jørgensen, H. (2017) Between marginality and universality: Present tensions and paradoxes in French colonial cultural heritage, civilizing mission, and citizenship in Puducherry, India. *Heritage & Society* 10(1), 45–67.

Jørgensen, H. (2022) 'Out through the door and in through the window?': Positioning colonial nostalgia for French India in Puducherry. *International Journal of Francophone Studies* 25(1–2), 59–92.

Lefebvre, H. (2012) From the production of space. In: *Theatre and Performance Design*. Routledge, pp. 81–84.

Little, W.E. (2000) Home as a place of exhibition and performance: Mayan household transformations in Guatemala. *Ethnology*, 163–181.

MacCannell, D. (1973) Staged authenticity: Arrangements of social space in tourist settings. *American Journal of Sociology* 79(3), 589–603.

Montero, C.G. (2020) *From Temporary Migrants to Permanent Attractions: Tourism, Cultural Heritage, and Afro-Antillean Identities in Panama*. University Alabama Press.

Namakkal, J. (2021) *Unsettling Utopia: The Making and Unmaking of French India*. Columbia University Press.

Reed, A. (2014) *Pilgrimage Tourism of Diaspora Africans to Ghana*. Routledge.

Reisinger, Y. and Steiner, C.J. (2006) Reconceptualizing object authenticity. *Annals of Tourism Research* 33(1), 65–86.

Sammells, C.A. (2016) Haute traditional cuisines: How UNESCO's list of intangible heritage links the cosmopolitan to the local. In: *Edible Identities: Food as Cultural Heritage*. Routledge, pp. 141–158.

Spears-Rico, G. (2015) Consuming the native other: Mestiza/o Melancholia and the performance of indigeneity in Michoacán (Doctoral dissertation, UC Berkeley).

Walter, P. (2017) Culinary tourism as living history: Staging, tourist performance and perceptions of authenticity in a Thai cooking school. *Journal of Heritage Tourism* 12(4), 365–379.

Section 6: Management and Policy

This section explores the impact of data and evidence on strategic planning and policymaking in the travel and tourism sectors. One case study examines 30 years of meteorological data to understand its implications for aviation and travel in Bali, while another analyses the development and outcomes of a hotel investment policy in Western Australia, highlighting the importance of evidence in policy development.

29 Meteorological Trends and the Travel Sector in Bali 1991–2020

Kadek Sumaja[1,2] and Tay T.R. Koo[2]
[1]I Gusti Ngurah Rai Meteorological Office, Indonesian Agency for Meteorology Climatology and Geophysics (BMKG), Bali, Indonesia; [2]School of Aviation, Faculty of Science, University of New South Wales (UNSW Sydney), Sydney, Australia

Summary
This case study describes 30 years of meteorological data for the purpose of sharpening the management, strategic and policy discussions on the potential impact of climate on the aviation and travel sectors in Bali. Findings show a mix of positive and negative meteorological trends that may have implications for developing regulatory, physical and economic strategies.

The Value and Interest of the Case Study

The study illustrates several of the important channels through which changes in the meteorological parameters can impact the travel sector in Bali, which is a significant tourism destination in the Asia-Pacific region. Specifically, three channels are described, focusing on inter-industry linkages, visitors and aviation. This is done from the perspective of a scientific weather body informing the medium to long-term plans of the destination management organization.

Background

Many studies have examined the impact of the aviation and tourism sector on climate change; for example, in the measurement and accounting of carbon emissions from travel activities as well as water and land used (Anger, 2010; Gössling and Peeters, 2015; Dube and Nhamo, 2019). However, only a limited amount of research has focused on analysing the impacts of changing climate and meteorological patterns on the international aviation and tourism sector in Bali (Furqan and Winandi, 2018; Ryley *et al.*, 2020). Addressing this gap is crucial to the understanding of the potential challenges faced by Bali's travel sector and in providing the knowledge base for sustainable destination management. For instance, climate highly correlates with the seasonality of Bali. Hence seasonal shifts, or their intensification, will require a mix of marketing (e.g. product and market diversification), regulatory and legislative (e.g. water access and land-use zoning), physical (e.g. flood infrastructures), educational (e.g. eco-edu products, visitor information, community disaster preparedness) and economic (e.g. land/housing price markets, foreign ownership) strategies. All of these are potentially relevant for sustainable tourism development in line with the UNWTO's sustainable development goals.

Against this background, the aim of this case study is to describe the meteorological trend of Bali over the past 30 years and discuss its implications for the tourism and aviation sectors. In doing so, three channels through which climate is expected to impact the travel sector in Bali are explained. The first issue describes the changes in the rainfall and groundwater reserve and how this creates competition for water between the tourism and agriculture industries. The second issue highlights whether, and how, the climate comfort index, which has potential implications for visitor satisfaction and experience in Bali, has changed over time. The third issue relates to the occurrences of extreme weather events in Bali. These events can cause unexpected flight delays and cancellations, as well as exacerbate the impact on aviation services of already pertinent natural events, such as volcanic eruptions in Bali. Therefore, understanding these phenomena is important for the efficiency and resilience of the operation of air service providers, which are the main modes of tourist arrival to the region, and tourism operators.

Data

For the analysis of the annual rainfall trend in Bali, data were collected from the Indonesia Agency for Meteorology Climatology and Geophysics (BMKG) and four Meteorological stations and twenty rain gauge points from 1991 to 2020. Also, water status and demand data in Bali were retrieved from the Ministry of Environment and Forestry (MoEF): Eco-region Development Control Center of Bali and Nusa Tenggara. La Nina/EL Nino event occurrences were retrieved from the Australian Bureau of Meteorology.

For the Holiday Climate Index (HCI) and humidity index (i.e. humidex), the 30 years of weather data (1991–2020) from 24 observation points were combined with the data from the European Union's climate data store (Available at: https://cds.climate.copernicus.eu/, accessed 30 July 2023) that comprises of maximum and average temperature (°C), relative humidity (%), cloud coverage (%), rainfall (mm/day) and wind speed (km/h). International tourist arrival data were obtained from the Center for Statistics Agency of Bali Province (BPS).

The extreme weather data were collected from I Gusti Ngurah Rai Meteorological Office's observation data with respect to rainfall, wind speed and temperature. The extreme category was derived using the following threshold: strong wind categorized as wind speed above 25 knots, heavy rain as rainfall that exceeds a threshold of 50 millimetres within a 24 hour period, and extreme air temperature as air temperature 3°C above the 30-year average. Volcanic ash coverage data were obtained from the Volcanic Ash Advisory Centre in Darwin, Australia (Available at: http://bom.gov.au/aviation/volcanic-ash/, accessed 30 July 2023).

Result and Discussion

Climate and water availability in Bali

Fig. 29.1 shows the rainfall data collected over the 24 rainfall observation points in Bali across 30 years. The average annual rainfall in Bali showed a minor fluctuation ranging from 1400 to 2800 mm/year. However, there is no clear discernible trend in rainfall in Bali as a whole. Given the slight decreasing trend in the maximum rainfall and generally stable average and minimum rainfall, a constant water supply is expected, reducing the water shortage probability. Most of the water sources in Bali are surface water supply, meaning the region is heavily reliant on the river (40%) and groundwater sources (49%), with the remaining water supply coming from springs (4%) and lakes/reservoirs (7%).

While there is no shortage of water availability in Bali as a whole, there is a significant spatial variation with respect to the expected balance of water supply and demand in Bali. As can be seen in Fig. 29.2, despite strong water availability as a whole, the three most popular tourism districts

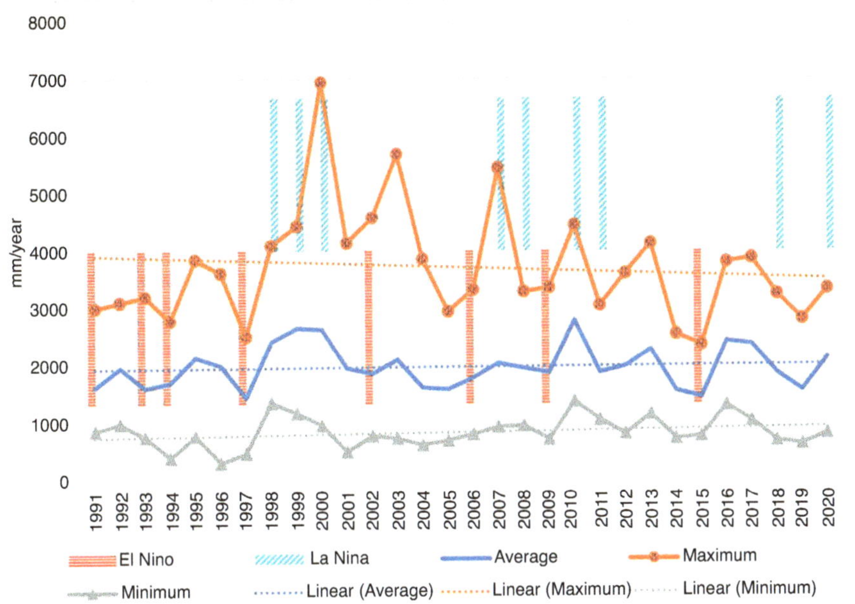

Fig. 29.1. Rainfall trend in Bali (source: Compiled from BMKG Observation Data and ENSO Outlook from Bureau of Meteorology Australia, 2023).

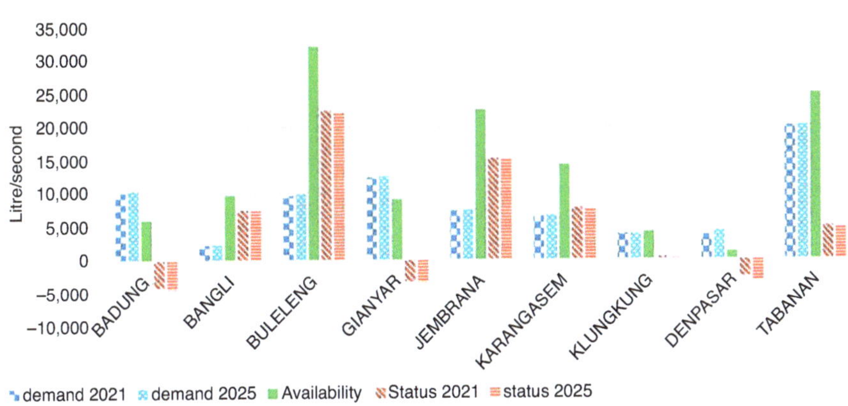

Fig. 29.2. Water status by region (source: Compiled from Ministry of Environment and Forestry (MoEF): Eco-region Development Control Centre of Bali and Nusa Tenggara).

(Badung, Gianyar and Denpasar) are experiencing water scarcity. Based on a report from Pusat Pengendalian Pembangunan Ekoregion Bali dan Nusa Tenggara (2021), Eco-region Development Control Center under the Ministry of Environment and Forestry, water scarcity in Badung and Gianyar is caused by high residential, wetland agricultural and tourism demand. Denpasar's water shortage is mostly attributable to the city's predominantly built-up areas, coupled with original fluvial landforms that are low in permeability and, therefore, high in run-off. In addition, due to the humidity and air temperature characteristics in Denpasar, an uncontrolled runoff will either become surface water runoff or, in flat terrains, lost through soil moisture or evaporation.

The above water scarcity plays a central role in the conflict between the international tourism and agriculture sectors. One may consider this as a displacement effect whereby the growth in one sector of an economy/destination has an adverse effect on another sector, through its impact on the competition for shared input. In this context, the key shared resource is water. The international tourism sector is typically an intensive user of water. The competition for water resources between the two sectors has feedback loops in that the agriculture usage is not only for local food supply but also for the preservation of UN Heritage-listed rice cultivation traditions called, Subak. This practice is one of the globally significant examples demonstrating how sustainable social-ecological systems emerged and self-organized in communities (Brooks *et al.*, 2018) and is of significant international visitor interest due to its culturally unique and visually striking display. Consequently, water scarcity will have the effect of reducing the net benefits of tourism to the local economy, further adding to the challenges of sustainable tourism in the destination.

As demonstrated in Fig. 29.1, rainfall usually correlates with the strength and frequency of El Nino and La Nina events. In years when La Niña takes place, there is an increase in rainfall, while during El Niño years, there is a decrease in rainfall. Thus, if climate change causes a change in the frequency and intensity of El Nino and La Nina, which the scientific evidence appears to suggest is possible (Ashok *et al.*, 2007; McPhaden *et al.*, 2011; NOAA, 2020), water supply issues may either be abated or aggravated, depending on whether it is an El Nino or a La Nina year. The key issue, therefore, is the uncertainty and the speed at which such changes may occur for the populous parts of the island heavily reliant on tourism.

Climate comfort for visitors

From the demand perspective, the climate is an important covariate of seasonal variation in tourism demand and influences the quality of the visitor's experience, which is associated with visitor spend and repeat visit intentions. This is especially the case for visitors arriving from countries during the home region's extreme climatic period. For example, there was the cancellation of flights between Darwin and Denpasar in October–November 2023 due to the Australian Defence Force's runway construction at the airport. This made it into the media in Australia (ABC, 2023)

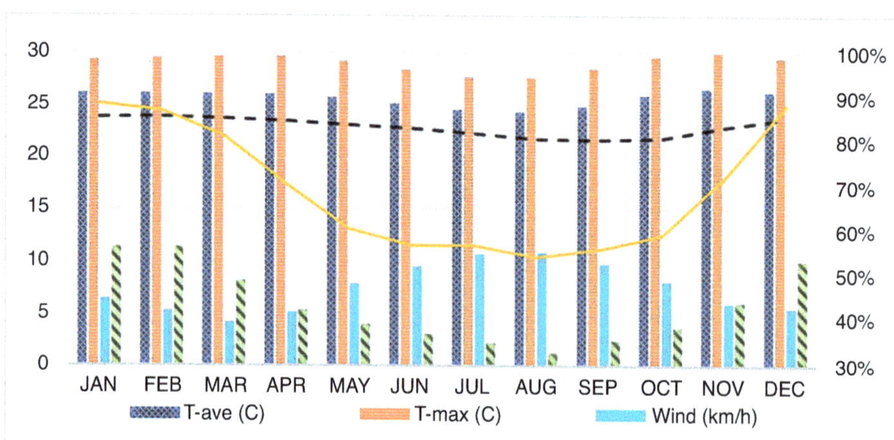

Fig. 29.3. Climate condition in Bali.

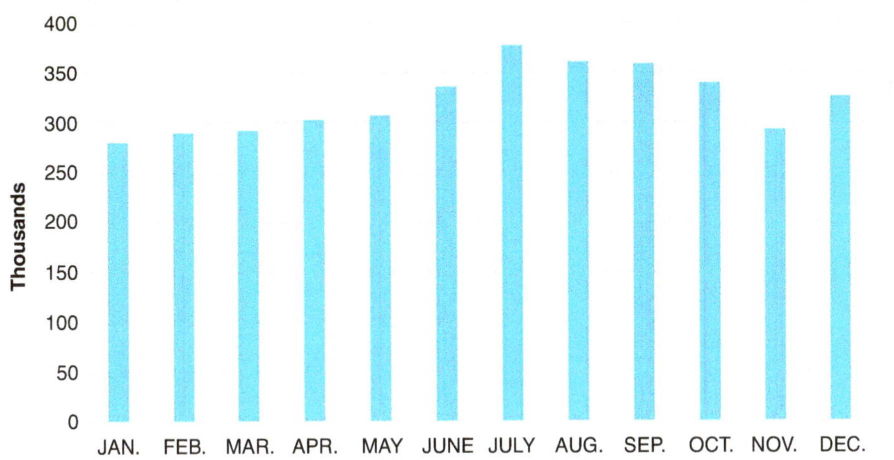

Fig. 29.4. Average monthly international tourist arrivals in Bali (2008–2019).

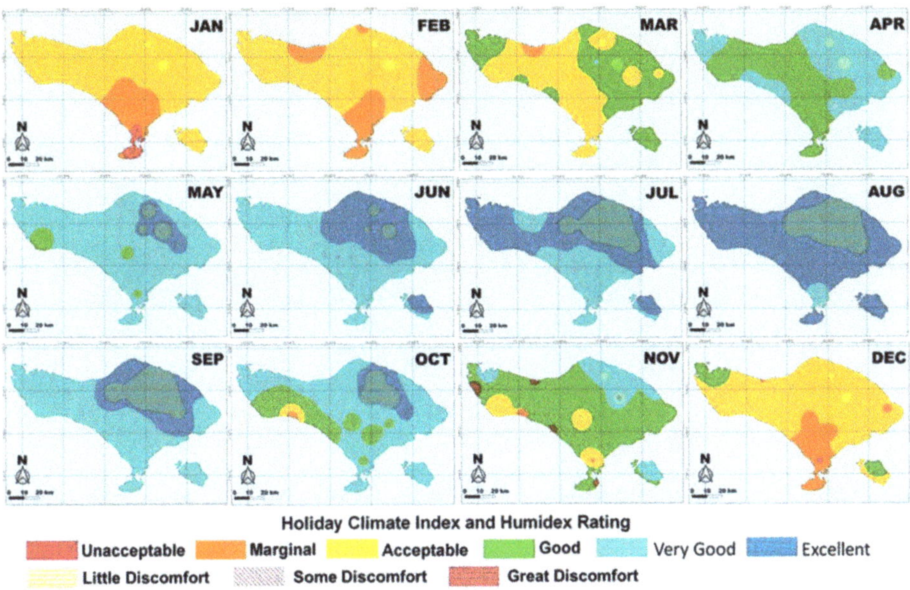

Fig. 29.5. Climate comfort pattern in Bali using HCI and Humidex (compiled by the authors).

because it caused difficulties for Darwin residents seeking to escape their 'build-up' season (build-up to the wet season that is typically humid and uncomfortable). Bali can be viewed as an all-year destination given the relatively narrow range between peaks of approximately 375,000 visitors per month and a trough of 280,000 per month (Fig. 29.4). Nonetheless, as can be seen in Figs 29.3 and 29.4, international tourism peak seasons coincide with northern hemisphere summer, southern hemisphere winter (Australia is the largest single inbound market for Bali), and the non-monsoon months (March–November),

Fig. 29.6. Climate comfort trend in Bali Holiday Climate Index (top) and Humidex (bottom).

which are arguably more suitable for tourism activities.

To identify any observable trends in climate comfort from the visitors' perspective, two well-established indices were used: the HCI and the Humidex. Fig. 29.5 is a visualization of the HCI. As can be seen, the Indonesian monsoon season (December to February) is not comfortable, and at times considered 'not acceptable'. For the majority of the year, however, Balinese weather is considered very favourable from a visitor perspective, and the highly comfortable climate coincides with the peak visitor seasons (see Fig. 29.6).

Between 1991 and 2020, the average HCI in Bali remained within the range of good and excellent, while the maximum HCI value ranged from very good to ideal. The minimum HCI value ranged from acceptable to very good. It appears the maximum and average HCI values experienced a slight decreasing trend, while the minimum HCI value showed a slight increasing trend (Fig. 29.6 top panel). This suggests a minor narrowing of the range of HCI towards generally favourable levels. As can be seen in Fig. 29.6 (bottom panel), the Humidity index shows a slightly increasing trend over the 30-year period. However, it remains within the bounds of some discomfort and great discomfort. Thus, overall, there is not a strong indication of deteriorating climate comfort. In fact, a slight improvement in climate comfort is observed.

The major variation remains across seasons, with the monsoon months showing significant discomfort. As climate change can influence all meteorological parameters used in the calculation of these indices, longitudinal monitoring of these indicators is critical for destination management. One of the climate-related risk factors from the visitor experience point of view would be if seasonal shifts in climate occur. However, at this point, the 30-year observation does not show evidence of important changes in seasonality.

Frequency and timing of extreme weather events at I Gusti Ngurah Rai Airport (DPS)

Based on the information provided by the Indonesian Agency for Meteorology, Climatology, and Geophysics (BMKG), the key meteorological indicators of interest concerning extreme weather events are wind, rain and air temperature. Strong wind is defined as that with a speed greater than 25 knots (or 46.3 km/h), heavy rain is that with greater than 50 mm within a 24 hour window and extreme temperature is defined as three or more degrees Celsius above the 30-year average.

Fig. 29.7 shows the relatively stable trend of extreme temperatures and rainfall occurrences. However, the extreme wind speed occurrences have increased in the past 10 years (2011–2020). Wind shear for the airport is a significant safety hazard, disrupting take-off and landing and resulting in flight delays and cancellations. The safety concerns can create operational inefficiencies and economic costs as alternate flight planning is required by the airlines. In the event of significant delays and cancellations, there are implications for insurance.

As can be seen in Fig. 29.8, extreme wind tends to occur during the wet season. Although this is considered an off-peak travel season, a significant number of passengers arrive during these months (over 300,000 per month – see Fig. 29.9). Therefore, the flow-on impact of flight disruption is potentially significant. Denpasar international airport runway has a bearing of 90 and 270, meaning it runs east–west. Therefore, the high wind, which mostly blows from the west, causes wind shear and tailwind. A very minor decreasing trend is observed for extreme rain events. From an aviation perspective, this is a positive as heavy rainfall impacts visibility and potential flooding risk at the airport, which is located adjacent to the sea at 6m above sea level.

Changes in the wind have the potential to significantly alter the impact of another type of well-known extreme event in Bali – the volcanic eruptions, highlighting the potential indirect impacts of climate change on aviation. These eruptions have a significant impact on aviation. For example, on 9 July 2015, the volcanic ash from the Raung volcano led to the cancellation of 250 flights. Similarly, on 3 November 2015, the ash from the eruption of Barujari volcano

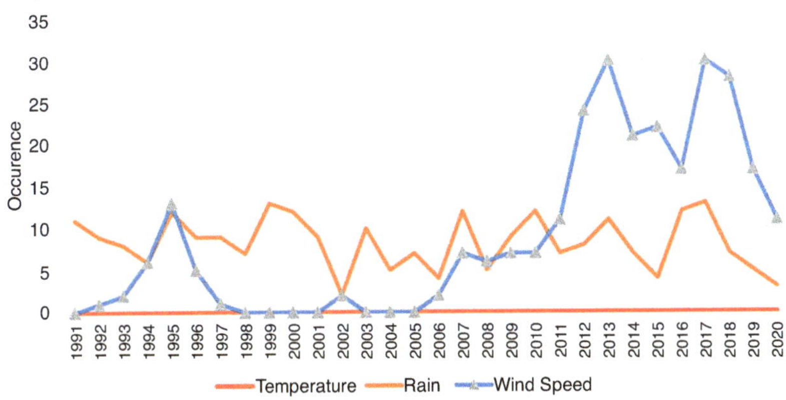

Fig. 29.7. Extreme weather occurrence in I Gusti Ngurah Rai Airport (DPS) for 30 years period.

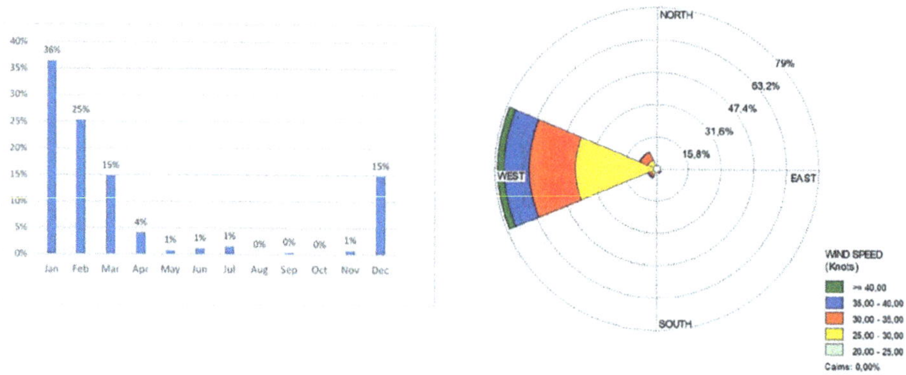

Fig. 29.8. Extreme wind occurrence (left) and extreme wind direction and speed (right) at I Gusti Ngurah Rai Airport (DPS).

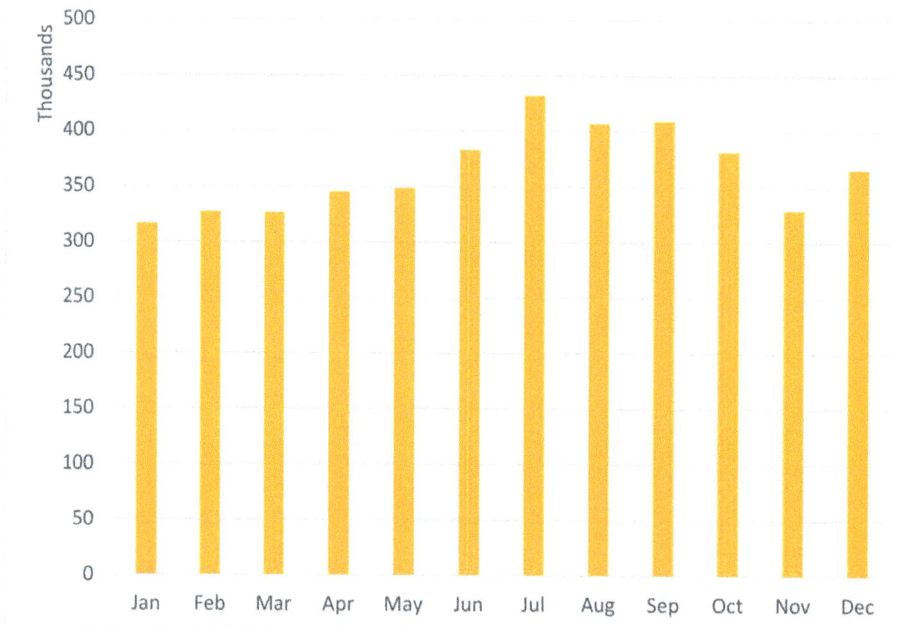

Fig. 29.9. Average monthly passenger arrival in I Gusti Ngurah Rai Airport (DPS) 2009–2019.

resulted in the cancellation of 692 flights. Then, on 27 November 2017, the ash from the eruption of the Agung Volcano led to 445 flight cancellations at Bali Airport. Wind changes can alter the direction of travel of volcanic ash, which can alter the impact on aviation.

Conclusion and Recommendations

From a tourism perspective, positive and negative meteorological trends co-exist. While stable trends can be observed with respect to rainfall and climate comfort levels, as demonstrated by the extreme events, the changes in the meteorological parameters can result in significant and geographically unpredictable impacts that compromise the safe operations of aviation and tourism activities. These changes can also exacerbate existing extreme events, such as the dispersal of volcanic ash, which also impacts the air travel sector.

One of the main concerns is the uncertainty regarding how frequently these extreme events will occur in the future. IPCC is predicting land and sea surface temperature to increase globally, and so is the Indonesian Agency for Meteorology, Climatology and Geophysics (Arora and Meehl, 2022; IPCC, 2022; BMKG, 2023). Although not always, an increase in temperature tends to be associated with changes in precipitation patterns and shifts in weather extremes, which can be associated with the severity and seasonal timing of eastern and western monsoons (Trenberth, 2011; Rummukainen, 2012; Loo *et al.*, 2015). Also, rainfall patterns in Bali correlate with the strength and frequency of El Niño and La Niña events, emphasizing the importance of understanding and preparing for these climate phenomena (Lestari *et al.*, 2018; Supari *et al.*, 2018).

Given the above, a recommendation for the future is to implement a mixed range of destination management strategies. An educational strategy can enhance the preparedness of the people in the coastal areas for storm surges, flash floods and landslides associated with extreme rainfall. This may be accompanied by physical/infrastructure strategies and expansion of early warning systems. Continuous monitoring of climatic trends, particularly those associated with El Nino and La Nina, as well as the creation of contingency plans (early warning and evacuation, shelter and food, medical care and infrastructure protection) for extreme weather occurrences, are key components of proactive climate resilience (Howe *et al.*, 2019; Nobre *et al.*, 2019). More weather observation points and equipment to understand with greater granularity the real-time effect of adverse events such as volcanic ash and how they interact with weather can help reduce unnecessary blanket bans on take-off and landing by identifying windows of very low safety risk. For the airport, the Bureau currently provides climate information for up to 3 months into the future. This information is used for planning on the ground and airside of the airport operations, as well as aiding airlines with their flight planning. As airlines typically set their schedule up to 6–9 months in advance, in the future, it would be advantageous for the airport to extend this period to up to 6 months. However, any increase in the time horizon will be coupled with increased forecast error.

An issue for sustainability in Bali is the difficult issue of balancing water supply and demand. The region's geographical heterogeneity has resulted in localized water scarcity challenges, indicating the necessity for specific, focused alternatives. Several approaches and techniques might be considered to solve these issues and maximize the potential given by Bali's climate and water resource dynamics. Water supply reliability may be improved by implementing sustainable water management practices such as rainwater collection and wastewater recycling in domestic, tourism and aviation sectors. Bali has two wastewater treatment plants (WWTP). Although the island generates around 642,000 cubic metres of wastewater per day, the treatment plants can only handle 61,000 cubic metres (Rochmadi *et al.*, 2010; Budiasa *et al.*, 2018). As a result, a large amount of untreated wastewater from houses, industry, hotels and agriculture pollutes the environment (Rai *et al.*, 2015; Budiasa *et al.*, 2018). Therefore, by addressing this weakness significant opportunity exists for Bali's water supply. About 70% of Bali's wastewater might be reclaimed and reused for purposes such as irrigation and manufacturing (Widianingtias *et al.*, 2023). This would help relieve pressure on current treatment facilities and alleviate environmental issues caused by untreated wastewater. With an average potential production of approximately 108.08 m^3 per year from a 200 m^2 catchment area, a rainwater harvesting system in the northern section of Denpasar City provides a steady source of water for home consumption (Sudiajeng *et al.*, 2020), while at Bali airport, there is the potential capacity to harvest an estimated annual rainfall of 1.2 million cubic metres (Sumaja *et al.*, 2021). To this end, collaboration between the tourism and agriculture industries is critical for optimizing water usage and reducing disputes (Deng *et al.*, 2022). The agricultural industry has practiced highly sustainable irrigation systems for many hundreds of years (through Subak). However, it has been experiencing continuous decline, as water and land resources generate greater economic gains through the tourism sector. Thus, enhanced wastewater management may have the double dividend of not only making the tourism industry more sustainable but also sustaining local heritage and contribute towards maintaining a diverse industrial base.

While there is no evidence of significant changes in rainfall and climate comfort over the past 30 years, the case study has illustrated that potential shifts in seasonality could pose a future risk. Therefore, counter-seasonal strategies are important pathways forward that will hedge against shifts in favourable climatic conditions while maintaining the economic sustainability of the tourism industry. Diversifying tourist seasons to attract visitors during the less-comfortable season can help to mitigate the impact of extreme weather such as high wind events during off-peak travel (Corluka *et al.*, 2013; Krabokoukis and Polyzos, 2023). This can be accomplished by promoting activities and attractions that are appealing during those seasons, offering discounts or special packages, and marketing the destination's unique features for that time of year. For example, during the rainy season, tourists can see fascinating waterfalls, and experience more exciting rafting because the water level and flow rate is higher. However, the success of such initiatives would be dependent on a variety of circumstances, including the destination's general appeal, marketing efficacy and tourists' willingness to visit during less popular times. Therefore, to develop these strategies, more understanding through research is required to disentangle the confounding influences of climatic variation, home-region seasonal/institutional factors, and market forces (e.g. prices and capacity) in determining the domestic and international tourism demand in Bali. This way appropriate and effective seasonal strategies can be developed by targeting marketing and/or policy-oriented variables (Li *et al.*, 2021).

Finally, understanding the fluctuations in rainfall, the spatial distribution of water scarcity challenges, changes in climate comfort, and the occurrence of extreme weather events is crucial for developing medium-to-long-term sustainable tourism strategies. While concerns about global warming and its possible future influence on climate are legitimate, it is important to note that the 30-year dataset utilized in this research shows no indication of substantial climatic changes throughout this time period. The yearly rainfall and climatic comfort statistics give a picture of the condition throughout this era. Further analyses, including using various climatic factors such as temperature, alternative analysis methodologies or longer-term datasets, will be required to acquire a more thorough knowledge of historical climate change and anticipated variations in climate patterns, such as those linked to rainfall, wind speed and extreme weather occurrences. Therefore, future research may focus on a more expansive and detailed study involving the global climate model and its linkages with Indonesian destinations.

References

Anger, A. (2010) Including aviation in the European emissions trading scheme: Impacts on the industry, CO2 emissions and macroeconomic activity in the EU. *Journal of Air Transport Management* 16(2), 100–105.

Arora, V.K. and Meehl, G.A. (2022) *The Observed and Projected Changes in the Climate System*. pp. 3–10.

Ashok, K., Guan, Z., Kumar, K., Yamagata, T. and Behera, S.K. (2007) The influence of climate change on El Niño and La Niña. *Nature* 443(7110), 691–696. DOI: 10.1038/nature05250.

Australian Broadcasting Corporation (ABC) (2023) Jetstar Suspends Bali Flights from Darwin Due to Defence Works. Available at: https://www.abc.net.au/news/2023-07-04/jetstar-suspends-bali-flights-from-darwin-due-to-defence-works/102561440 (accessed 31 July 2023).

Badan Meteorologi, Klimatologi, dan Geofisika (BMKG) (2023) Proyeksi perubahan iklim Indonesia 2022–2050. BMKG, Jakarta.

Brooks, J., Reyes-García, V. and Burnside, W. (2018) Re-examining Balinese subaks through the lens of cultural multilevel selection. *Sustainability Science* 13, 35–47. DOI: 10.1007/s11625-017-0453-1.

Budiasa, I.W., Santosa, I.G.R., Sunarta, I.N., Suada, I.K., Rai, I.N. *et al.* (2018) The potential use of Bali wastewater for crop production based on Moscow region experience. *Water Resources* 45, c138–c147.

Bureau of Meteorology (2023) ENSO Outlook. Available at: http://www.bom.gov.au/climate/enso/outlook/ (accessed 31 July 2023).

Corluka, G., Radi, M. and Geić, S. (2013) Selective forms of tourism – The way of extending the summer season. *Economy Transdisciplinarity Cognition* 16(2), 88–96.

Deng, L., Guo, S., Yin, J., Zeng, Y. and Chen, K. (2022) Multi-objective optimization of water resources allocation in Han River basin (China) integrating efficiency, equity and sustainability. *Scientific Reports* 12(1), 798.

Dube, K. and Nhamo, G. (2019) Climate change and the aviation sector: A focus on the Victoria Falls tourism route. *Environmental Development* 29, 5–15.

Furqan, A. and Winandi, F.I. (2018) The impact of climate change on coastal tourism destination: Case of Kuta Beach, Bali, Indonesia. *ASEAN Journal on Hospitality and Tourism* 16(2), 84–98.

Gössling, S. and Peeters, P. (2015) Assessing tourism's global environmental impact 1900–2050. *Journal of Sustainable Tourism* 23(5), 639–659.

Howe, P.D., Marlon, J.R., Mildenberger, M. and Shield, B.S. (2019) How will climate change shape climate opinion? *Environmental Research Letters* 14(11), 113001.

IPCC (2022) Climate Change 2022: Impacts, Adaptation and Vulnerability. In: *Contribution of Working Group II to the Sixth Assessment Report of the Intergovernmental Panel on Climate Change*. IPCC, Geneva.

Krabokoukis, T. and Polyzos, S. (2023) Analyzing the tourism seasonality for the Mediterranean countries. *Journal of the Knowledge Economy* 14(2), 1–24.

Lestari, D.O., Sutriyono, E. and Iskandar, I. (2018) Respective influences of Indian Ocean Dipole and El Niño-Southern Oscillation on Indonesian Precipitation. *Journal of Mathematical & Fundamental Sciences* 50(3), 257–272.

Li, X., Hossein Rashidi, T. and Koo, T.T.R. (2021) Exploring the temporal travel choices: A joint modelling of how long to travel and when. *Current Issues in Tourism* 24, 2532–2553.

Loo, Y.Y., Billa, L. and Singh, A. (2015) Effect of climate change on seasonal monsoon in Asia and its impact on the variability of monsoon rainfall in Southeast Asia. *Geoscience Frontiers* 6(6), 817–823.

McPhaden, M.J., Zebiak, S.E. and McPhaden, J.C. (2011) The role of climate change in the recent El Niño-Southern Oscillation. *Nature Climate Change* 1(1), 22–27. DOI: 10.1038/nclimate1023.

NOAA Research (2020) How Will Climate Change El Niño and La Niña? Available at: https://research.noaa.gov/News/ArtMID/451/ArticleID/2685/mediaid/2040?mediaid=2040%2C2041 (accessed 1 August 2023).

Nobre, G.G., Muis, S., Veldkamp, T.I. and Ward, P.J. (2019) Achieving the reduction of disaster risk by better predicting impacts of El Niño and La Niña. *Progress in Disaster Science* 2, 100022.

Pusat Pengendalian Pembangunan Ekoregion Bali dan Nusa Tenggara (2021) Dokumen Status Daya Dukung Air Pulau Bali. Kementerian Lingkungan Hidup Dan Kehutanan – Republik Indonesia.

Rai, I.N., Shoba, S., Shchegolkova, N., Dzhamalov, R., Venitsianov, E. *et al.* (2015) Analysis of the specifics of water resources management in regions with rapidly growing population under different climate conditions: Case study of Bali Island and the Moscow Region. *Water Resources* 42, 735–746.

Rochmadi, R., Ciptaraharja, I. and Setiadi, T. (2010) Evaluation of the decentralized wastewater treatment plants in four provinces in Indonesia. *Water Practice and Technology* 5(4), wpt2010091.

Rummukainen, M. (2012) Changes in climate and weather extremes in the 21st century. *Wiley Interdisciplinary Reviews: Climate Change* 3(2), 115–129.

Ryley, T., Baumeister, S. and Coulter, L. (2020) Climate change influences on aviation: A literature review. *Transport Policy* 92, 55–64.

Sudiajeng, L., Wiraga, I., Mudhina, M., Waisnawa, I.G.N.S. and Sudiarsa, I.M. (2020) Assessment of the Effectiveness on Domestic Rainwater-harvesting Wells (SPAHUDO) in the Northern Area of Denpasar City–Bali Indonesia Through Ergo-Hydrogeology Approach. *Journal of Physics: Conference Series* 1569(4), 042098.

Sumaja, K., Satriyabawa, I.K.M., Dewi, T.P.P. and Fadianika, A. (2021) Potential of rainwater harvesting (RWH) to overcome the problem of water scarcity at the airports: A case study in I Gusti Ngurah Rai Airport, Bali. *IOP Conference Series: Earth and Environmental Science* 893(1), 012079.

Supari Tangang, F., Salimun, E., Aldrian, E., Sopaheluwakan, A. and Juneng, L. (2018) ENSO modulation of seasonal rainfall and extremes in Indonesia. *Climate Dynamics* 51, 2559–2580.

Trenberth, K.E. (2011) Changes in precipitation with climate change. *Climate Research* 47(1–2), 123–138.
Widianingtias, M., Kazama, S., Benyapa, S. and Takizawa, S. (2023) Assessment of water reclamation and reuse potential in Bali Province Indonesia. *Water* 15(14), 2642.

30 Evidence Use in Policymaking: The Case of a Hotel Policy in Western Australia

Carolina I. Garcia, Christof Pforr and Michael Volgger
School of Management and Marketing (Tourism Research Cluster), Faculty of Business and Law, Curtin University, Western Australia, Australia

Summary

This case study explores the development of a policy launched in 2011 to incentivize hotel investments in Western Australia, especially in its capital city, Perth. To understand the policy process, we studied relevant documents, including policy plans, promotional material, annual reports, industry reports and records of Parliament debates. This approach allowed us to assess the use of evidence throughout three policy phases: (i) the process of building a case for the development of hotel incentives (the *why*); (ii) the design of hotel incentives as part of a broader tourism strategy (the *what*); and (iii) the implementation of this policy and the outcomes it achieved (the *how*).

The Value and Interest of the Case Study

The case study portrays an example of a wide range of evidence used throughout three policy phases. It also shows the complexities and non-linearity of both developing policies and assessing their outcomes in dynamic settings with concurrently occurring external events.

Background

Covering around a third of its landmass, Western Australia (WA) is Australia's largest state. Agriculture and extractive industries are the traditional backbone of WA's export economy, with the mining sector accounting for 46% of its gross state product (GSP) in 2021–2022 (Government of WA, 2023). Tourism is a growing industry, representing 5.6% of the state's employment and 2.2% of its GSP (Government of WA, 2023). While visitor numbers are still recovering from the COVID-19 pandemic, in the year ending March 2023 visitor expenditure recuperated strongly, even being ahead of 2019 figures by 18% (Tourism WA, 2023b).

Tourism WA (TWA) is the statutory authority in charge of tourism in the state. It was established by the Western Australian Tourism Commission Act 1983 and is governed by a Board of Commissioners. Its areas of focus are marketing (including working with airlines, travel wholesalers, retail and online travel agents), major events (including developing, attracting and promoting them) and improving access, accommodation and tourism experiences (working with government and industry).

Western Australia is divided into five tourism regions (Fig. 30.1). These are

Fig. 30.1. Western Australia's five tourism regions.
Source: Tourism WA (n.d.).

'Destination Perth', 'Australia's Coral Coast', 'Australia's Golden Outback', 'Australia's North West' and 'Australia's South West'.

Perth is the state's capital and forms part of the 'Destination Perth' tourism region, which features major events, corporate tourism and attractions in its six tourism precincts: 'Perth City', 'Fremantle and Rottnest Island', 'Sunset Coast', 'Peel and Rockingham', 'Swan Valley and Darling Range' and 'Avon Valley'.

In 2019, Destination Perth received close to 5.7 million overnight visitors, most from WA (56%) or other Australian states and territories (27%). Considering data of 2018 and 2019, the main reasons for these visits were holidays and visiting friends/relatives for intrastate and international visitors. Meanwhile, the main reason behind visits from other states was business, accounting for 47% of these stays (Tourism WA, 2023a).

Perth has long played a crucial role as an important destination in WA, with 68% of international visitors and 62% of interstate visitors only spending time in the 'Destination Perth' region when visiting the state in 2019 (Tourism WA, 2023b). Furthermore, it acts as a gateway to the other regions and, as a result, connectivity and infrastructure tend to be part of the tourism agenda. Another key focus is the availability and characteristics of the accommodation on offer, particularly in the Perth City tourism precinct.

As mentioned earlier, Western Australia's rich natural resources have traditionally provided a solid foundation for the state's economic development and the height of a mining boom around 2010 saw peak occupancies and rising room rates in Perth. Perth City was leading all capital cities in Australia in terms of occupancy (84%), average room rate (AU$192.42), room yield (AU$163.73) and room yield growth (15.8%), according to data from STR in September 2011 (Government of Western Australia, 2011). This limited spare hotel capacity threatened tourism development in the state and prompted the design of a policy initiative to incentivize investments and broaden the supply of tourism accommodation: The Hotel Development Incentives Policy.

Aim and Methodology

This case study explores the development of a policy launched in 2011 to incentivize hotel investments in Western Australia, particularly in the Perth City tourism precinct. By unravelling the program, this contribution shows how different types of evidence were used throughout the policy process. To identify and understand the evidence used, we studied a range of relevant documents, including policy plans, promotional material, annual reports, industry reports and records of Parliament debates.

Hotel Policy Process in Western Australia

The Hotel Development Incentives Policy was an initiative by Tourism WA focused on the state's capital, Perth, particularly the Perth City tourism precinct. The policy was officially introduced in 2011 but spanned over several years, especially from 2007 to 2016.

We study the use of evidence throughout three policy phases (Fig. 30.2): (i) the process of building a case for the development of hotel incentives (the *why*); (ii) the design of the hotel incentives as part of a broader tourism strategy (the *what*); and (iii) the implementation of this policy and the outcomes it achieved (the *how*).

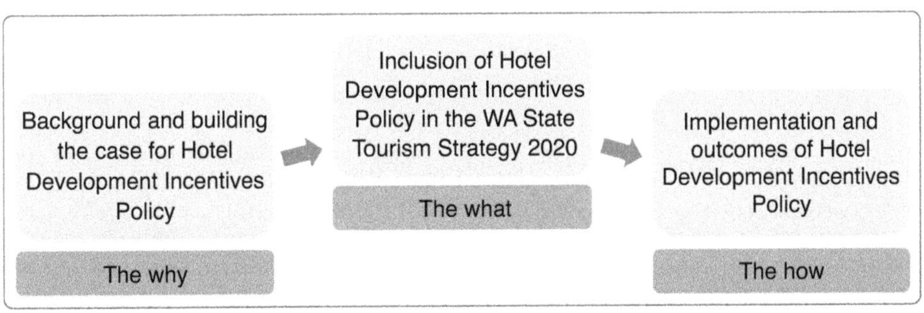

Fig. 30.2. Three phases of the hotel development incentives policy.

The *why*: Building the case for a hotel policy

In order to unravel how the Hotel Development Incentives Policy came to life, we look at two stages. In the first stage, the need for the development of the hotel sector became part of the 'political agenda'. In the second stage, this topic was further explored and characterized.

Stage 1 – Problem identification

As mentioned before, 'Destination Perth' is a key destination in Western Australia. As the gateway to the other regions in the state, Perth's connectivity and infrastructure were a focal point in the lead-up phase to the policy. References to a need for hotel incentives in the Perth City tourism precinct can be identified from 2005 to early 2008.

In 2005, hotel capacity in the city was questioned in relation to the Australian Tourism Exchange (ATE), a key industry annual event, being hosted in Perth City. Attempts to secure new hotels were back then mentioned in Parliament by the WA Minister for Tourism, who emphasized that 'the hotel sector in Western Australia is very professional, and we are looking to secure new hotels. (...) The state's capacity is improving all the time' (Hansard, 2005, p. 9). A year later, with another Minister in charge of the tourism portfolio, this need was more explicitly recognized during a discussion in Parliament:

> The other long-term strategy that we must address is access to hotel accommodation (...). We are conscious that at the moment, particularly with conventions of that size, there is a shortage of hotel accommodation. Because we are in a great economic boom, there is a lot of commercial and business use of our hotels. It is ironic, but it is putting pressure on availability of accommodation. Therefore, we need to be looking at how we can secure more hotel accommodation. (Hansard, 2006, p. 7)

The Tourism WA's Annual Report 2007–2008 also highlighted this supply problem. The public agency had a *Tourism Accommodation Development Register* to monitor the progress of tourism accommodation projects in Western Australia's five tourism regions. But in 2007, it engaged a consultancy firm to conduct specific research on hotel investment in Perth on the basis that

> Notwithstanding the very strong performance of the Western Australian economy over the past few years and with the Perth hotel market reaching historically high levels of occupancy, little evidence exists of willingness by developers or investors to facilitate new hotel development in Perth.
>
> The existing inventory of hotels in Perth is ageing and very limited construction of traditional full service hotels has occurred over the past twenty years. Additions since that time have been primarily conversions of office buildings, and serviced apartments, both limited and full service offerings.
>
> A small number of new properties have opened in the past 2 years but looking forward, there are currently very few developments under active planning. (Transocean Group, 2007, p. 1)

In its annual report, TWA highlighted some findings from this study among current and emerging issues and trends in the state. Capacity issues were raised and substantiated by making reference to hotel occupancy rates in Perth being above 80% and accommodation demand growing in 2005 and 2006 at its highest rate compared to the previous 18 years. Such high demand growth was attributed to both increasing numbers of international leisure visitors and the mining and corporate sectors. At the same time, the state's growing ability to attract business events meant a need to accommodate delegates in this constrained scenario. Furthermore, increasing construction costs challenged investment in accommodation.

Against this backdrop, one of the public agency's tourism industry development plans for the financial year 2008–2009 was to 'address the shortfall in hotel room supply through national and international promotion of existing building conversion opportunities and the construction of new hotels in Perth and surrounds' (Tourism WA, 2008, p. 36). In fact, already in 2007–2008 TWA had participated in seven hotel investment events within Australia and overseas in Asia and the Middle East (Tourism WA, 2008).

Stage 2 – In-depth understanding of hotel capacity concerns

The identification of hotel capacity concerns was followed by attempts to better understand this problem. Such efforts were characterized by additional evidence-gathering and the commissioning of specialized studies from 2008 to 2010.

In 2008, TWA commissioned a study conducted by the Property Council of Australia (2008). The report reviewed the potential for hotel property conversions in Perth City and provided a list of 24 buildings with high potential for such conversion. It pointed out that, while there was interest in hotel investment, some factors prevented action. Barriers included higher returns of other property uses, high construction costs, and business risks of having one tenant in a hotel vs multiple tenants in an office building.

It is worth noting that reports from both Transocean Group (2007) and Property Council of Australia (2008) included recommendations that closely aligned with the Hotel Development Incentives Policy which was later adopted. Transocean Group (2007) suggested as actions: (i) hotel development as a precondition for developing large-scale sites, (ii) provision of well-located properties for conversion, (iii) taxes and other charges relief, (iv) tourism zoning and planning flexibility. In line with the aforementioned second action item, the Property Council of Australia concluded that 'the best case scenario for a viable hotel conversion in Perth today would involve the government making available a suitable existing commercial property on a long term lease, to a developer and hotel operator' (Property Council of Australia, 2008, p. 1).

In 2008, the then Minister for Tourism cited a report by Colliers International during a parliamentary discussion (Hansard, 2008). He stated that, according to the study, Perth's hotel market showed the strongest growth in Australia and could lead hotel investment opportunities in the country. He also indicated that the government was trying to attract such investments.

> The government is working extremely hard to attract hotel investors of international reputation to consider Perth and Western Australia as a strong investment opportunity to meet the future demand. It is pleasing to see these conditions for investment in Western Australia and we will be very aggressively seeking high-quality hotel investors to make sure that we meet the continuing growth in our wonderful state. (Hansard, 2008, p. 1)

Shortly after, TWA tasked Access Economics with the *Perth Hotel Economic Impact Study* (2010). This study included updated and comparative figures, highlighting that Perth's hotels were close to capacity, with average occupancy rates above 80% in the previous financial years 2006–2007 and 2008–2009. The study also found that capacity constraints came at a significant economic cost for the State, forecasted to reach AU$435 million in gross state product in the 10 years to 2019–2020 (Table 30.1). In parallel, TWA also worked with Jones Lang LaSalle Hotels on a *Review of Tourist Accommodation Development in WA Report* (Tourism WA, 2010), which offered further insights into necessary planning, finance and investment. Further forecasting was developed by AEC Consulting, as highlighted in the Tourism WA, 2020: 'Perth will need an additional 1900 hotel rooms by 2020 to keep occupancy at viable and sustainable levels from a demand and supply perspective, far in excess of currently known investments' (AEC consulting, 2011, in Tourism WA, 2012b, p. 13).

Table 30.1. Forecasted impact of hotel shortages, from 2009 to 2010 to 2019–2020.

	Perth	Rest of WA	WA
Gross regional/state product ($m)	–391.3	–44	–435.3
Real house consumption ($m)	–336.5	–5.7	–342.2
Employment (full-time equivalent)	–406	–35	–442

Notes: The table shows the general-equilibrium modelling impacts of hotel shortages in Australian dollars in net present value terms (2010 value) for the decade 2009–2010 to 2019–2020.
Source: Based on Access Economics (2010)

In 2009 and 2010, concrete action to incentivize hotel investment from an institutional point of view could already be noted with the restructuring of Tourism WA, when the division *Industry Development and Visitor Servicing* was replaced by *Infrastructure & Investment*. Further, TWA wrote a guide for foreign investors together with Blake Dawson and Jones La Salle (Blake Dawson, Tourism Western Australia and Jones Lang Lasalle Hotels, 2009). This guide introduced legal and regulatory matters for commercial and hotel real estate investment in the state. It also presented Perth-specific information, indicating that 'Perth is the fast growing capital of Western Australia and major gateway into the State' (Blake Dawson *et al.*, p. 13).

The *what*: Hotel development incentives policy

Specifically, the government's Hotel Development Incentives Policy consisted of an incentives package for expanding and upgrading the hotel sector in WA, especially in the Perth City tourism precinct. The policy was an initiative of Tourism WA in collaboration with LandCorp (a government agency managing land use)[1], the Metropolitan Redevelopment Agency (MRA) and the City of Perth (the local government). The incentives were announced in October 2011 and included three main components: provision of State Government land, funding for infrastructure upgrades, and plot ratio (floor space) bonuses. Box 30.1 presents some additional details about these incentives, which were granted on a case-by-case basis and overseen by a Government Incentives Steering Group.

The Hotel Development Incentives Policy was not a standalone initiative, but an integral part of the broader *Strategy for Tourism in Western Australia* released in 2012 (Tourism WA, 2020). Its overarching goal was to double tourist spending in the state by 2020—from AU$6 billion in 2010 to AU$12 billion in 2020. To achieve this goal, seven strategic pillars for growth were determined: Brand, Indigenous Tourism, Infrastructure, Ongoing Operations, Asia, Business Travel, Regional Travel and Events. The Hotel Development Incentives Policy was a core component within the infrastructure pillar, which focused on capacity expansion in three distinct areas: hotels, aviation and tourism workforce (Fig. 30.3). Importantly, the strategy was informed by a 3-month consultation and submission process that involved consultative workshops and key stakeholder consultation (Tourism WA, 2009). Participation was encouraged via emails, media and website promotion. The consultative workshops were held in 13 locations throughout WA with the tourism industry, government agencies and the local community. The

Box 30.1. Government actions to provide hotel incentives in the Perth City tourism precinct.

- Release Crown land for hotel development through expressions of interests
- Grant longer term Crown leases for short stay accommodation developments where appropriate
- Determine concessional lease payments for Crown land on the Valuer General's advice that specifically promote hotel accommodation developments
- Discount Crown land values for hotel developments on a 'case by case' basis in consultation with the Department of Regional Development and Lands, and other relevant agencies
- Consider leasing Crown land for hotel use at a peppercorn rate*
- Designate the use of land under Government control, Crown land, properties and/or buildings for future hotel use or conversion
- Allocate funds for the provision of infrastructure upgrades to facilitate hotel accommodation development on a 'case by case' basis
- Provide flexible plot ratio bonuses for hotel developments

*Peppercorn rate refers to a token or nominal rent.
Source: Government of Western Australia (2011, p. 4).

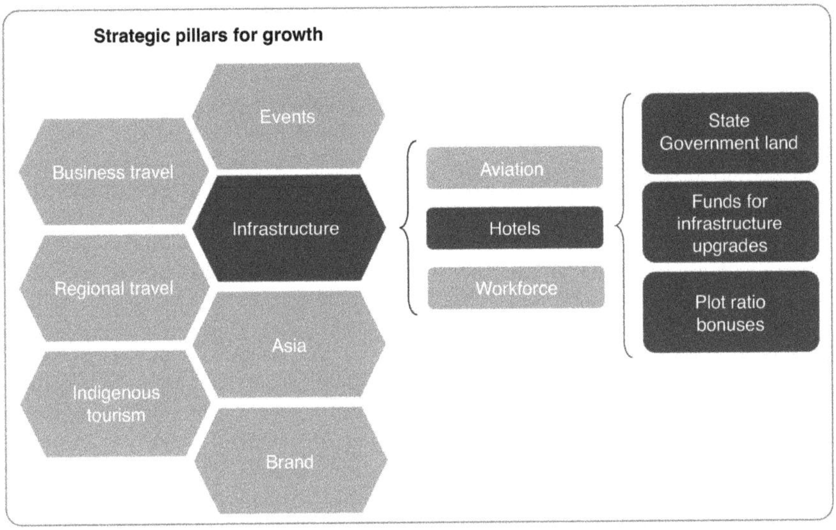

Fig. 30.3. Hotel development incentives policy as part of the Tourism WA (2020).
Source: Based on Tourism WA (2012b).

workshops aimed at presenting the draft vision and generating discussions and additional ideas for tourism in the state.

Two government strategies in this tourism plan focused on the development of the hotel sector in WA:

1. Develop policies that encourage investment: 'Government must work across agencies to develop a policy environment that increases the relative attractiveness of investment in hotels to address the room shortfall. A 'whole of State' view must be taken to understand the downstream benefits of encouraging hotel investment' (Tourism WA, 2012b, p. 19).
2. Facilitate access to land and tenure for the development of tourism: 'The importance of tourism as a strategic land use must be communicated and understood across government agencies to ensure that tourism is properly represented in new development opportunities' (Tourism WA, 2012b, p. 19).

The above government strategies were accompanied by an industry stakeholder engagement strategy:

Engage with industry stakeholders to assist the government to guide investment attraction: 'Industry operators have an unmatched perspective of market trends and future opportunities. By working together to incorporate this perspective, more accurate, relevant and aligned long-term policy can be developed' (Tourism WA, 2012b, p. 20).

From an institutional point of view, the importance of infrastructure development led to a restructuring of Tourism WA in 2010, when the division *Industry Development and Visitor Servicing* was replaced by *Infrastructure & Investment*. The new division focused on facilitating infrastructure projects in the state, including investment in tourism assets and the development of major tourism projects. In particular, the division concentrated its efforts on aviation access, accommodation (especially Perth City hotel developments), caravan/camping and cruising infrastructure (Tourism WA, 2012a). To achieve its objectives, TWA collaborated with the private sector and other government agencies.

It is worth noting that the Hotel Development Incentives Policy was not a standalone measure in Perth City as other important infrastructure projects were also taking place at the time. These included the construction of Elizabeth Quay and the City Link, an upgrade to the city's Riverside, as well as the building of a new Perth Stadium and the Perth Arena (Tourism WA, 2012a).

The *how*: The policy in action

To understand the Hotel Development Incentives Policy in action, we consider two stages. In the first stage, we present the measures undertaken as part of the release of the hotel incentives policy. In the second stage, we turn to the outcomes of the policy initiative.

Hotel incentives and investment promotion

The Hotel Development Incentives Policy was announced in October 2011 and, as previously indicated, focused on three main components: provision of State Government land, funding for infrastructure upgrades, and plot ratio (floor space) bonuses. Rolling out the Hotel Incentives Development Policy involved the identification of potential sites and calls for expressions of interest. As highlighted by the Minister for Tourism in 2012, 'a committee in Tourism WA has been looking for opportunities to redevelop existing buildings or to use government land that is no longer required for hotel development' (Hansard, 2012, p. 1). One of the sites identified was the Fire and Emergency Services Authority (FESA)'s headquarters, whose staff had been relocated to another building.

> Different groups have expressed quite a bit of interest in developing that site. We will ask for expressions of interest, leaving it open for developers to come up with options for the site. We will not dictate what we want, other than to say that there has to be hotels on the site with a mixed use of cafes, restaurants and shops. They would be needed to make the development viable. (Hansard, 2012, p. 2)

At the same time, strong efforts were made to communicate opportunities for hotel investment in WA and the Perth City tourism precinct in particular. As mentioned above, in 2007–2008 TWA had already participated in several hotel investment events within Australia and overseas in Asia and the Middle East (Tourism WA, 2008). After the launch of the Hotel Development Incentives Policy, an expansion of these initiatives via partnerships with Tourism Australia and Austrade, the leading tourism government institutions at the Australian federal level, could be noted. WA's investment opportunities were promoted in Australia, Asia, the Middle East, Europe and North America (Tourism WA, 2013). To this end, promotional documents were launched, explaining the Hotel Development Incentives Policy and marketing specific hotel opportunities in Perth City in English and Chinese.

Outcomes of hotel development efforts

These efforts led to the development of a number of new hotels which greatly expanded the accommodation capacity in the Perth City tourism precinct. However, this achievement was not necessarily linked specifically to the Hotel Development Incentives Policy. Indeed, in 2015 the Minister for Tourism was questioned in Parliament about projects under the Incentive Policy and confirmed that the FESA site had been the only one released as part of the initiative, while some development projects in Perth City had applied for the plot ratio benefit (Hansard, 2015). A Westin hotel was developed at the former FESA site, in partnership with LandCorp, the Department of Fire and Emergency Services, and the Department of Health (Tourism, 2014).

The communication efforts undertaken as part of the policy had a notable impact on hotel investments in the Perth City tourism precinct. According to TWA's Annual Report 2012–2013 (2013), 'following the promotion of opportunities in Perth by Tourism WA, the Intercontinental Hotel Group (IHG) announced a proposal (...)' (p. 24). TWA also 'provided introductions that ultimately led to the acquisition of the Esplanade Suites by one of China's top 10 hotel groups (...)' (p. 24). The year after, other facilitation efforts were also highlighted: 'Alongside the Metropolitan Redevelopment Authority (MRA), Tourism WA facilitated the selection of the luxury Ritz-Carlton hotel proposal as a cornerstone of the Elizabeth Quay development, and is actively introducing investors to another hotel opportunity to be situated adjacent to Perth Arena in the CityLink precinct' (Tourism, 2014, p. 19).

The 'active support for hotel investors and developers' led to commitments for the development of 1300 rooms in Perth City in

2013–2014 (Tourism WA, 2014, p. 19), with the target of securing 1900 additional hotel room commitments by 2020. Already in the next financial year, this target was reached, with more than 2000 rooms committed by 2019 (Tourism WA, 2015). At this stage, it was highlighted that the achievement was the result of efforts 'both directly and through partners such as the Metropolitan Redevelopment Authority, LandCorp and City of Perth. Any further hotel development in Perth is being driven by market forces' (Tourism WA, 2015, p. 22).

Having reached and surpassed the target, in 2014–2015 TWA's focus changed towards destination marketing and events to attract more tourists to Perth (Tourism WA, 2015). However, new hotels continued to open. A 2020 report shows a growing hotel supply and, at a slower pace, demand in the period 2016–2019 – with a consequently decreasing occupation rate – in the Perth City tourism precinct and its surroundings (Tourism WA, 2020).

Table 30.2 lists the hotels that opened from 2012 to 2020. Most hotels opened in 2015, 2016 and 2019, with most new rooms coming to market in 2016 and 2019. The 1,900 rooms target was greatly surpassed, with 3,859 additional rooms in 24 hotels in the CBD itself and 1,523 new rooms in 17 hotels in the Greater Perth Metro area becoming available

Table 30.2. New hotels in the Perth CBD.

Project name	Opening	Scale	Rooms
Fraser Suites Perth	2012	Luxury	236
Baileys Serviced Apartments	2014	Independents	25
The Nest on Newcastle	2015	Independents	20
Alex Hotel	2015	Independents	74
COMO The Treasury	2015	Luxury	48
Attika Hotel	2015	Independents	22
Citylights Hotel	2016	Independents	46
Quest Kings Park	2016	Upscale	71
Sage Hotel West Perth	2016	Independents	101
Quest Mounts Bay Road	2016	Upscale	71
Quest East Perth	2016	Upscale	130
Peppers Kings Square Hotel	2016	Upper Scale	120
Crown Towers Perth	2016	Luxury	500
Tribe Perth	2017	Independents	126
The Westin Perth (FESA House)	2018	Upper Scale	368
QT Perth	2018	Upper Scale	184
The Sebel West Perth Aire Apartments	2018	Upper Scale	64
Hostel G Perth	2019	Midscale	77
DoubleTree by Hilton, Northbridge	2019	Upscale	206
Ibis Styles East Perth	2019	Midscale	252
Adnate Art Series	2019	Upscale	250
Ritz Carlton	2019	Luxury	204
Novotel Perth Murray Street	2019	Upscale	435
DoubleTree by Hilton, Perth Waterfront	2020	Upscale	229
Total completed			3859

Source: Based on Tourism WA (2020).

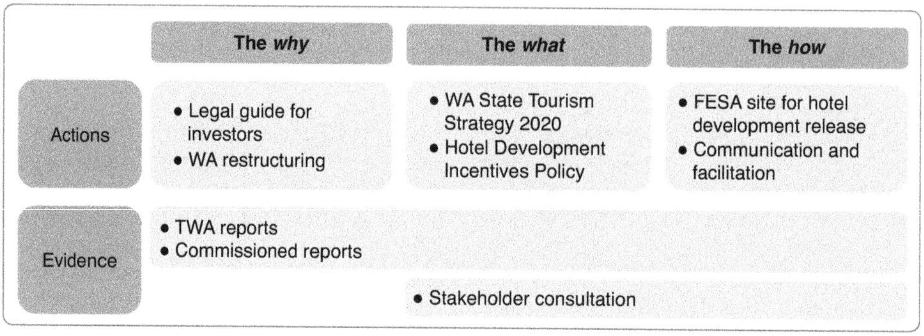

Fig. 30.4. Evidence and actions in the three phases of the hotel development incentives policy.

by December 2020. Furthermore, by 2020 five hotels were either renovated or refurbished in the Perth CBD and three were expanded in the Greater Perth Metro area.

What, *how* and *why*: Reflecting on the hotel development incentives policy

Reflecting on the Hotel Development Incentives Policy in Perth, some key learnings stand out. These learnings are related to (i) the evidence used in the three policy phases (the *why*, the *what* and the *how*) and (ii) the policy context for evidence use in the policy.

Evidence use in three policy phases

In relation to evidence use, the case study outlined the range of evidence the government considered throughout the policy process (Fig. 30.4). This evidence included internal and external reports, commissioned specialized research and stakeholder consultation. Evidence to support policy development was collected over several years and was a relevant input factor in the Hotel Development Incentives Policy, not only in terms of justification and design but also for creating promotional material to attract investors.

Relevant considerations relate to the type of evidence used. Different mixes of evidence may be selected and can vary according to the jurisdiction, institution or policy under development. In the case of the Hotel Development Incentives Policy, commissioned studies played an important role. These were mainly elaborated by private consultancies and advocacy groups with potential interests in relation to hotel development in Perth.

Meanwhile, stakeholder consultation was present during the policy process, which can be a means to incorporate a range of perspectives to inform policy. We found that more comprehensive participatory approaches (involving industry, government agencies and the community) were present in the development of the *Tourism Strategy 2020* when infrastructure concerns were initially identified. However, when designing and implementing the specific Hotel Development Incentives Policy, stakeholder participation seems to have been more limited in scope. TWA worked on the hotel policy development with specialized agencies such as LandCorp, the MRA and the City of Perth. However, we did not find any records indicating that the local community was consulted more extensively for the development of the specific hotel policy. This might have been relevant and beneficial, as the hotel policy has an impact on the type and speed of urban development, with a range of potential impacts on local (non-tourism) actors.

Policy context for evidence use

Additional reflections are possible when turning to the policy context for the use of evidence in the Hotel Development Incentives Policy. It has already been mentioned that this policy was formulated at a time when Perth was experiencing accelerated growth as a result of a mining boom and the development of important projects, like the Elizabeth Quay and Perth Stadium.

The Hotel Development Incentives Policy appears aligned with this context of accelerated urban development in Perth. Indeed, the policy's three main components (i.e. provision of State Government land, funding for infrastructure upgrades, and plot ratio or floor space bonuses) are intertwined with the city growth paradigm. The policy development started because capacity constraints were identified. As Perth's hotel accommodation was close to full capacity, the argument was that more hotels were needed. But, of course, this is not the only imaginable course of action, considering that a mining boom with its particular patterns of mobility demand and timeframes was a main driver behind the capacity issues at Perth hotels. Whether alternatives were assessed is unknown to the authors. The commissioned specialized reports appear focused on assessing hotel development impacts and opportunities, which already implies the assumption that the key solution was building new hotels (or extending existing ones).

Moving forward to the outcomes of the Hotel Development Incentives Policy, it is difficult to comprehensively and accurately assess its impacts. The Hotel Development Incentives Policy was not the only trigger for the expansion of hotel capacity in the City of Perth. Strictly speaking, only one site was released under the policy while the plot ratio benefit offered as part of the scheme was demanded by more hotels. Related activities also generated interest and investment beyond the specific Incentive Policy measures. These primarily included facilitation, promotion and communication actions via distribution of material for investors and presentations at specialized events in Australia and abroad. Figures show that the number of new rooms built by 2020 exceeded the original target. This building of numerous hotels in a relatively short timeframe resulted in market consequences in the short term and may unleash different dynamics in the longer term. For example, it is likely that the downward trend in occupancy rates and room prices observed in the aftermath of the process can be at least partially attributed to the expansion of hotel capacity. Other contemporaneous events such as the COVID-19 pandemic would have exacerbated this situation, making causal attributions challenging. However, it seems warranted to assess that the policy had impacts on the local hotel and construction markets. It would be important to conduct specific policy impact assessments to try to determine the (socio-spatial) effects of the initiative in a fine-grained and comprehensive fashion. Counter-factual analysis may help to investigate whether a similar development would have occurred in the absence of the policy.

Concluding Remarks

Overall, the learnings in this case study point towards the complexities and non-linearity of developing tourism policies and the utilization of evidence during the different stages of the policy generation and implementation process. The case study also highlights challenges in properly assessing tourism policy outcomes. This is especially the case in dynamic settings, where events and initiatives other than the policy itself take place concurrently and might affect the overall outcome.

The Hotel Development Incentives Policy was created as a response to a mining boom and the limited hotel capacity in the Perth City tourism precinct. This was a particular point in time for WA and its capital city, where several infrastructure initiatives were taking place at the same time, like the now emblematic Elizabeth Quay. The tourism sector was also at a turning point, with the WA State Tourism Strategy 2020 under construction. These features mark the context in which the Hotel Development Incentives Policy became part of the policy agenda and came to life using a variety of evidence along the way.

Note

[1] In 2019, Land Corp merged with the Metropolitan Redevelopment Authority (MRA) to form Development WA, the State Government's land development agency.

References

Access Economics. (2010) Perth Hotel Economic Impact Study, February 2010.

Blake Dawson, Tourism Western Australia and Jones Lang Lasalle Hotels (2009) Western Australia Real Estate: A Legal Guide for Foreign Investors.

Government of Western Australia (2011) Hotel Investment Incentive Packages Western Australia.

Government of Western Australia (2023) WA Economic Profile – August 2023. Available at: https://www.wa.gov.au/government/publications/western-australias-economy-and-international-trade (accessed 1 September 2023).

Hansard (2005) Extract from Hansard [Assembly – Wednesday, 8 June 2005]. pp. 276a-289a. Available at: https://www.parliament.wa.gov.au/Hansard/hansard.nsf/0/9bfb2ef1296e96abc82575720029e77e/$FILE/A37%20S1%2020050608%20p276a-289a.pdf (accessed 25 August 2023).

Hansard (2006) Extract from Hansard [ASSEMBLY – Thursday, 25 May 2006]. pp. 518b–527a. Available at: https://www.parliament.wa.gov.au/Hansard/hansard.nsf/0/07ee2ad85796c17ec8257573008084a4/$FILE/A37%20S1%2020060525%20p518b-527a.pdf (accessed 25 August 2023).

Hansard (2008) Extract from Hansard [ASSEMBLY – Wednesday, 9 April 2008]. Available at: https://www.parliament.wa.gov.au/Hansard/hansard.nsf/0/12dfea04bb07f40cc8257570000f8038/$FILE/A37%20S1%2020080409%20p2078a-2078a.pdf (accessed 25 August 2023).

Hansard (2012) Extract from Hansard [ASSEMBLY – Wednesday, 30 May 2012]. pp. 218c–228a. Available at: https://www.parliament.wa.gov.au/Hansard/hansard.nsf/0/3f3f2a7156bec76848257a210030b844/$FILE/A38+S1+20120530+p218c-228a.pdf (accessed 25 August 2023).

Hansard (2015) Extract from Hansard [ASSEMBLY – Tuesday, 5 May 2015]. pp. 3079c–3079c. Available at: https://parliament.wa.gov.au/Hansard/hansard.nsf/0/ebdc096cfa833ee648257eb2001889cc/$FILE/A39%20S1%2020150505%20p3079c-3079c.pdf (accessed 25 August 2023).

Property Council of Australia (2008) Perth CBD Hotel Property Conversion Study. October 2008.

Tourism WA (2008) Annual Report 2007–2008.

Tourism WA (2009) Tourism 2020 A Vision for the Tourism Industry, Process Update. May 2009.

Tourism WA (2010) Annual Report 2009–2010.

Tourism WA (2012a) Annual Report 2011–2012.

Tourism WA (2012b) Strategy for Tourism in Western Australia 2020.

Tourism WA (2013) Annual Report 2012–2013.

Tourism WA (2014) Annual Report 2013–2014.

Tourism WA (2015) Annual Report 2014–2015.

Tourism WA (2020) Perth Hotel Development Pipeline Report. Available at: https://www.tourism.wa.gov.au/Publications%20Library/Markets%20and%20research/2021/Ad%20hoc/Hotel-Development-Pipeline-Dec-2020.pdf (accessed 25 August 2023).

Tourism WA (2023a) Destination Perth Overnight Visitor Factsheet 2022. Available at: https://www.tourism.wa.gov.au/Publications%20Library/Markets%20and%20research/2022/Visitor%20Factsheets%20-%20Regions/Regional%20Factsheet%202022%20-DP.pdf (accessed 14 September 2023).

Tourism WA (2023b) WA Tourism Snapshot, Visitors and Spend in Year Ending March 2023. Available at: https://www.tourism.wa.gov.au/Publications%20Library/Markets%20and%20research/2023/IVS%20NVS/Overview%20of%20Visitation%20to%20WA%20-%20YE%20Mar%2023.pdf (accessed 14 September 2023).

Tourism WA (n.d.) Tourism Regions of Western Australia. Available at: https://www.tourism.wa.gov.au/Publications%20Library/Industry%20support%20and%20events/Regional-Tourism-Organisation-map.pdf (accessed 15 September 2023).

Transocean Group (2007) Perth Hotel Investment Study, June 2007.

Section 7: Emerging Trends in Tourism

This final section showcases new trends in the industry. Cases from China explore topics like surfing tourism in Hainan and educational tourism in Mount Danxiashan, demonstrating how tourism evolves to meet changing demands and interests while bringing positive economic impacts to local communities.

31 Surfing in Hainan Island: An Emerging Surfing Tourism Destination in China

Yana Wengel[1,2], Longrui Han[3] and Yunxin Hu[1]
[1]HNU-ASU Joint International Tourism College, Hainan University, Haikou, China; [2]Borneo Tourism Research Centre, Faculty of Business, Economics and Accountancy, Universiti Malaysia Sabah, Kota Kinabalu, Malaysia; [3]The Development Planning Unit, University College London, London, UK

Summary
Surfing has gained significant scholarly attention recently, yet Hainan Island has received less attention. Hainan Island is an emerging destination for the global surfing community. This case study provides a historical overview of surfing in Hainan and discusses the current state of challenges and opportunities related to the surfing industry on the island. The case study also discusses the cultural and economic impacts of surfing on the island, including the emergence of surf schools, the hosting of surfing competitions, and the development of related infrastructure in Riyue Bay. We also discuss the development of surfing after this sport became an Olympic discipline. Additionally, the case study highlights some key surf spots in Hainan, such as Riyue Bay in Wanning, and discusses the unique characteristics that make these spots ideal for surfing. Overall, the case study provides a comprehensive overview of the development and current state of surfing in Hainan Island, shedding light on the increasing significance of surfing in the region's tourism industry and its potential for future growth.

The Value and Interest of the Case Study

The value of this case study lies in its exploration of a relatively unexamined location of the surfing world: Hainan Island. By providing an in-depth overview of the history, current status, challenges, and opportunities of surfing on the island, the case study sheds light on the emergence of Hainan as a global surfing destination. It delves into the cultural and economic impacts of this growth, including the rise of surf schools, competitions, and the sport's potential as an Olympic discipline, making it a valuable resource for researchers, surf enthusiasts, and stakeholders in the tourism industry.

Background

Surfing originated in Polynesia, where it was a central part of the culture and lifestyle. European explorers and missionaries documented the practice in the late 18th century (Lemarié, 2020). In the early 20th century, Hawaiian surfers like Duke Kahanamoku popularized the sport, spreading it globally (Lemarié, 2018). Following World War II, there was a surge in

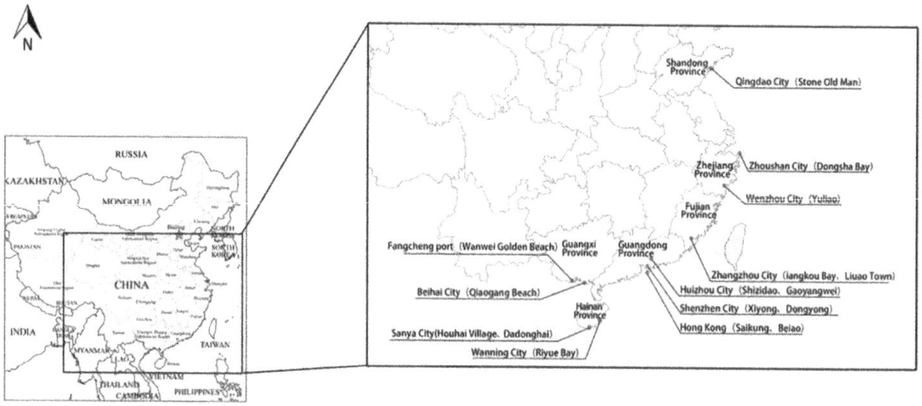

Fig. 31.1. Surfing locations in China.

consumerism, which particularly impacted young people and their desire to consume leisure activities such as movies, driving, and sports, including new activities like surfing (Ormrod, 2005). The 1960s witnessed a surfing surge due to the development of modern surfboards, while surfing's image became a symbol of freedom and rebellion (Warshaw, 2010). Today, surfing is a widely practised sport and lifestyle, with a global community and professional competitions, yet Hainan Island is a new spot in the surfing world (Zhang and Qi, 2022). Contemporary Chinese surfing diffused from Hong Kong (where international surfing events were held in 1997), while Hainan Island is an emerging destination for global surf tourism (Fig. 31.1).

One of the surfing pioneers in Hainan's surfing industry was Peter Drouyn (1948–2019), a professional Australian surfer recruited by the government with the goal of training a team of young Chinese surfers back in 1985. Regrettably, this endeavour did not succeed, but it also did not prevent other international surfers from exploring Hainan during the 1980s and 1990s (Zanella, 2019).

Another outstanding surfing person in the history of Hainan's surfing is Japanese surfer and entrepreneur Hiroshi Yonekawa, who set up a Monran surfboard production factory in Haikou in the mid-1990s and inspired a group of locals to explore surfing. At the edge of the millennium, surfing in Hainan was popular among a few locals and foreigners living on the island. In 2008, the emerging surfing scene in Hainan was ready for the first annual competition created by Brendan Sheridan,[1] the Surfing Hanan Open. The first official surfing event organized by the government was the Wanning International Surfing Festival in 2010, and more events followed (Zanella, 2019).

Despite all the efforts to promote surfing, it was not a popular sport in China until 2016, which marks a transformative moment for the sport in the country. In 2016, the International Olympic Committee voted to include surfing as a new discipline for the Tokyo 2020 Olympic Games (International Olympic Committee , 2016). Before 2016, the number of active surfers, most of whom were foreign residents of Hainan, remained limited. Following the inclusion of surfing into the Olympic games, China began to regard surfing as a legitimate sport, instituting national championships and official teams. The landscape shifted as young athletes from various backgrounds were identified and enlisted into the national team. In China, Hainan Island is one of the best places for surfing. Hence, it was chosen as the training base for the national surfing team. The inclusion of surfing in the Olympics was a significant milestone for the sport and illustrated its global popularity and recognition as a competitive discipline. Incorporating surfing into the Olympic programme led to the wider popularity of surfing's appeal and accessibility in Hainan. Hence, domestic and international athletes and tourists travelled to Hainan, which, in turn, contributed to its status as a popular surfing destination.

The natural geographical advantages of Hainan Island, governmental support of surfing development, and international exposure through events like the World Surf League (WSL) competitions brought global attention to Hainan's surfing opportunities, further popularizing the sport in the region. Due to the fast growth of the popularity of surfing in Hainan, infrastructure and surfing-related businesses developed rapidly. Many experienced surfers have set up surfing schools and clubs.

As part of a larger study on Hainan's surfing scene, we collected data for this case study in Riyue Bay from February to July 2023, making five visits with over 20 overnight stays. Methods of data collection included participant observation, immersive ethnographic engagement by one of the researchers who is familiar with surfing and interviews. We conducted formal interviews and casual discussions with surf community members, including coaches, shop owners, employees, and volunteers. The case study explores how surfing has grown in popularity in Hainan over the past decade as more surfers have discovered the island's surfing spots.

The majority of surfing tourists to the island are trying surfing for the first time in Riyue Bay. They are predominantly male, with higher than average education and income, and they have more free time (Zhao *et al.*, 2020). Interestingly, despite the male dominance noted by Zhao *et al.* (2020), our discussions with surf coaches in 2023 suggest that females often drive adventurous surf experiences, influencing their male counterparts to join in.

Apart from surfing tourism, Riyue Bay serves as the location for China's national surfing team academy, hosting seven professional provincial-level surfing teams from Hainan, Shanxi, Henan, Hebei, Zhejiang, Shandong, and Liaoning for training. In the private sector, the Wanning area currently hosts about 50 surf clubs, with 35 of them situated in Riyue Bay and employing 229 individuals (Hainan International Media Centre, 2023a). The nearby Tianxin Village (田新村), formerly reliant on crop cultivation and animal husbandry, has been transformed by opening tourism businesses. As such, a surge in new accommodations (bed and breakfasts, homestays, and hostels) and restaurants and shops (convenience stores, clothes, souvenirs, and surfing gear) has occurred. By late 2022, the village had 34 lodgings featuring 308 guest rooms (Hainan International Media Centre, 2023b). Some providers established campgrounds with glamping accommodations.

Benefits in the Advancement of the Surfing Industry in Hainan

As surfing and tourism synergistically evolve, Hainan Island has witnessed a proportional surge in revenue from surfing tourism. The growing popularity of the surfing culture has introduced a novel cultural dimension and intrinsic value to Hainan. This case study, focusing on Riyue Bay, assesses the advantageous influence of surfing in this location.

Riyue Bay is a picturesque coastal area in Wanning City, Hainan Island, China. The half-moon-shaped bay (Riyue Bay) is 168 km from Haikou and 87 km from Sanya. It is renowned for its stunning natural beauty, featuring pristine sandy beaches, crystal-clear waters, and lush forests. The name 'Riyue Bay' translates to 'Sun and Moon Bay', alluding to the breathtaking sunrises and sunsets that can be observed from the bay. This scenic location offers a tranquil retreat for beachgoers, with its serene atmosphere and opportunities for various water activities. Riyue Bay has gained recognition as a prominent surfing destination, attracting surf enthusiasts from around the world due to its consistent waves and favourable surfing conditions. The coastline of Riyue Bay is long and narrow due to the unique underwater topography of Wanning City and monsoons and typhoons. Riyue Bay is known for its high-quality wave patterns, point breaks, and beach breaks. The most famous wave is the longest left-handed wave in Asia, which is 2.8 km long. The average wave height is 1.5–2m all year round (Tian, 2020). These unique wave conditions have attracted many surf professionals and national team competitors to train and compete. The winter months (October to March) tend to have larger and more consistent waves, making it more suitable for experienced or professional surfers. On the other hand, May to September brings smaller and more manageable waves, providing an ideal environment for beginners to learn and practice.

This blend of natural beauty, cultural significance, and recreational opportunities has solidified Riyue Bay's reputation as a sought-after destination for both surfers and tourists seeking a tranquil coastal experience. As the popularity of Riyue Bay's natural advantages grows, the area regularly hosts a variety of surfing competitions and events, including international-level competitions and local contests. These events attract many competitors and spectators, generating significant revenue for the local economy.

The development of surfing tourism in Hainan has had a multifaceted and positive impact on Riyue Bay. One essential facet of this impact is economic growth, driven by the increasing number of tourists drawn to this characteristic sport, which is renowned for its strong demand. As Yang (2019) highlighted, the rise of surf tourism in Riyue Bay has led to the growth of local industries, including hotels, catering, and retail, creating employment opportunities and boosting tax revenues. In 2021, the authorities and the private sector agreed to invest in and enhance rural revitalization and industrial integration by approving the Riyue Bay Surf Town plan.

Simultaneously, this surge in tourism has stimulated entrepreneurial opportunities and jobs within the community, especially in the Tianxin Village. The development of surfing tourism in the area has created substantial employment opportunities for local youth and women, providing them with above-average monthly income. For example, 16 surf clubs are run by young locals, and over 60 registered surfing coaches are from Riyue Bay (Hainan International Media Centre, 2023b). Residents have capitalized on the growth of surf culture by establishing surfing schools, shops, accommodations, and restaurants, contributing substantially to their income.

Beyond the economic sphere, the influence of surfing tourism extends to the local community and culture. The arrival of domestic and foreign tourists who engage in surfing activities has fostered a sense of community within Riyue Bay. Residents have actively embraced surfing culture, resulting in a closely-knit network. However, it is essential to note that some businesses within the community may compete with one another as they cater to the same tourist segment.

In the spirit of surf culture, an emphasis on the protection and sustainable development of the marine environment has emerged, albeit still in its early stages. While this emerging awareness is promising, there is a growing need for further efforts to raise awareness and enact measures that prioritize the preservation of the coastal and marine ecosystems. Surfing enthusiasts and surf schools have engaged in beach and ocean clean-up activities, actively raising awareness about environmental preservation. Some residents have even chosen to become volunteers, dedicated to protecting the marine environment and the propagation of surfing culture. Others have opened environmentally friendly businesses, such as producing environmentally friendly coloured sunscreen, which has become part of Riyue Bay's surfing culture (Fig. 31.2).

While surfing can be associated with extreme conditions and is often considered an extreme sport (e.g. big wave surfing and tow-in surfing), not all forms of surfing necessarily fall under the category of 'extreme.' Surfing contributes to lifelong physical, social, and psychological well-being (Wheaton, 2017; Lemarié, 2022). The rise of surfing in Riyue Bay encourages outdoor sports and promotes a healthy lifestyle among locals, reflecting the positive impact of surf culture on Hainan Island. Since December 2010, when the International Surf Festival was held in Hainan, various surfing events held locally include the WSL Women's Longboard Surfing Championship, WSL Men's Longboard Surfing Championship, ISA China Cup Men's and Women's Shortboard Competition, ASP Men's Shortboard Hainan Classic, WSL Women's Shortboard Championship Trials and Surfing Hainan Open, among others.

Challenges Within the Surfing Industry in Hainan

While the surfing industry has brought numerous advantages to Hainan Province, certain deficiencies have emerged. First, the absence of a unified market regulator has resulted in

Fig. 31.2. Coloured environmentally friendly sunscreen (left top and left bottom) and volunteers collecting rubbish on the beach (right), courtesy of Hu Yang.

pricing disparities for surfing lessons between formal surf clubs and independent instructors known as 'Beach Boys'. Beach Boys, as surf instructors operating outside established schools or clubs, can offer lessons at significantly lower rates due to their exemption from shop rental fees. This price difference can lead to disparities in the quality of instruction provided. Notably, some freelance instructors have professional education as surfing coaches, while others lack the required qualifications. This situation has partly arisen due to the surging demand for surfing lessons (Lai, 2021). Nevertheless, Beach Boys, who are often local residents, express their gratitude towards surf clubs for bringing the surfing industry to Wanning, as it has generated employment opportunities for them.

Long-time Riyue Bay surfers call themselves 'làngrén' (浪人[2]), which can be translated as 'wave rider' or Ronin.[3] These surfers come from all over the world, and by chance, they discovered that Riyue Bay has excellent wave points. In this way, the surfing culture was brought to Riyue Bay. The surfers often come here purely for the love of surfing, not to make a profit. Therefore, they see Riyue Bay as a surfing Eden where they can relax and develop their hobbies. Riyue Bay's streets demonstrate the elements of surfing, skateboarding and hip-hop culture. The connection between skateboarding, surfing, and hip-hop is rooted in their shared countercultural and rebellious origins and their influence on youth culture and self-expression. The streets display graffiti (Fig. 31.3), and several small skate spots are in the area.

Riyue Bay is a popular destination among surfing lovers, tourists, and business people looking for new opportunities. The area has attracted opportunistic entrepreneurs, raising concerns among the local surfing community. The previously undeveloped area has now been transformed into a bustling commercial and tourism hub in Riyue Bay, featuring numerous hair braiding shops, high-end bars, restaurants, nightclubs, and boutique bed and breakfast operators. This commercialization trend is gradually overshadowing the emerging authentic surf culture as profit motives take precedence. Consequently, the cost of living for dedicated surf enthusiasts, làngrén (浪人), who engage in surfing purely for its intrinsic value, is rising. In particular development zones, làngrén (浪人) have resorted to graffiti to express disagreement with current tourism planning and investment initiatives (see Fig. 31.4). Consequently, some have opted to seek alternative surfing havens, departing from Riyue Bay in search of a more suitable and less commercialized environment.

Fig. 31.3. Hiphop graffiti.

Fig. 31.4. Protest graffiti '不懂冲浪文化, 你搞个毛, 垃圾!' (Translation: If you do not understand surfing culture, just do not blindly demolish and build something; just do not fool around, loser!).

The influx of surfers to Hainan, some drawn by the excellent waves in Riyue Bay and others seeing surfing as a business opportunity, creates a unique socio-cultural challenge. While long-time surfers (wave riders) value Riyue Bay as a surfing paradise, prioritizing their love for the sport over profit, the increased integration of surfing with skateboarding and hip-hop culture in the area reflects a broader cultural and economic shift. Navigating the divide between long-time surfers (who primarily value Riyue Bay as a surfing paradise) and surfing as commercial product poses a socio-cultural challenge as the communal space transforms, incorporating leisure-driven surfers and those looking for commercial prospects.

Surfers have introduced a thriving surfing culture to Riyue Bay, and the Hainan Provincial Government recognizes the economic potential of surfing tourism. As part of infrastructure enhancements, the government has unveiled plans to construct the 'Surfers Paradise' station along the Hainan Coastal Scenic Highway,[4] which links numerous scenic locations throughout Hainan's 12 coastal cities and counties, including Riyue Bay. Beyond typical highway services like food, petrol, and opportunities to rest, the station will be developed as a themed hub, emphasizing its focus on providing visitors with a comprehensive and immersive surfing-themed experience.

These initiatives align with the government's overarching commitment to invest in and develop Riyue Bay's infrastructure and increase residents' quality of life, enhancing the region's overall appeal. As a further step, the government intends to allocate approximately 400 million yuan to two key projects – the Surf Centre and a surfing wave pool (Hainan International Media Centre, 2023a). Through school activities that promote youth involvement in surfing, authorities aspire to elevate sports offerings and enhance the overall quality of life in the region.

The unregulated tourism-driven expansion leads to traffic congestion and limited parking, especially during peak season. However, after some spontaneous tourism development, authorities are working on strategic plans for tourism development, rural revitalization and modernization of the region. As such, Wanning's Rural Revitalization Bureau plans to invest two million yuan in building parking areas to solve traffic problems (Hainan International Media Centre, 2023b). Overall, the development in Riyue Bay is marked by widespread, often unregulated or spontaneous construction, the opening of stalls and restaurants, and inadequate waste management, which may lead to severe environmental degradation within the region (Li *et al.*, 2023).

At the same time, the increase in the number of tourists leads to crowding (which deteriorates the visitor experience) and environmental problems. The increase in tourism and surfing activities can stress the local environment, leading to beach erosion, litter, and damage to the marine ecosystem. Maintaining the delicate balance between tourism and environmental preservation is crucial. Among other challenges are weak infrastructure and an absence of lifeguards in the bay. While development is essential, it must be balanced to ensure that it benefits the local community and is sustainable in the long term. The rapid influx of tourists and surfers may lead to changes in the local community and lifestyle. Balancing the needs and interests of the local community with those of tourists and surfers can be challenging.

Conclusion

Contemporary tourists seek authentic experiences in outdoor settings, and sports tourism is becoming increasingly important. The distinctive appeal of Riyue Bay, characterized by its remarkable wave points and vibrant surf culture, attracts individuals to this location. This cultural phenomenon exerts a positive influence across several dimensions, notably encompassing economic expansion, the advancement of tourism, the propagation of environmental preservation, and the promotion of salubrious lifestyles, thereby encapsulating the manifold benefits of surf culture for both Hainan Island and Riyue Bay. Surfing tourism in Hainan's Riyue Bay has become one of the region's defining characteristics, drawing global attention through surfing competitions and events and bolstering the island's reputation as a tourist destination. However, there is still room for improvement in managing and avoiding excessive commercialization in Hainan's surf tourism industry. We acknowledge the need for a more nuanced exploration of challenges, particularly excessive commercialization in Hainan's surf tourism industry. To address this, future research could look at the impact of commercialization on the authentic surf experience, local businesses, and environmental sustainability. Additionally, we suggest exploring sustainable development strategies and presenting successful models from other surf destinations to inspire Hainan's stakeholders.

Overall, this study benefits students and researchers exploring surfing beyond established surf locations. It also provides insights for tourism stakeholders, including local businesses, surf schools, and competition organizers, by examining the cultural and economic impacts of surfing on Hainan Island. Additionally, it appeals to a broader audience interested in the cultural and economic dimensions of surfing, contributing to a deeper understanding of how the sport shapes the region's identity and tourism sector. The study informs and guides stakeholders to understand Hainan as an emerging surfing destination.

Notes

[1] Brendan Sheridan is a Californian surfer who relocated to Hainan in 2004. Brendan played a pioneers' role in establishing the surfing culture in the region. In 2006, he founded the Hainan Surfing Open contest, contributing significantly to the promotion of surfing in Riyue Bay (Zanella, 2019).
[2] làngrén 浪人- waverider (浪- wave, 人- person).
[3] A ronin was a warrior, a samurai without a master, who travelled the country offering his service to anyone in need of a sword to hire (Breithaupt, 2004).
[4] The Hainan Coastal Scenic Highway is a 988.2 km highway loop that connects 12 coastal cities and counties and features 84 scenic landscape areas, 22 coastal headlands, 25 lighthouses, 68 bays, 26 coastal lagoons, 16 mangrove forests, 40 courier stations, 31 A-level scenic spots, 21 tourist resort areas, and 261 coastal scenic spots and historical sites (Quinn, 2021).

References

Breithaupt, H. (2004) Ronin, PhD. *EMBO Reports* 5(9), 831. DOI: 10.1038/sj.embor.7400233.
Hainan International Media Centre (2023a) Introduction to Riyue Bay (Rosenberg, M. trans). Press Junket Background Information (released at the 2023 Boao Forum For Asia).
Hainan International Media Centre (2023b) Introduction to Tianxin Village (Rosenberg, M. trans). Press Junket Background Information (released at the 2023 Boao Forum For Asia).
International Olympic Committee (2016) IOC Annual Report 2016. Available at: https://stillmed.olympic.org/media/Document%20Library/OlympicOrg/Documents/IOC-Annual-Report/IOC-Annual-Report-2016.pdf (accessed 18 January 2024).
Lai, S. (2021) Surfing to the masses, where are the qualified trainers? Workers' daily. DOI: 10.28277/n.cnki.ngrrb.2021.005073.
Lemarié, J. (2018) Surfing: A History of Gliding | Surf: Une histoire de la glisse.
Lemarié, J. (2020) Le processus de sportisation du surf à Hawaï, en Australie et en Californie de 1778 à 2015. In: Guibert, C. (ed.) *Les mondes du surf: Transformations historiques, trajectoires sociales, bifurcations technologiques*. Maison des Sciences de l'Homme d'Aquitaine (MSHA), pp. 37–62.
Lemarié, J. (2022) 'Surf Tourism'. In: *Encyclopedia of Tourism Management and Marketing*. Edward Elgar Publishing, Cheltenham, UK. DOI: 10.4337/9781800377486.surf.tourism.
Li, F., You, H. and Qin, J. (2023) Hainan Riyue Bay Surfing test design. *Secondary School Geography Teaching* (15), 67–68.
Ormrod, J. (2005) 'Endless Summer (1964): consuming waves and surfing the frontier. *Film and History: An Interdisciplinary Journal of Film and Television Studies* 35(1), 39–51.
Quinn, P. (2021) Construction of Hainan's 'No.1 National Coast Scenic loop highway' is underway! Tropical Hainan. Available at: https://www.tropicalhainan.com/construction-of-hainans-no-1-national-coast-scenic-loop-highway-is-underway/ (accessed 18 January 2024).
Tian, Y. (2020) *Research on the Current Situation and Development Strategies of Surfing Sports in Wanning City*. Wuhan Sports University, Wuhan, China.
Warshaw, M. (2010) *The History of Surfing*. Chronicle Books, San Francisco, CA.
Wheaton, B. (2017) Surfing through the life-course: silver surfers' negotiation of ageing. *Annals of Leisure Research* 20(1), 96–116. DOI: 10.1080/11745398.2016.1167610.
Yang, Z. (2019) *Research on the Impact of Surfing Events on Tourism in Wanning*. Hainan Tropical Ocean University, Sanya, China.

Zanella, N. (2019) *Children of the Tide. An Exploration of Surfing in Dynastic China*. Gardners, Eastbourne, UK.

Zhang, Q. and Qi, Y. (2022) Research on the current situation and countermeasures of surfing tourism development in Sanya. *Tourism and Photography* 09, 32–34.

Zhao, H., Liang, R. and Ma, F. (2020) Reflections on the international development model of the Hainan surfing event industry. *Sports Boutique* 2, 48–50.

32 Learning by Doing: Educational Tourism at Mount Danxiashan

Guiqing Li and Ding Xu
School of Tourism Management, Sun Yat-sen University, China

Summary

Educational tourism is a fast-growing market in China. Since Mount Danxiashan was listed as a Global Geopark and World Heritage Site in 2010, educational tourism has rapidly grown in the local community, becoming a flagship tourism product. This makes Mount Danxiashan an informative case for understanding educational tourism development and community participation. The present case study has two foci. First, we articulate the stakeholder collaboration mode at Mount Danxiashan. Various stakeholders, including but not limited to the administrative authority, research and education institutes, and local communities, co-determined to develop educational tourism as the best practice for local resources. The collaboration mode and the educational focus are critical in sustaining local development. Second, we depict the learning trajectory of the local community. Led by local intellectual elites, residents acquired both tourism- and nature-related knowledge and participated in educational tourism. This enabled them to design and deliver nature and heritage education curricula for tourists. Moreover, this process encouraged and empowered villagers to co-create new knowledge.

The Value and Interest of the Case Study

The critical interest of this case is how educational tourism can be developed, and how locals, through participation in educational tourism operations, can improve economic income and achieve personal growth.

Background

Educational tourism in China's heritage sites

Educational tourism can be generally conceptualized as learning activities in the traveling context (Li and Liang, 2020). It epitomizes the ideas of experiential and informal learning, in which participants achieve learning outcomes while traveling and visiting destinations (Xu and Ho, 2021). Often the participants are students of various age groups. Exchange programs and summer camps are common forms of educational tourism (Chau, 2021). There is a recent trend in China, in which educators and tourism operators collaborate to provide diverse and accessible educational tourism products (Wen *et al.*, 2023; Zhou and Liu, 2023). This involves many types of tourism contexts, such as cultural tourism, ecotourism, and rural tourism. The learning activities can take many forms, including self-learning programs, popular science lectures, and on-site observation. Importantly, the term popular science refers to the communication of scientific information and concepts to a general audience in an accessible and engaging manner. Some like Daum (2009) consider it as 'science "lite"', in which scientific knowledge is made understandable to non-experts, often through simplified explanations, visual aids, and storytelling techniques.

Heritage tourism in China is witnessing an increase in integrating education and learning components (Li and Zhang, 2023). For tourists, it enriches their on-site experiences and improves the meaningfulness of leisure travel. Operating educational tourism can benefit the locals intellectually and economically. Running educational tours encourages them to continuously learn new knowledge, and allows more profits than common tourism services as people tend to pay a premium for knowledge (Wen *et al.*, 2023). For destination managers, educational tourism helps promote heritage conservation and sustainable use of resources and is useful for building community capacity for both knowledge acquisition and tourism operation (Zhou and Liu, 2023).

However, successfully developing educational tourism in heritage sites is not an easy task. A critical consideration is to balance the interests of various stakeholders. Their different understandings of heritage values and practices lead to institutional conflicts in ecology and cultural resource utilization (Li *et al.*, 2023). Furthermore, local communities are often marginalized in institutional designs (Zhang, 2022), which threatens the sustainable development of heritage sites. Thus, a significant concern lies in the transmission of heritage values into the communities, and the personal growth of locals. Considering the increasing popularity of educational tourism and the critical importance of learning in heritage tourism, a case where educational tourism grows steadily can offer insights for tourism researchers and practitioners. Thus, the objective of this case study is to present Mount Danxiashan and how it has effectively united stakeholders to develop educational tourism.

Mount Danxiashan is one of the first World Heritage Sites in China to develop educational tourism and has formed a mature system to cultivate educational tourism. Different stakeholders – heritage site management agencies, community residents, heritage site enterprises, tourists, and students – have come together in this process. This cooperative system places each stakeholder in a suitable role, which jointly promotes the branding of Mount Danxiashan educational tourism. Importantly, this system enables the growth of several local (village) knowledge experts, who are role models for other community members, encouraging them to participate in the project.

The context of the case

Mount Danxiashan is in Shaoguan City, Guangdong Province, China, with a total area of 292 km^2. It is famous for its unique Danxia landform and rich biodiversity resources and was listed as a UNESCO Global Geopark in 2004 and a World Natural Heritage site in 2010 (UNESCO, 2023a, b). Mass tourism in Mount Danxiashan began in the 1980s, and educational tourism has developed since 2004. Mount Danxiashan has attracted many operators, both locals and immigrants, to carry out educational tourism programs. Mount Danxiashan has been awarded multiple national titles (Table 32.1). The educational tourism program design was also identified as one of the *Top 10 Innovative Cases of World Heritage Education* by the World Heritage Institute of Training and Research for Asia and the Pacific Region under the auspices of UNESCO (WHITR-AP, 2023). These titles demonstrate the wide recognition of Mount Danxiashan's educational tourism development.

Research Method

This case study is based on data collected from fieldwork that lasted from January 2020 to August 2021. The aim was to connect with the community, perform chronological participatory observation, interview various stakeholders, and collect secondary data (publications, web texts, government reports, and planning texts). We have acquired consent from the participants that their personally identifiable information can be used as references to increase the credibility of our work. Importantly, one author has volunteered at Danxiashan since 2017. This author was deeply involved in Danxiashan's educational tourism development by formulating the community participation mode in educational tourism, composing educational tourism courses, designing education tourism

Table 32.1. National titles of Danxiashan related to educational tourism.

No.	Titles	Time	Issuing departments
1	Science popularization base of homeland resources	2009	Ministry of land and resources
2	National ecotourism example area	2013	Department of protection, ministry of environment, national tourism administration
3	The national science popularization and education base	2015	China association for science and technology (CAST)
4	National primary and middle school students study and practice education base	2017	Ministry of education of the people's Republic of China (MOE)
5	The national natural education base	2019	Chinese academy of forestry
6	Long-term scientific research base of the national forestry and grassland administration	2020	The national forestry and grassland administration
7	The national science popularization and education base	2022	China association for science and technology (CAST)
8	National forest and grass popular science popularization base	2023	National forest and grass popularization science popularization base

Data source: The Danxiashan Management Committee (2023).

Table 32.2. Key informants in the case study.

No.	Surname	Title and roles
1	Chen	Deputy head of the Danxiashan Management Committee
2	Gu	Local, educational tour designer
3	Guo	Local, expert on plant taxonomy
4	Liu	Local, owner of Airbnb *Shanyuanse*
5	Zhu	Local, expert on dendrobium farming

festivals and events, and training community educational tourism practitioners. He was recognized as a key helper by the community and provided critical data in drafting the present case study. The case study uses material mostly retrieved from interviews with five key informants (Table 32.2).

Stakeholder Collaboration for Educational Tourism

Unifying one goal for educational tourism

The UNESCO enlistment urges tourism transformation at Mount Danxiashan. The local community at Mount Danxiashan relied heavily on the tourism business. Since enlistment in 2010, human activities have been restricted by a series of regulations. Mass tourism products were constrained by limited routes and accessibility. Mount Danxiashan needs new types of tourism operations to ensure local businesses can be sustained.

The Danxiashan Management Committee is a governmental organization subordinated to the Shaoguan Municipal Government. The committee is designed to implement and supervise UNESCO and governmental regulations about Mount Danxiashan, and moderate issues, conversations, and conflicts between various stakeholders. The committee first realized the potential of educational tourism in addressing the loss of the mass tourism market and helping maintain local tourism businesses. Moreover, Chen (key informant 1) revealed that they believe educational tours enable heritage values to be better circulated among tourists, which fulfills UNESCO's expectations to engage in science literacy education as a Global Geo-park.

From Chen, we realize how the committee utilized its resources to promote education tourism development. First, they established partnerships with research and education institutions to explore the heritage values of Mount Danxiashan. These institutions helped translate their academic writings about Mount Danxiashan into popular science messages. Second, the committee circulated these messages to different stakeholders. This enhanced their understanding of the geological features, history, culture, and natural ecological values of Mount Danxiashan. Last, the experts from the research and education institutions, including geographers, ecologists, and tourism scholars, helped design educational tourism (Li *et al.*, 2023). Under the auspices of UNESCO (WHITR-AP, 2023), they collaborated with the committee to organize activities for educational tourism, such as the 'Wild Plant Identification Competition of Mount Danxiashan' (Fig. 32.1). These titles (Table 32.1) demonstrate the wide recognition of Mount Danxiashan's educational tourism development.

After 6 years of promoting educational tourism, various stakeholders finally aligned. This was because in 2016, as the Chinese Central Government imposed national policies to encourage educational tourism in nature reserves, the committee took the chance to unify the region for educational tourism development. The moderation of the Danxiashan Management Committee laid the foundation for the unification of such developmental goals, which fuelled the rapid growth of educational tourism in this region (Li *et al.*, 2023).

Educating stakeholders for promoting educational tourism

There are mainly five stakeholder groups in the educational tourism development of Mount Danxia. These include internal stakeholders – heritage and nature reserve management

Fig. 32.1. Wild plant identification competition of Mount Danxiashan.

authorities, local communities, and local enterprises, and external ones such as mass tourists and student tourists (Li and Zhang, 2023). Commonly in developing heritage tourism or ecotourism, the emphasis is given to the tourist experiences or marketing of local tourism products, or how to share benefits by satisfying the community's needs. Such practices neglect the need for proficient tourism operations and may lead to failures in heritage value communication.

The committee of Mount Danxiashan spent considerable effort in lobbying various stakeholders, mostly the residents and tourism businesses, about educational tourism development. The committee acted as a coordinator and made use of the research and education institutions to first improve the scientific literacy of managerial personnel, community residents, and local entrepreneurs. Chen suggested the committee organized reading clubs, workshops, seminars, and special interest groups to facilitate internal stakeholders' learning journey. These activities communicated geological knowledge, natural ecology, history, culture, and educational tourism skills. As a result, many local individuals are now capable of undertaking various roles in running an educational tour, such as nature education rangers and bird-watching guides (Fig. 32.2).

Since the Chinese Central Government endorsed educational tourism, Mount Danxiashan has become a popular education tourism destination for school tours. The specification of market orientation pushed changes in Mount Danxiashan's educational tourism supply capacity. Different from the previous actions that were top-down and led by the Management Committee, the residents became more self-organized. Local enterprises and community residents realized the importance of the school markets and strengthened their communication skills to adapt to the needs of younger consumers without alerts from the Danxiashan Management Committee. Now heritage values and ecotourism concepts are delivered more understandably and interestingly.

Happy residents, happy tourists

Balancing the interests of different stakeholders is necessary to achieve sustainability. The primary needs of internal stakeholders (residents and businesses) include income generation and diverse livelihood strategies. Educational tourism, compared to farming or conventional B&B businesses, ensured their

Fig. 32.2. Local-led birdwatching.

needs were satisfied. Moreover, through learning scientific knowledge, they better understood the importance of conservation and believed that educational tourism is a solution for the sustainable use of resources.

The essential request of external stakeholders (students, parents, and schools) is to enjoy quality educational tourism. The guidance of the Management Committee and assistance of the research and education institutions has cultivated competent education tourism products for tourists to consume. For parents, educational tourism is a leisure trip to improve the parent-children relationship. For schools, it is engaging in extra-curricular programs that fit both the expectations of the students and the parents. An important learning outcome is children's increased awareness of pro-environmentalism and science literacy. This makes educational tourism at Mount Danxiashan a meaningful experience to be promoted in the consumer markets.

The main interest of the Danxiashan Management Committee is to implement UNESCO and national regulations, particularly in the conservation and utilization of heritage resources. By uniting different stakeholders in the joint development of educational tourism, they have achieved this goal. They introduced educational tourism to promote sustainable and profitable livelihood strategies, and effectively communicate heritage site values while preserving heritage resources.

Community Growth in Developing Educational Tourism

The transformative growth of local people

Residents in Mount Danxiashan experienced several stages of transformation. In 2004, Danxiashan was listed as a UNESCO Global Geopark, which started the development of educational tourism. In the early stage of this development, some local villagers, despite not going through systematic school education, worked as field assistants with external knowledge experts due to their local knowledge. They devoted themselves to learning geological features and the ecological environment.

Gu (key informant 2) originally worked as a tour guide and was introduced into educational tourism as an on-site assistant managing school children. After undertaking several education tourism programs since 2004, she has built connections with various experts, which has helped her learn geology and plant taxology. Through a process of continuous knowledge accumulation and practice, she is now a local expert in the field of popular science – making scientific knowledge more understandable and interesting to non-experts through interpretation and storytelling. She now shares solid connections with the local tourism board and research institutes and has helped draft and promoted a series of standard procedures in guiding educational tours related to educational tourism. Her case illustrates how social capital can be gained from educational tourism and helps its further development. The connections with knowledge experts and governmental officials can be invaluable for the regional population as the communities were often underprivileged. Engagement in educational tourism can help cultivate such connections.

An increasing number of knowledge experts, university researchers mostly, have been invited to Mount Danxiashan to aid community education since 2016. They have provided diversified learning channels, such as community lectures on specific topics, at no charge. In exchange, the committee allowed the researchers to access the core zone of Mount Danxiashan for plant, insect, and ecological fieldwork. Additionally, the locals are encouraged to join researchers' field investigations. Under professional supervision, the villages can better acquire scientific knowledge. These opportunities enhanced the local people's ability to facilitate educational tourism. The Management Committee initiated the certification of 'Popular Science Influencer.' The committee certifies individuals who complete required courses and are capable of producing and disseminating popular science content to educational tourists.

Developing educational tourism encourages local residents to value their own culture. Taking Liu (key informant 3) as an example, he was originally an Airbnb host, which provided him with his necessary income. With the rise of the educational tourism market, he collaborated with the Management Committee and transformed his hostel into a base with the functions of both accommodation and local intangible cultural education. He repositioned his Airbnb to serve as a local cultural preservation center – a public space for hosting traditional events and showcasing local cultural artifacts. The local villages are dominated by Hakka culture, which is famous for its distinct language, customs, food, beliefs, and traditional festivals. However, these cultural components were not fully presented to tourists in the mass tourism era. The community members, especially elderly elites, highly appreciated Liu's conduct in preserving their culture, which in turn improved his local networks and community status.

In 2019, the Management Committee established the initiative of Danxiashan Science Popularization Town. The initiative aimed at developing educational tourism through a top-down management approach, and carefully planned for buildings and management of educational tourism (Zhou and Liu, 2023). An important measure was to make certified popular science influencers as role models for other villagers. The demands for educational tourism services – accommodation providers, nature education rangers, and food suppliers – keep increasing. A large number of locals and even immigrants have joined the educational tourism businesses. The locals may use local knowledge in developing place-specific educational products, and the immigrants exploit their knowledge to provide more customized products. Mount Danxiashan witnessed the emergence of local experts in the fields of geology, plants, birds, insects, and Hakka culture (Fig. 32.3).

From local to scientific knowledge production

An interesting contribution of the locals is the production of scientific knowledge. Since many researchers enter Danxiashan for field investigation, locals have picked up scientific knowledge and research skills from interactions with the researchers. Guo (key informant 4) is one example. He used to learn Chinese medicine

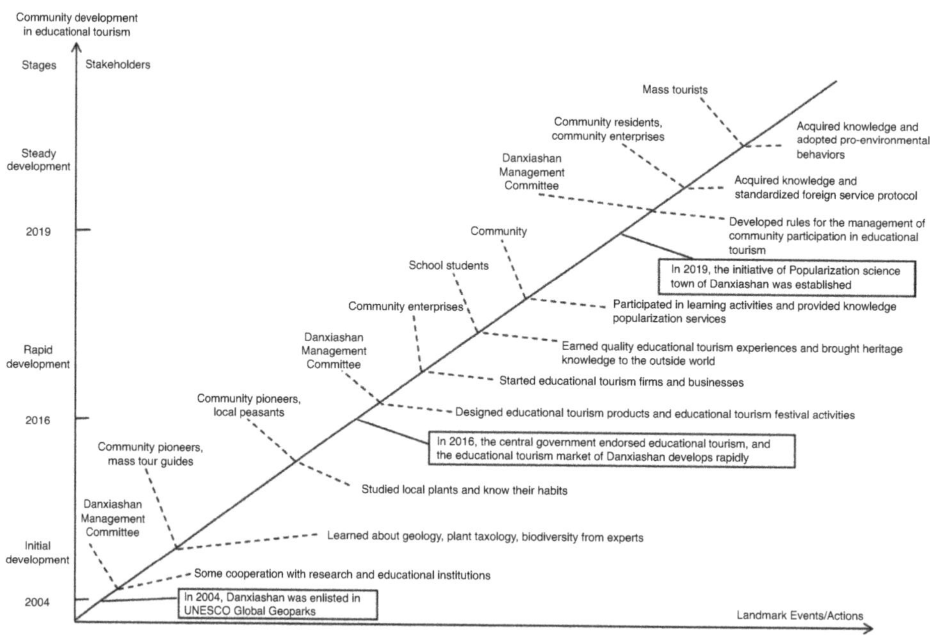

Fig. 32.3. Educational tourism development of Mount Danxiashan.

and acquire traditional knowledge about local plants. In collaboration with biologists conducting plant surveys in Mount Danxiashan, he developed an interest and competencies in plant taxonomy. He helped with the discovery and naming of various species in Mount Danxia, such as *Semiaquilegia danxiashanensis* (Zhou *et al.*, 2019), *Diospyros penghuae* (Zhao *et al.*, 2018), *Primulina huangjiniana (Gesneriaceae)* (Huang *et al.*, 2020), and *Asplenium danxiaense nov. (Aspleniaceae, Aspleniineae)* (Xu *et al.*, 2022), and authored various academic publications. He is now recognized as a home-grown expert by the provisional forestry department. Local experts like Guo are transferring local wisdom into scientific knowledge, and benefit local and non-local societies with useful insights from Mount Danxiashan.

Beyond academic knowledge, the locals also transfer traditional wisdom into practical knowledge, particularly in farming dendrobium. Local villagers, represented by Zhu (key informant 5), accumulated traditional skills in growing dendrobium. Dendrobium is a large family of lithophytic rock orchids. It is an expensive medical ingredient with anti-inflammatory effects. However, wild dendrobium often grows on the exposed edges of Mount Danxiashan's red rocks, which makes it difficult to harvest. The locals, through decades of farming, gradually developed a method to home-grow dendrobium (Fig. 32.4). In 2007, some medical researchers came to Mount Danxiashan. Zhu approached them by chance and demonstrated the local methods of home-growing dendrobium. These methods were found to be more effective than contemporary techniques and are now written into curriculum material for teaching medical university students.

In addition, to enriching the educational tourism experiences in Mount Danxiashan, local villagers developed diverse extra-curricular courses from local knowledge. They carefully revisited the history of ancient villages, farming methods, and local cuisines, and connected the information with contemporary scientific explanations. Such content was further designed as educational tourism products aimed at tourist segments traveling with school-age children. Creating new educational tourism products from local knowledge benefited the locals through higher incomes, communicated heritage values, and branded Mount Danxiashan with diverse product lines.

Fig. 32.4. Zhu and his dendrobium farm.

Conclusions

As an eco-friendly approach, educational tourism plays an important role in Mount Danxiashan. It alleviates the conflicts between conservation and utilization in this World Heritage Site and builds community capacity in terms of livelihood and local knowledge. A key to success is the Danxiashan Management Committee. First, they took a leading role in planning and fostering educational tourism. This released the community from the heavy pressure of designing, lobbying, and sometimes investing in educational tourism. Thus, people engage in tourism development as 'participants' instead of 'drivers,' which lowers their cost of success. Second, the committee coordinated conflicts and provided resources to help individuals progress, which sustained stakeholder cooperation and catalyzed capacity building within the community.

It is worth noting that the educational tourism community differs from asset-based tourism communities in two ways. The first is the essence of the tourism products. The former emphasizes the educational function and learning experience of the product, while the latter focuses on leisure and entertainment (Li *et al.*, 2023). Second, there are different requirements regarding the abilities of community participants. The former requires participants to continuously accumulate local knowledge and convert it into products, while the latter requires them to master tourism service skills.

As the representative agent of UNESCO and national governments, the management committees of heritage sites are responsible for conservation, community growth, and sustainable development. However, across the globe, there is a dearth of discussion about how such management committees can effectively help with community growth and the sustainable development of heritage sites. We hope that this case study offers insights into the roles and actions of heritage management committees. Moreover, the case sheds light on why educational tourism should be introduced in heritage sites and nature reserves, and how local communities can participate and contribute to the sustainable development of this type of tourism. However, it is worth noting that more evidence is needed to provide stronger implications from Mount Danxiashan, such as the share of educational tourism contributing to the overall economy, and the effectiveness of university experts and scholars in helping regional residents with educational tourism.

References

Chau, S. (2021) Antecedents and outcomes of educational travel in higher education. *Journal of Hospitality, Leisure, Sport & Tourism Education* 29, 1–13.

Danxiashan Management Committee (2023) Mount Danxiashan. Available at: http://dxs.sg.gov.cn/xwzx/jqyw/index.html (accessed 5 October 2023).

Daum, A.W. (2009) Varieties of popular science and the transformations of public knowledge: Some historical reflections. *Isis* 100, 319–332.

Huang, C., Meng, K., Guo, J., Chen, F., Liao, W. and Fan, Q. (2020) Primulina huangjiniana (Gesneriaceae), a new species from Guangdong, China. *Guihaia* 40, 1429–1437.

Li, P. and Liang, H. (2020) Factors influencing learning effectiveness of educational travel: A case study in China. *Journal of Hospitality and Tourism Management* 42, 141–152.

Li, G. and Zhang, C. (2023) Stakeholder cooperation mode and formation mechanism of world heritage education. *Study on Natural and Cultural Heritage* 8, 30–36.

Li, G., Xie, J. and Liu, J. (2023) A study on the mechanism of intellectual elites' development in the knowledge-oriented community tourism. *Tourism Tribune* 38, 109–121.

UNESCO (2023a) China Danxia. Available at: https://whc.unesco.org/en/list/1335 (accessed 3 July 2023).

UNESCO (2023b) Danxiashan UNESCO Global Geopark (China). Available at: https://en.unesco.org/global-geoparks/danxiashan (accessed 3 July 2023).

Wen, S., Chen, Y. and Liao, J. (2023) The evolution of animal landscape and practice of nature education in urban zoos: A case study of Guangzhou Zoo. *Journal of Chinese Ecotourism* 13(3), 526–539.

WHITR-AP (2023) Top 10 Innovative Cases of World Heritage Education Showcase. Available at: http://www.whitr-ap.org/index.php?classid=1518&newsid=3277&t=show (accessed 10 July 2023).

Xu, J. and Ho, P.S.Y. (2021) Whether educational tourism can help destination marketing? An investigation of college students' study tour experiences, destination associations and revisit intentions. *Tourism Recreation Research*, 1–16.

Xu, K.-W., Lin, C.-X., Guo, J.-Q., Zhou, X.-X., Liao, W.-B. *et al.* (2022) *Asplenium danxiaense* sp. nov. (Aspleniaceae, Aspleniineae). A new tetraploid fern species from Guangdong, China, based on morphological and molecular data. *European Journal of Taxonomy* 798, 162–173.

Zhang, C. (2022) Review and prospect: Relationship between world heritage conservation and tourism development in the 50 years. *Chinese Cultural Heritage* 2022, 38–42.

Zhao, W., Fan, Q., Chen, F., Huang, C., Chen, Z., Hou, R. and Liao, W. (2018) Diospyros penghuae, a new species of Ebenaceae from Danxiashan mountain in Guangdong, China. *Acta Scientiarum Naturalium Universitatis Sunyatseni* 57, 98–103.

Zhou, T. and Liu, J. (2023) Research on production mechanism of educational tourism products in nature reserves: A case study of Danxia Mountain, Guangdong Province. *Journal of Chinese Ecotourism* 13(3), 466–482.

Zhou, J.-J., Huang, Z.-P., Li, J.-H., Hodges, S., Deng, W.-S., Wu, L.E.I. and Zhang, Q. (2019) Semiaquilegia danxiashanensis (Ranunculaceae), a new species from Danxia Shan in Guangdong, southern China. *Phytotaxa* 405(1), 1–14.

Conclusion

Taufik Abdullah[1] and Reni Polus[2]
[1]Faculty of Social Science Education at Universitas Pendidikan Indonesia; [2]Business School at Edinburgh Napier University, Scotland

Charting the Path Forward: Final Thoughts and Future Outlook for Asia-Pacific Tourism

Reflecting the journey

As we wrap up *Tourism Case Studies from the Asia-Pacific Region*, it is time to reflect on the rich tapestry of stories and insights woven throughout these sections. This volume has offered a vibrant exploration of how tourism is not just a catalyst for economic growth but a powerful force for cultural preservation, community empowerment and sustainable development across the Asia-Pacific. Each case study has contributed to a deeper understanding of the dynamic role that tourism plays in this diverse region, highlighting its importance in the global tourism landscape.

From the outset, we have seen how innovative environmental practices are reshaping tourism. The green trails initiative by Indiahikes in India and the sea turtle conservation efforts in Malaysia stand as shining examples of how commitment to environmental stewardship can harmonize with tourism. These stories highlight that true sustainability is not a one-size-fits-all approach but requires a nuanced understanding of each destination's unique characteristics. In exploring cultural sustainability, we have discovered how tourism can be a custodian of heritage. The preservation of traditional arts in Cambodia and the integration of cultural heritage in Lau Pa Sat, Singapore illustrate that tourism can indeed safeguard cultural identities while driving economic growth. These cases offer a compelling narrative of how respecting and promoting cultural heritage can coexist with modern tourism demands.

Our journey through community empowerment showcased how tourism can be a transformative force for local populations. The stories from the Philippines, Maldives, India and Thailand reveal how community-based tourism initiatives can ignite social change and economic opportunities. By involving locals in tourism planning and implementation, these projects foster a sense of pride and ownership, proving that tourism, when done right, can create lasting positive impacts. Resilience and recovery emerged as crucial themes, especially in the wake of crises like the COVID-19 pandemic. The remarkable rebound of destinations such as Miso Walai Homestay in Malaysia and Bojo Aloguinsan Ecotourism Initiative in the Philippines illustrate how tourism can be a cornerstone for recovery and sustainability. These cases highlight that resilient tourism is built on adaptive strategies and strong community ties, setting inspiring examples for future recovery efforts.

The ethical dimensions of tourism were laid bare in our examination of indigenous rights and socio-economic disparities. From wartime legacies in the Philippines to indigenous tourism ventures in Malaysia, these studies challenge us to think critically about inclusivity and fairness in tourism practices. They advocate for a tourism ethos that respects local cultures and addresses socio-economic imbalances. Furthermore, the ethical dilemmas surrounding elephant-based tourism in Nepal highlight the complex balance between animal welfare and tourism benefits, highlighting the need for compassionate and equitable practices across all aspects of the industry.

Innovative management and policy approaches were also spotlighted, emphasizing the role of evidence-based decision-making in fostering sustainable tourism. The insights from meteorological trends in Bali, Indonesia and hotel investment policies in Western Australia show that leveraging data and research can drive smarter, more sustainable tourism policies. Looking to the horizon, emerging trends in China such as surfing tourism in Hainan and educational tourism at Mount Danxiashan reveal exciting new opportunities and challenges. These trends offer a glimpse into how tourism consumption is evolving and how destinations are adapting to meet new demands while pursuing sustainability.

In wrapping up, this volume bridges theoretical frameworks with practical realities, advocating for interdisciplinary and culturally sensitive approaches in tourism. By weaving together concepts of sustainability, cultural identity and community empowerment with real-world contexts, it deepens our understanding and paves the way for more ethical tourism practices. Serving as a valuable guide for both higher and vocational education in tourism, the volume offers crucial insights for instructors and students alike. Whether you seek to enhance your teaching practices or aspire to succeed in the tourism industry, *Tourism Case Studies from the Asia-Pacific Region* is your essential companion on this enlightening journey. It equips you with the tools to navigate the complex landscapes of tourism, deepening your understanding of the unique characteristics of people and places in the Asia-Pacific region.

Additionally, this volume highlights the importance of integrating evidence-based decision-making and data-driven policies to achieve sustainable outcomes. For practitioners, it provides actionable insights into balancing economic growth with social and environmental stewardship. The case studies highlight the significance of building robust partnerships, preserving cultural authenticity, and enhancing visitor experiences while protecting natural and cultural resources. In this context, this volume not only illuminates the diverse challenges and opportunities within tourism but also advocates for a thoughtful, inclusive approach to crafting policies and practices. It stands as a vital resource for advancing tourism that is both responsible and rewarding for communities and travellers alike.

Emerging themes and future directions

Several key themes have emerged from this volume, charting a path for the future of tourism in the Asia-Pacific region and beyond. Across all sections, the critical role of collaboration among stakeholders – government, industry, local communities and tourists – was a recurring theme. Effective tourism development and management require the alignment of diverse interests and concerted efforts towards shared goals. Equally important is the recognition that the diversity of the Asia-Pacific region necessitates tailored approaches that consider the unique cultural, environmental and social contexts of each destination. One-size-fits-all solutions are inadequate. Instead, adaptive and context-specific approaches are essential for sustainable tourism development.

Sustainability as a core principle is another key theme. The shift towards sustainable tourism is not merely a trend but a necessity. Whether addressing environmental conservation, cultural preservation or community empowerment, sustainability must be at the heart of tourism planning and practice. This principle is closely tied to the theme of resilience and adaptability. The tourism sector must be resilient and adaptable to challenges such as global pandemics, environmental changes and socio-economic shifts. Building resilience through sustainable practices and community engagement can help destinations navigate uncertainties and emerge stronger. Lastly, education and capacity building are vital for the long-term success of tourism initiatives. Training programmes, knowledge sharing, and access to resources can equip communities with the skills needed to manage and benefit from tourism sustainably.

Future directions and ongoing research

Building on the emerging themes, the ongoing and forthcoming research highlighted by contributors offers an insightful vision for the future of tourism in the Asia-Pacific region. Notable ongoing and future research includes:

- **Sustainable Hiking Through the Green Trails: Examples of Exceptional Responsible Practices** by Anjusha P.P. and Toney K. Thomas will benefit from conducting in-depth interviews with trekkers and community members. This enriched approach aims to uncover the transformative experiences of trekkers and the broader benefits of eco-friendly practices. Future research will explore the changes and challenges faced while implementing community engagement initiatives in remote Indian villages, providing deeper insights into sustainable trekking practices.
- **Sustainable Tourism and Stakeholders' Satisfaction in Batulayang Tourism Village, Bogor, Indonesia** by Ade Suherlan will expand to include comprehensive stakeholder perceptions and the impacts of tourism on community welfare, employment, and social solidarity. Future research will address the challenges and successes in maintaining cultural values and local wisdom amid growing tourism, using both qualitative and quantitative insights to provide a holistic view of sustainable development.
- **Overtourism and Tourism Sustainable Management in the Komodo National Park, Indonesia** by Reiza Miftah Wirakusuma, Sarah Gardiner and Alexandra Coghlan will develop a Causal Loop Diagram (CLD) to map stakeholder conflicts and explore potential system weaknesses. By validating the CLD with stakeholders and analysing conflict-driving archetypes, future research aims to offer effective solutions for managing overtourism and reducing reliance on visitor numbers as a primary tourism strategy.
- **Improving Cultural Sustainable Tourism: Evidence of Saba Budaya Baduy from the Baduy Tribe, Indonesia** by I. Putu Gede Eka Praptika, Amanda Natasanti, Della Ade Saputri, Muhammad Ibnu Azzulfa, Brigita Talentiana Mahen and Fahmi Prihantoro will further investigate how the Baduy Indigenous people in West Java use the concept of 'Saba Budaya Baduy' to preserve their cultural identity and promote sustainable tourism. Future research will focus on community resilience and effective resource management, providing insights into how cultural preservation and tourism can mutually reinforce each other.
- **Hawker Centres, Cultural Heritage and Tourism – A Case Study of Lau Pa Sat, Singapore** by Lan Wang will continue to examine the transformation of Lau Pa Sat. Future research will include comparative studies with similar global heritage sites to identify best practices in adaptive reuse. Longitudinal studies will monitor the transformation's effects on the local economy, community engagement and cultural preservation, contributing to a comprehensive conservation model that balances historical preservation with contemporary needs.
- **Experiencing Cambodia's Living Arts Through Tourism** by Bobbie Chew Bigby, Madura Thivanka Pathirana and Sarin Chhuon will document the ongoing efforts of Cambodian Living Arts (CLA) to revitalize Cambodian cultural heritage through tourism. Despite the cessation of their tourism programme, future research will focus on how CLA's initiatives continue to influence cultural heritage, ensuring their contributions to cultural preservation are recognized alongside academic perspectives.
- **Lived Religious Practice and Tourism Perceptions of Southern Buddhist Monks in an Ethnic Tourism Community in Xishuangbanna, China** by Yu Li will further explore the impact of religious tourism on the lived religion of community residents, thereby enriching the understanding of cultural sustainable development in religious tourism destinations.
- **Community-Based Ecotourism and Conservation in Tboli, Philippines** by Ramon Benedicto A. Alampay will benefit from personal visits and interviews with key community actors. This enriched approach will provide an in-depth assessment of how the community-based tourism model adapts

to post-COVID growth. Future research will also explore the challenges and opportunities in sustaining this model amid increasing market demand, offering valuable insights into its long-term viability and success.

- **Business Strategy and Development Towards Sustainability: A Story of Ban Bo Chet Luk Community Entrepreneurship, Thailand** by Morakot Ditta-Apichai will investigate how community-based tourism enterprises in Thailand integrate sustainability principles. Future research will assess capacity-building programmes for tourism micro-entrepreneurs and evaluate marketing strategies to attract environmentally conscious tourists, contributing to a more sustainable tourism sector.
- **Baan Khoa Lak: A Case Study on Sustainable Rural Development Through Community-Based Tourism and Natural Resource Conservation** by Krittawit Krittayaruangroj and Suparak Suriyankietkaew aim to replicate the community-based tourism model of Baan Khoa Lak in other rural Thai areas. Their ongoing research will explore how this model can drive regional development and provide insights into rural sustainability through tourism.
- **Green Recovery and Regenerative Tourism: The Success Story of Miso Walai Homestay in Building a Resilient Rural Tourism Destination** by Nadzirah Hosen, Isabel Paulino and Amran Hamzah will compare the Miso Walai Homestay's green recovery model with other community-based tourism initiatives in Southeast Asia. This comparative analysis will enhance the Miso Walai model and investigate its potential as a blueprint for other community-based tourism projects, identifying best practices and challenges.
- **The Bojo Algouinsan Ecotourism Initiative: Crisis Mitigation Through Community Tourism Development** by Zainub Ibrahim and Paul Gemmell will further investigate the Bojo Aloguinsan Ecotourism Initiative in the Philippines, focusing on its multidimensional impacts on the community. Future longitudinal studies will assess the project's evolution, cultural, environmental and economic impacts, providing a comprehensive understanding of its long-term sustainability.
- **The Growth of Tourism and 'Development' in an Indigenous Setting: Exploring a Case of the Chittagong Hill Tracts, Bangladesh** by Sabiha Yeasmin Rosy will explore recent changes in tourism development for Indigenous peoples in the Chittagong Hill Tracts of Bangladesh, focusing on the sustainability of ethnic crafts and indigenous resilience against uneven tourism impacts. Future research will offer insights into how Indigenous communities adapt and thrive amidst evolving tourism dynamics.
- **Equal Opportunities in Rural Tourism Entrepreneurship: Challenges in Malaysia** by Joo-Ee Gan will assess the long-term impacts of tourism on protected areas in Malaysia, particularly under the National Ecotourism Plan 2016–2025. This ongoing study will evaluate the socio-economic impacts of tourism and contribute to the development of sustainable tourism practices in protected areas.
- **Elephant-Based Tourism in Nepal: The Key to Reducing Human Poverty or a Tool to Commodify Endangered Species?** by Michelle Szydlowski will continue to document the health and welfare of captive elephants in Nepal, with plans to expand research on the interactions between elephants and humans. Future research will focus on the well-being of elephants and mahouts, and explore strategies to improve conservation and cultural appreciation.
- **Haunting Memory of Forgotten History: A Case Study of World War II 'Comfort Stations' in the Philippines as a Tourism Resource Featured in Selected Filipino Films** by Aileen C. Ani will expand on the history and narratives of World War II "comfort stations" in the Philippines. Future research will aim to capitalize on local histories and media complexities, providing a deeper understanding of these sites and exploring their potential as high-interest tourist destinations.
- **'We Invite You for Lunch': Post/Colonial Subjectivities and Performances of Tourism in Puducherry, India** by Aindrila Chakraborty will advance research

on heritage tourism in Puducherry, India, by examining how heritage politics influence local, national and international heritage-making processes. Ongoing doctoral research will investigate the interplay between colonial legacies and contemporary nationalism in heritage tourism.

- **Evidence Use in Policymaking: The Case of a Hotel Policy in Western Australia** by Carolina I. Garcia, Christof Pforr, and Michael Volgger have developed a new CABI case study exploring fiscal incentives for hotels in Buenos Aires, Argentina. This new study complements their current case and future research will involve comparing findings to enhance understanding of evidence use in tourism policy and develop effective strategies for incentivizing hotel investments.
- **Surfing in Hainan Island: An Emerging Surfing Tourism Destination in China** by Yana Wengel, Longrui Han, and Yunxin Hu will explore the future of surfing tourism in Hainan, China, focusing on its growing role in the global surfing community. Future research will address the authenticity of surfing experiences, infrastructure issues and environmental impacts, while assessing the potential for sustainable development in surfing tourism.
- **Learning by Doing: Educational Tourism at Mount Danxiashan** by Guiqing Li and Ding Xu have published a significant article on educational tourism and its role in balancing conservation and community development at Mount Danxiashan. This article provides valuable insights into how knowledge transfer can resolve conflicts between conservation and development, offering a model for other protected areas to follow.

Final thoughts and call to action

As we consider the future, it is clear that the Asia-Pacific region will continue to play an important role in shaping global tourism. The lessons from these case studies – from environmental stewardship and cultural preservation to community resilience – are invaluable. They challenge us to rethink our travel habits, destination management strategies, and our role in fostering a sustainable and equitable tourism industry. The rich insights provided in this collection and the future research plans serve as both a celebration of tourism's transformative power and a call to action. They urge us to embrace sustainable practices and work collaboratively towards a more inclusive and culturally respectful tourism industry. Let us embrace these insights and strive to make tourism a cornerstone of sustainable development in the Asia-Pacific region and beyond.

As we close this volume, we extend our deepest gratitude to the contributors whose expertise has illuminated the complexities of tourism across the Asia-Pacific region. Their work inspires a future where tourism not only enriches our experiences but also contributes positively to the communities and environments we visit. *Tourism Case Studies from the Asia-Pacific Region* stands as a testament to the collaborative efforts and shared vision of advancing tourism knowledge and practice. As we move forward, let's continue to explore, learn and innovate, forging a path towards a more sustainable and vibrant future for global tourism.

List of Original *Tourism Cases*

Ajoon, E.J. and Hayes, S. (2024) Resilience and loyalty in the face of adversity: A case of local employees in Kerala's backwater tourism. *Tourism Cases*, 2024, DOI: 10.1079/tourism.2024.0021.

Alampay, R.B.A. (2024) Community-based ecotourism and conservation in Tboli, Philippines. *Tourism Cases*, 2024, DOI: 10.1079/tourism.2024.0006.

Ani, A.C. (2024) Haunting memory of forgotten history: A case study of WWII 'comfort stations' in the Philippines as a tourism resource featured in selected Filipino films. *Tourism Cases*, 2024, DOI: 10.1079/tourism.2024.0001.

Anjusha, P.P. and Thomas, T.K. (2024) Sustainable hiking through the green trails: Examples of exceptional responsible practices. *Tourism Cases*, 2024, DOI: 10.1079/tourism.2024.0011.

Bigby, B.C., Pathirana, M.T. and Chhuon, S. (2024) Experiencing Cambodia's living arts through tourism. *Tourism Cases*, 2024, DOI: 10.1079/tourism.2024.0012.

Chakraborty, A. (2023) 'We invite you for lunch': Post/colonial subjectivities and performances of tourism in Puducherry, India. *Tourism Cases*, 2023, DOI: 10.1079/tourism.2023.0048.

Chan, J.K.L. (2023) Adopting sustainable practices at an Island ecotourism destination in Sabah through a sea turtle conservation project. *Tourism Cases*, 2023, DOI: 10.1079/tourism.2023.0046.

Clissold, R., Westoby, R., McNamara, K.E. and Obed, V. (2024) The realities of peripheral handicraft livelihoods in the face of disasters in Vanuatu. *Tourism Cases*, 2024, DOI: 10.1079/tourism.2024.0005.

Delevska, A., Osorio, S.C. and Nguyen, V.K. (2023) 'Not all fun and games': The case of the Victoria 2026 Commonwealth Games. *Tourism Cases*, 2023, DOI: 10.1079/tourism.2023.0045.

Ditta-Apichai, M. (2023) Business strategy and development toward sustainability: A story of Ban Bo Chet Luk community entrepreneurship, Thailand. *Tourism Cases*, 2023, DOI: 10.1079/tourism.2023.0050.

Gan, J.E. (2024) Equal opportunities in rural tourism entrepreneurship: Challenges in Malaysia. *Tourism Cases*, 2024, DOI: 10.1079/tourism.2024.0047.

Garcia, C.I., Pforr, C. and Volgger, M. (2024) Evidence use in policymaking: The case of a hotel policy in Western Australia. *Tourism Cases*, 2024, DOI: 10.1079/tourism.2024.0048.

Gogoi, P. and Maheshwari, A. (2024) Community-based tourism as a sustainable approach to tourism development in Jibhi Valley, Himachal Pradesh, India. *Tourism Cases*, 2024, *DOI: 10.1079/tourism.2024.0066*.

Hosen, N., Paulino, I. and Hamzah, A. (2023) Green recovery and regenerative tourism: The success story of Miso Walai Homestay in building a resilient rural tourism destination. *Tourism Cases*, 2023, DOI: 10.1079/tourism.2023.0049.

Ibrahim, Z. and Gemmell, P. (2023) The Bojo Algouinsan ecotourism initiative: Crisis mitigation through community tourism development. *Tourism Cases*, 2023, DOI: 10.1079/tourism.2023.0028.

Ito, E. (2023) Aikido tourism development in Tanabe, Wakayama, Japan. *Tourism Cases*, 2023, DOI: 10.1079/tourism.2023.0042.

Krittayaruangroj, K. and Suriyankietkaew, S. (2024) Baan khoa lak: A case study on sustainable rural development through community-based tourism and natural resource conservation. *Tourism Cases*, 2024, DOI: 10.1079/tourism.2024.0067.

Li, G. and Xu, D. (2023) Learning by doing: Educational tourism at Mount Danxiashan. *Tourism Cases*, 2023, DOI: 10.1079/tourism.2023.0043.

Li, Y. (2024) Lived religious practice and tourism perceptions of southern Buddhist monks in an ethnic tourism community in Xishuangbanna, China. *Tourism Cases*, 2024, DOI: 10.1079/tourism.2024.0075.

Nguyen, D.N. and Cordoba, G.F. (2023) Responses from businesses in rural communities recovering from the 2011 Tohoku Disasters during the 2020 Tokyo Olympics. *Tourism Cases*, 2023, DOI: 10.1079/tourism.2023.0053.

Pathirana, M., Bigby, B., Scott, N. and Gardiner, S. (2024) Tourism development opportunities in post war: A case of Jaffna, Sri Lanka. *Tourism Cases*, 2024, DOI: 10.1079/tourism.2024.0069.

Porya, A., Mishra, S., Sunayana and Chowdhary, N. (2024) Forging social identity: The making of a destination – A case of the Bullet Baba Shrine in Marwar, Rajasthan, India. *Tourism Cases*, 2024, DOI: 10.1079/tourism.2024.0009.

Praptika, I.P.G.E., Natasanti, A., Saputri, D.A., Azzulfa, M.I., Mahen, B.T. *et al.* (2024) Improving cultural sustainable tourism: Evidence of Saba Budaya Baduy from Baduy Tribe, Indonesia. *Tourism Cases*, 2024, DOI: 10.1079/tourism.2024.0020.

Rosy, S.Y. (2024) The growth of tourism and 'development' in an indigenous setting: Exploring a case of the Chittagong Hill Tracts, Bangladesh. *Tourism Cases*, 2024, DOI: 10.1079/tourism.2024.0022.

Shenaan, M. and Schänzel, H. (2024) The guesthouse phenomenon in the Maldives – development and issues. *Tourism Cases*, 2024, DOI: 10.1079/tourism.2024.0010.

Suherlan, A. (2024) Sustainable tourism and stakeholders' satisfaction in Batulayang Tourism Village Bogor Indonesia. *Tourism Cases*, 2024, DOI: 10.1079/tourism.2024.0007.

Sumaja, K. and Koo, T.T. (2024) Meteorological trends and the travel sector in Bali 1991–2020. *Tourism Cases*, 2024, DOI: 10.1079/tourism.2024.0068.

Szydlowski, M. (2024) Elephant-based tourism in Nepal: The key to reducing human poverty or a tool to commodify endangered species? *Tourism Cases*, 2024, DOI: 10.1079/tourism.2024.0003.

Tiwari, R. (2023) Breaking barriers, building futures: The transformative power of tourism in empowering women in Munsiyari, India. *Tourism Cases*, 2023, DOI: 10.1079/tourism.2023.0040.

Wang, L. (2024) Hawker centres, cultural heritage and tourism – a case study of Lau Pa Sat, Singapore. *Tourism Cases*, 2024, DOI: 10.1079/tourism.2024.0004.

Wengel, Y., Han, L. and Hu, Y. (2024) Surfing in Hainan Island: An emerging surfing tourism destination in China. *Tourism Cases*, 2024, DOI: 10.1079/tourism.2024.0014.

Wirakusuma, R.M., Gardiner, S. and Coghlan, A. (2024) Overtourism and tourism sustainable management in the Komodo National Park, Indonesia. *Tourism Cases*, 2024, DOI: 10.1079/tourism.2024.0026.

Index

CABI – who we are and what we do

This book is published by **CABI**, an international not-for-profit organisation that improves people's lives worldwide by providing information and applying scientific expertise to solve problems in agriculture and the environment.

CABI is also a global publisher producing key scientific publications, including world renowned databases, as well as compendia, books, ebooks and full text electronic resources. We publish content in a wide range of subject areas including: agriculture and crop science / animal and veterinary sciences / ecology and conservation / environmental science / horticulture and plant sciences / human health, food science and nutrition / international development / leisure and tourism.

The profits from CABI's publishing activities enable us to work with farming communities around the world, supporting them as they battle with poor soil, invasive species and pests and diseases, to improve their livelihoods and help provide food for an ever growing population.

CABI is an international intergovernmental organisation, and we gratefully acknowledge the core financial support from our member countries (and lead agencies) including:

UKaid from the British people

Ministry of Agriculture
People's Republic of China

Australian Government
Australian Centre for International Agricultural Research

Agriculture and Agri-Food Canada

Ministry of Foreign Affairs of the Netherlands

Schweizerische Eidgenossenschaft
Confédération suisse
Confederazione Svizzera
Confederaziun svizra

Swiss Agency for Development and Cooperation SDC

Discover more

To read more about CABI's work, please visit: **www.cabi.org**

Browse our books or explore our online products at:
https://www.cabidigitallibrary.org

Interested in writing for CABI? Find our author guidelines here:
https://www.cabidigitallibrary.org/books/authors

www.ingramcontent.com/pod-product-compliance
Lightning Source LLC
LaVergne TN
LVHW071335210926
848486LV00003B/154